HISTORICAL FOUNDATIONS OF COGNITIVE SCIENCE

PHILOSOPHICAL STUDIES SERIES

VOLUME 46

HISTORICAL FOUNDATIONS OF COGNITIVE SCIENCE

Edited by

J-C. SMITH

Youngstown State University

KLUWER ACADEMIC PUBLISHERS

DORDRECHT / BOSTON / LONDON

Library of Congress Cataloging in Publication Data

Historical foundations of cognitive science / edited by J.-C. Smith.
p. cm. -- (Philosophical studies series ; v. 46)
Includes bibliographical references.
ISBN 0-7923-0451-9
1. Knowledge, Theory of--History. 2. Philosophy of mind--History.
3. Cognitive science--History. I. Smith, J.-C., 1946- II. Series.
BD161.H537 1989
128'.2'09--dc20 89-38208

ISBN 0–7923–0451–9

Published by Kluwer Academic Publishers,
P.O. Box 17, 3300 AA Dordrecht, The Netherlands.

Kluwer Academic Publishers incorporates
the publishing programmes of
D. Reidel, Martinus Nijhoff, Dr W. Junk and MTP Press.

Sold and distributed in the U.S.A. and Canada
by Kluwer Academic Publishers,
101 Philip Drive, Norwell, MA 02061, U.S.A.

In all other countries, sold and distributed
by Kluwer Academic Publishers Group,
P.O. Box 322, 3300 AH Dordrecht, The Netherlands.

Printed on acid-free paper

Printed in the Netherlands

CONTENTS

PREFACE

My interest in gathering together a collection of this sort was generated by a fortuitous combination of historical studies under Professor Keith Lehrer and studies in cognitive science under Professor R. Michael Harnish at the University of Arizona. Work on the volume began there while I was an instructor in the Department of Linguistics and was greatly encouraged by participants in the Faculty Seminar on Cognitive Science chaired by Professor Lance J. Rips. I wish to express my appreciation to all of these and to many other individuals with whom I discussed the possibility of contribution to this work.

I am especially grateful to the authors of the essays included here, as they showed more patience than I could have hoped for in seeing me through a number of uncertain stages in development of the project. My thanks are also due to my colleague Charles Reid for assistance in reviewing submissions, to Tim McFadden for computer resources, and again, to Keith Lehrer for continuing advice in arrangements for publication. Financial support for manuscript preparation was provided in part under University Research Grant No. 617 from the University Research Council, Youngstown State University.

THE EDITOR

J-C. SMITH

INTRODUCTION: PROBLEMS AT THE ROOTS

Whenever a new discipline threatens to emerge from the concentrated attempt at solving an enduring set of problems, there is a need to search out the historical foundations upon which its investigations can rest. Such continuity is desirable not only for the intellectual respectability the new discipline may thereby gain but also as a protection against the repetition of errors and the loss of pretheoretical insight. The development of a dominant computational methodology and substantial theories of information processing in the cognate areas of what has come to be known as "cognitive science" is a case in point. Philosophical perspectives on specific research projects in cognitive psychology, theoretical linguistics, neurobiology, and computer science modeling of artificial intelligence have led to systematically new approaches to a broad range of traditionally intractable problems. Yet what we find here is often that technology is outstripping our theoretical understanding and that these varied proposals lack firm grounding within the historical tradition which is the heritage of any scientific pursuit. This amounts to accepting an intellectual orphan of uncertain origins, whereas establishing historical continuity may be a way to secure achievements and point the direction for growth.

The purpose of this collection of essays is to show that particular historical treatments of problems in philosophy of mind and philosophical psychology provide relevant bases for theoretical development in the special area of modern cognitive science. These investigations address both ontological and epistemological questions concerning the compositional nature of mental particulars, their typical organization, and their relations to physical contexts such as the body and its natural environment. The historical views brought to bear on these issues include those of the ancient Greeks, the medieval scholastics, early modern philosophers from Descartes to Hume and Reid, Kant and the 19th century phenomenologists, and the founders of 20th century behaviorism and analytic philosophy. While these lacked most of the detailed empirical evidence of current research, their approaches contain the first steps toward a complete account of the mind and provide striking parallels with much of the work in modern cognitive science. Indeed, there emerges here rather clear support for the old saw, "There is nothing new under the sun." The articles presenting and interpreting these views not only clarify the historical background of general cognitive studies but also elucidate many of the central difficulties which remain problematic for the fledgling special field of cognitive science.

I think it would be useful here to review some of the more significant of these problems with which both the traditional and modern approaches to the study of cognition have been confronted, albeit not always in the same terms. It is progress toward the resolution of such problems during this century that supports the claim that a new discipline is being hatched. Reviewing some of them will hopefully provide a focus for study of the works in this volume, while reminding us of the need for deeper, more unified investigation in the future. I hasten to add, however, that what I shall have to say should not be taken in any way as a complete story about the challenges that have been met nor as a complete litany of the obstacles yet to be surmounted.

Before proceeding, we might adopt a working definition of the subject at issue as follows:

> *cognitive science* $=_{df}$ the systematic scientific study of cognition and related psychological phenomena evidenced by the structuring, storage, and utilization of knowledge in intentional behavior.

Needless to say, such an attempt at definition is bound to be fraught with controversy. In particular, what comes to count as observational evidence of such phenomena will be crucially suspect if the explanations are indeed to be scientific. Yet this in itself suggests an important difference between the modern orientation and most traditional approaches toward the same subject. The results of modern investigations in the area are generally to be gained *a posteriori*, in some manner as a consequence of experimental testing and research, rather than through any process of purely *a priori* reasoning as in the past. Indeed, 'no armchair psychology' has been a rallying call of the evolving science. The problem before was always that nothing could be seen as providing evidence 'hard' enough in the area of cognitive phenomena to support lawlike generalizations of the sort we expect from science. Of course, to some extent this obstacle is still with us—and some argue that it always will be for any attempt at *scientific* psychological explanation—but we seem to have made at least some significant inroads across this landscape.

Specifically, to many there have appeared to be significantly new and interesting proposals that provide, if not solutions, at least plausible strategies for resolving problems in the following areas. As the articles in this collection demonstrate, these problems lie at the roots of cognitive science, in the intellectual traditions where its study first took hold. I have tried here to recommend some of the more relevant examples of this.*

Belief formation/fixation: the problem of explaining references to beliefs in the folk or scientific psychological explanation of intentional behavior. What is belief and how is it acquired? Is there *really* such a thing at all? What is the import of statements expressing propositional attitudes like 'Sam believes that grass is green'? This seems to be a two-place relation '*s*B*g*' where 'that grass is green' is a singular term picking out some proposition *p*. But how is *that* relation determined? Is there some corresponding relation 'xΨr' such that Ψ is a psychological relation of believing which person *x* stands in and *r* is some mental representation of *x*'s which is indexed by the proposition *p*? These sorts of questions are possibly the most central in philosophy of mind, language, and psychology and are addressed in various ways by every article here. They are of special interest in the article by Kitcher, presenting Kant's theory of belief formation, and that by Smith, concerned with a category of belief fixation. For background, consult Fodor (1975), Stich (1983), and Garfield (1988).

Black Box Problem (of mentalism): the problem of inferring internal states and processes on the basis of observed external behavior alone. This notorious difficulty derives from the fact that observed behavior, especially of intentional agents, can proceed from indefinitely many internal circumstances. Not only are we unable to definitively arrive at the particular sequences of internal states or processes, since the program will always depend upon the architecture of the internal mechanism or programming language. But we also seem incapable of demonstrating the need for a programming (or functionalist) analysis, as opposed to a purely causal explanation, at all. The articles by Kitcher and Cummins provide examples of how this problem arises when the observable behavior is verbal, inferential, and introspective. Cf. Pylyshyn (1984), Cummins (1983), and Dennett (1978, 1987).

Chauvinism/liberalism: the concern that our psychological explanations either will be so narrowly tied to our physiology as to needlessly exclude, for example, the possibility of alien lifeforms with cognitive abilities, or else will be so broadly functionalist as to include obviously non-psychological realizations, such as the chance execution of a program by

*The reader should consult the index for page numbers of topics in individual articles; some general references to introductory and other relevant works in the field also accompany each description.

random sub-nuclear events (or by the coordination of activities of the entire population of China). This concern is closely linked to the problem of the periphery as that is the point at which specification must seemingly turn non-functional and species-restrictive or remain so abstract that all manner of realizations are allowed. The general question of realizability or implementation is thus also relevant here. In the article by Shields an Aristotelian approach toward human physiology seems to invite liberalism, and the Meerbote article defends a liberalist interpretation of Kantian "anomalous monism." But the position taken in Searle's article on the biological basis of mentality is presumably chauvinistic. Cf. Block (1978, 1980) and Searle (1980).

Concept acquisition: the problem of explaining how conceptual categorization proceeds, especially in response to perceptual input. What is a concept, anyway? Is it the same as a concep*tion*? Is it an exemplar or prototype? Is there a rule that is followed in acquiring one? How could one acquire a new conceptual category if one did not already have concepts of the features defining the category? Clearly a wide range of problems concerning the nature of cognitive processes, perception, and representation will confront views about how concepts are acquired. The article by Tweedale recounts the major debate in medieval philosophy over the difference between universal and particular conception, while the Meerbote and Kitcher articles describe the elements of Kant's revolutionary theory of conceptual development based on the innate structure of mind. Cf. Smith and Medin (1981), Fodor (1981), and Putnam (1981).

Consciousness: the subject of either (i) the awareness of one's own mental operations, as represented by concepts and beliefs concerning them, or (ii) the 'raw feeling' of intentional existence. What is it, in either case? Is it functionally explainable; i.e., in terms of (causal) relations among mental states, inputs, and outputs? The second concern is closely tied to the problem of qualia, while the first raises the issues of introspectionism. The second is meant to capture the sense of 'what it is to be like' a sentient creature, and the first, the sense of what it is to have 'self-knowledge'. The Smith article undertakes construction of a specific account of the latter based on Thomas Reid's distinction between consciousness and reflection. The article by Tweedale is also relevant to this perspective on consciousness as a form of apprehension. Consideration of the more general, phenomenological 'feel' of *being* conscious is included in the McIntyre article. Cf. Ornstein (1972, 1973), Sayre (1969), Dennett (1978).

Fregean Presumption: the understanding that the structure of thought is revealed in the structure of natural language, as presupposed in linguistic analysis. Issues concerning representational relations and specific features such as the mental copula will surface with this presumption. But is this as obvious or plausible a theoretical plank as most philosophers of mind and language have thought? Are there mental words? Is there mental grammar? There is no *a priori* reasoning which necessitates this connection between language and thought, since the claim that thought has syntactic structure at all is itself in need of evidential support. Surely it is possible that complex syntactic structures of natural language map onto formally simple but unique mental events that provide the appropriate semantic relations. The article by Normore, on the problems of synonymy and definition in mental language, and the McIntyre article, on the syntactic characterization of "noemata" or mental meanings, present both sides of this issue. Cf. Frege (1892), Fodor (1975), Brand (1984), and Garfield (1988).

Functions: an area of questions about the relations among internal states, inputs, and outputs, as the basis for attribution of mental properties; e.g., whether or not such relations are purely causal or syntactic, what their connection to semantic interpretation or representational content is, and the extent to which they are teleological (or forward or backward looking). This is a particularly confusing area since there are numerous distinct senses of 'function' appealed to in logic, mathematics, and science. Is functional explanation the paradigm for psychological explanation? The view probably most responsible for renewed interest in cognitive studies is Functionalism, which holds that mental states, roles, or properties are identical with functional states, roles, or properties. This ontologically neutral alternative to (type-type) physicalist Identity Theory opens the door for multiple realizability of cognitive processes and thus for the minds and machines analogy in general. See the article by Shields for a useful introduction to functionalist concerns, including the distinction between versions according to their "topic neutrality." The Tweedale, Cook, and Biro articles also demonstrate early functionalist leanings in medieval and modern philosophy. Cf. Nelson (1976), Block (1978, 1980), Cummins (1983), and Bigelow and Pargetter (1987),.

Homunculi/Hume's Problem: the alleged impossibility of explaining intentional behavior in terms of increasingly complex interactions of decreasingly intelligent devices or internal 'homunculi'; i.e., explanation without appeal to components having capacities similar to those under investigation. What if explanation without "undischarged homunculi" is impossible?

See the article by Biro in particular for a clear analysis of what is at stake with this problem and whether it is appropriately handled by Hume. Related discussion can also be found in the Smith and McIntyre articles. Cf. Dennett (1978, 1987) and Searle (1980, 1983).

Image/objectual vs. propositional/discursive representation: basically the imagery debate over the possibility or likelihood of a pictorial format for mental representation, but also includes concerns over the non-reducibility of 'knows-how' knowledge in skills or direct acquaintance to 'knows-that' knowledge of the sort verbally communicable. Traditional Idea Theory has a stake in the outcome of this debate, as well as the present-day attempt to account for the 'novice-expert shift'. Is all behavior the result of discursive rule-following, or is some, like swimming or riding a bicycle, produced by a more homeostatic process of analogical connections? See the Dreyfus article for discussion of this last question as leading to the need for modern "connectionist" treatments of sophisticated skills. The Biro, Tweedale, and Kitcher articles can be consulted for perspectives on the imagery issues of this problem area. Cf. Block (1978), Kosslyn (1980), Pylyshyn (1981).

Intentionality: the problem of explaining the "inherent aboutness" of mental states that includes the ability to represent non-existent objects or objects whose representations will never figure into input-output relations. Is this a natural/realistic or merely functional/theoretical property (assuming such distinctions)? How is it related to the property of intelligence? Is it to be equated with the inten*sion*ality exhibited by transparent linguistic contexts (i.e., those expressing *de re* beliefs)? Of course, one could not consider this topic without raising the issues of phenomenology. The article by McIntyre provides a thorough response to the claim, especially as made by Hubert Dreyfus, that Husserl's work on intentionality was a central step toward cognitive science. Spinoza's earlier claims on the intrinsically representational character of the mental are considered in the Cook article. Cf. Cummins (1983), Searle (1983).

Introspectionism: the view that the experience and testimony of some forms of self-awareness provide sufficient or at least significant evidence for theoretical claims concerning the nature of mental operations. If there is such a thing as introspection, is it the same as consciousness, 'inner' perception, or some other capacity or faculty? The Smith article directly addresses this question in the context of the doctrine of privileged access, which holds the evidence of introspection to be incorrigible. Spinoza's rejection of Cartesian reliance on such evidence becomes an issue in the

Cook article, and how David Hume sought to integrate this view within his system is discussed in the Biro article. In addition, the Cummins article focuses on an early twentieth century attempt to make the view scientific in psychology. Cf. Smith and Jones (1986) and Audi (1988), for an epistemological introduction to these issues.

Meaning/content: a subject related to the problem of intentionality and to the Fregean Presumption and constituting the central area of concern with linguistic and mental representation. Specific considerations involve the relation between syntactic and semantic properties of language. Is mental language the meaning or content for spoken language? The Meerbote article considers perhaps the first systematic claim that it is. Whether meaning or some narrow content is determined just by what is in the head, a possibility currently accepted under Individualism, comes up in the Biro article. For consideration of meaning as intentional content, both the Smith and McIntyre articles review the claims of phenomenology. Cf. Dennett (1987), Fodor (1987), Garfield (1988), Haugeland (1985), and Schiffer (1987).

Mechanism: the view that something like the science of Newtonian mechanics—perhaps expanded to include all later pre-Einsteinian physics—is sufficient for explanation and prediction of not only inanimate but also all animate and intentional behavior. For example, human verbal behavior may be considered explainable by ordinary mechanical physics. The Chomsky article addresses the Cartesian denial that any such explanation could be forthcoming, due to the creativity of speech, and then further reviews early failed attempts by linguists to provide one. In the Searle article, the more modern claims of 20th century functionalists about possible mechanical realizations of thoughts is analyzed. Cf. Dennett (1978), Nelson (1976), and Searle (1980).

Mental copula: the problem of explaining what exactly it is that associates a representation of the subject with a representation of the predicate in the mental expression of a proposition; i.e., along the lines of the Fregean Presumption, the problem of what exactly is the mental correspondent of the 'is' or 'are' in categorical statements. Is the mental copula a further *concept* or some syncategorematic mechanism for combining concepts? How this issue emerged in medieval reworking of Aristotle's views is discussed in the Normore article. The problem as it relates to modern work on the interpretation of language is treated in the McIntyre article. Cf. Brand (1984) and Castañeda (1975).

Mind/Body Problem: the traditional question of the ontological status of mind and body or of the explanatory roles of theoretical references to them and any possible interactions between them. Is this a real problem or merely a pseudo-problem created by the conceptual presuppositions of a certain theoretical framework? Although minds and bodies are surely not *observably* independent and interactive, is there an *explanatory* sense in which they are distinct from each other? Or is that just a matter of the logical consequences of rather arbitrary choices among possible systems of categorization? Leibniz's parallelist resolution of this problem forms an early basis for cognitive science from the perspective of the Kreiling article. But the Searle article seeks more to defuse the problem altogether, as merely an excuse for ignoring the actual biological nature of mentality. Every article in this collection can be taken to bear on this issue in one way or another. Cf. Smith and Jones (1986).

Modularity/faculty thesis: the psychological thesis that the brain/mind is somehow constituted by systematically independent components, modules, organs, or faculties. If they exist, are their processes or operations mental or cognitive, conscious or unconscious, inferential or direct? Are there both input systems and output systems? What are their inputs *from* and their outputs *to*? Should all mental capacities be viewed as emanating from such closed systems, or should some be taken as general features of open-ended cognition. The Normore and Kitcher articles provide versions of Kantian reasoning about mental faculties, while the Smith article focuses on the modular nature of attentive reflection but *not* consciousness under Reid's faculty psychology. Finally, the article by Rips provides a careful analysis of ordinary beliefs about the component relations (e.g., part-of and kind-of relations) among a wide range of cognitive abilities. Cf. Brand (1984) and Fodor (1983).

Naturalism/materialism vs. mentalism/idealism: the distinction between (i) those views accepting that reality consists entirely in the nomologically explainable system of material, spatio-temporal objects and events and (ii) those views holding that reality is constituted by, or is explainable in some degree as the product of, regular mental or ideational phenomena (perceptions, beliefs, etc.). The Kreiling article addresses this contrast in terms of the major problem for Leibniz of reconciling physicalist analyses of natural processes with his views on logic and metaphysics. Consequences of 'bracketing' the 'natural standpoint' are also discussed in the McIntyre article's interpretation of Husserlian mentalism. Cf. Putnam (1981) and Churchland (1984).

Problem of Old Evidence: the problem raised by the use of evidence that is part of background knowledge to confirm a new hypothesis, when the positive relevance definition of confirmation impossibly requires that the probability of the hypothesis given the evidence be greater than the probability of the evidence given background knowledge which entails it. This complication of Bayesian philosophy of science finds its representative in theories of concept acquisition and change, where the question of reliably codifying new concepts on the basis of old ones arises. The Kitcher article attributes to Kant a provision for "automatic continuous updating" of concept-rules that offers a more holistic solution to this theoretical problem. The Tweedale article also raises such issues in discussion of the cause/effect relations over time of background intellectual habits and new acts of intellectual apprehension. Cf. Glymour (1980) and Fodor (1981).

Problem of Perception: the traditional problem of explaining how the stimulation of sensory organs leads to perceptual knowledge about the external world. Is perception modularly organized? Are its processes best viewed as inferentially based or as unmediated and direct? How do perceptual beliefs about the world acquire justification? The Meerbote article contains an account of responses to this problem from Descartes to Locke and then in greater depth Kant's notion of a process of synthesis by the understanding. Cf. Smith and Jones (1986).

Problem of the Periphery: the difficulty in functionalism of providing a topic neutral characterization of inputs and outputs in the specification of the functional roles of internal states. This is the underlying problem leading to chauvinism or liberalism in the attribution of mental properties and the central issue in theories of input systems and output systems. If internal states are to be characterized only in terms of causal or formal relations to other states, inputs, and outputs, then two questions arise: (i) How should the inputs and outputs be named in that characterization? (ii) How are the inputs and outputs to be characterized themselves? If functionalism is to remain ontologically neutral, then the peripheral items must be referenced in some neutral vocabulary. Yet those items themselves cannot be *functionalistically* characterized, as seemingly required to keep the vocabulary neutral, without the same problem resurfacing in characterization of *their* peripheral relations. The Shields article runs onto this dilemma in the course of distinguishing weak from strong functionalism in Aristotle. The articles by Normore and Meerbote both encounter the problem in terms of what is *given* in the sensory impressions that are the peripheral points of perceptual contact with the world. Lastly, Searle's article addresses consequences of the failure of functionalists to resolve this issue. Cf. Block (1978).

Psychological explanation: the general area of theoretical development concerned with predicting the intelligent and intentional behavior of animate entities. Can this be systematically scientific or is it only informal, 'folk' theory at best? Does it require causal transition laws, as with chemistry or physics, or is it fundamentally a form of functional explanation? If the latter, as most cognitive scientists seem to think, then is computational modeling adequate for experimental results? This last question is, of course, a recurrent concern in the Searle article. As for other interests, the Kitcher article responds to methodological attacks on Kantian "transcendental psychology," and the Husserlian view on psychology as an "eidetic science" is discussed in the McIntyre article. Other approaches toward psychological explanation that are considered here include such different assessments of a "science of the mind" as Spinoza's in the Cook article and Hume's in the Biro article, as well as the Titchnerian structuralist account reviewed in the Cummins article. Cf. Fodor (1980), Putnam (1981), Cummins (1983), and Dennett (1987).

Qualia Problem: the obstacle to complete functional explanation presented by the 'raw feel' or qualitative aspect that makes it *what it is like* to have perceptual or other experiences or to be an experiencer of them. Why does this aspect of mental life seem not to be functionally characterizable? Does it matter if it is not? Is it tied to the particular realization or implementation of the process constituting the experience? One perspective on this problem is that it marks the boundary between adequate applications of (type-type) Identity Theory (on the qualitative level) and necessitated (token-token) functionalist applications (on the discrete/discursive level). An exceptionally clear analysis of this well-known problem appears in the Biro article. The Smith article also contains a relevant discussion on the qualitative aspect of consciousness. Cf. Nagel (1974), Shoemaker (1975), Block (1978), and Putnam (1981).

Realizability/implementation: allowing that there are mental or cognitive events, this is the matter of how and what token material processes count as their occurrences; the computational question for cognitive science. Are there distinctly different types of realizing processes for the different types of realized processes, or is this only a token-token relation? Much of what has appealed to cognitive scientists about Functionalism is derived from its implication of the multiple material realizability of mental particulars. The Shields article makes much of Aristotle's similar claim that anything performing the *function* of a body part could, in principle, be used to replace it *regardless of material composition*. In the Cook article, the issue is raised by Spinoza's requirement of an ultimate reduction to natural laws for a final explanation of mind. As for the present-day computationalist maxim that content relations must be reliably preserved by specific formal relations, the McIntyre article provides a concise account of how a comparable claim gained consideration in earlier phenomenological debate. Cf. Fodor (1980) and Cummins (1983).

Representational relations: an area of investigation into the capacity of internal states for encoding content, especially the truth-revealing connection between perceptual processes and the external world. Does thought consist in the manipulation of mental representations? What is the relation between representation and intention? While the traditional Representative Realism of Locke did not survive Berkeley's attack on knowledge of representational relations, the search for an alternative basis for mental representation has continued through the modern period. In cognitive science this search has been focused on what is known as the Representational Theory of Mind. The Biro article presents a useful overview of this theory, although elements of it can be found here in the accounts of much earlier history as well. The McIntyre and Kitcher articles also provide 'transcendental' perspectives on mental representation. Cf. Fodor (1975, 1987) and Schiffer (1987).

Rules vs. Connections: the contrast between (i) views holding that all mental processes can be explained as rule-governed processes or executions of programs and (ii) views denying that such explanation is possible, at least for some processes that are instead the results of either non-discrete summing of strength vectors or homeostatic settling into stability. But is this just a question of implementation rather than a real difference between competing computational models of cognition? The issue of knowledge-*that* vs. knowledge-*how* is also clearly relevant here. Whether all skills can be accounted for in terms of rule-following behavior is an ancient question, and the Dreyfus article traces it back to positions held by Socrates and Plato while showing that it is at the cutting edge of current research. In the Kitcher article a "two-tiered theory of belief" attributed to Kant seems to allow for low-level connectionist processing in combination with rule-governed inferential associations. Cf. Rumelhart et al. (1986) and Fodor and Pylyshyn (1988).

Teleology: the attribution of purpose or intended result to natural processes, including the characterization of functional roles in terms of what consequences are typically produced by natural events. Can the Aristotelian notion of final cause be entirely eliminated from functional theory? The Shields article offers a critical look at this prospect. On elimination of the teleological element, the Cook article presents Spinoza's solution, which was to take "persevering in being" as the goal of the body. But on the other side the Kreiling article describes Leibniz's defense of teleology as coherent convergence toward an ideal. The modern tendency toward teleology in the form of explanatory instrumentalism is also illustrated in the Cummins article. Cf. Smith and Jones (1986), Bigelow and Pargetter (1987), and Dennett (1978, 1987).

Unconscious mental processes: the concern of that level of psychological explanation at which references to unreportable but experimentally demonstrated information-processing events are made; especially relevant in addressing questions about the nature of processes within encapsulated modules or faculties. Are attributions of perceptual hypothesis formation and inferences that a subject is unaware of appropriate or necessary for explaining cognition? How is the line drawn between implicit reasoning that is merely an abstract development of attitudes and actions and explicit reasoning that the subject really experiences but simply does not attend to sufficiently to report? As the Dreyfus article holds up for inspection, cognitiv*ism* is historically committed to content ascription on unconscious levels. The contrast between this view and medieval scholastic thought is brought out in the Tweedale article, which also contradicts claims in the Normore article about Ockham's position on precisely this issue. And of course Kant's architectonic approach opens up similar questions about unconscious structure that are discussed in the Meerbote and Kitcher articles. Cf. Pylyshyn (1984) and Garfield (1988).

Verificationism: the positivist view that the meanings of statements (and perhaps their terms) are the conditions of verification of their truth. As the prevailing view of mid-twentieth century analytic philosophy, it suffers from well-known defects of self-refuting inconsistency and unjustified presupposition of criterial standards. Similar difficulties plagued its behaviorist representatives in psychology. Yet how can these scientific programs be dismissed without falling prey to historical relativism and cultural and psychological indeterminacy? The Kitcher article concludes with a surprising contention about the famous attack by Thomas Kuhn on the verificationist distinction between observational and theoretical statements. An alternative method is tried in the Cummins article, and the Searle article discusses the overall role of verificationism in shaping present-day cognitive science. Cf. Fodor (1975) and Putnam (1981).

No doubt there are other problems and areas of concern that are touched upon in the following historical studies. I have merely tried to isolate some of those that seemed to me to carry forward into current studies in cognitive science. Of course, the rapid rate of developments in this field means that such a correlation is tentative at best. But one gets the feeling that some significant subset of these problems will endure for the foreseeable future.

How that future will read these problems is as unlikely to be within the same conceptual framework employed in the present as our readings are to capture the perspectives of the past. There can be no doubt that exercises in historical interpretation of the sort found here are at best just glimpses of previous world pictures "through a glass darkly." In the attempt to garner support for our current designs we perhaps inevitably rewrite history to suit our purposes, if only through use of the given vocabulary in our entrenched outlook. As Robert D'Amico (1989, p. 55) observes at one point in his review of historicist debate:

> A research program is not simply *there* in the history of science as a documented fact, but is more like a Weberian ideal type. It is the idea of the program or tradition that tells us what to look for in history.

In retrospective there may seem to us to be a chain of development down through the ages tied to the same collection of intractable problems, and viewing the past in this way may be very useful to us. Yet we should not deceive ourselves about our inability to know how those problems actually appeared to previous thinkers, any more than we can expect to know what subset of them will emerge as perspicuous links to understanding for future thinkers.

In fact the failure to have knowledge of either sort may not be crucial at all. If recent history is any indication, technology and experimentation will proceed to reveal much of the subject without the sought-after assistance of overarching theoretical solutions. It is not so much what the philosophical attitudes of the time really were that matters, as it is what they conspired to produce of importance to all other times. Still, a settled perspective on the theoretical level would surely be a more promising position for pursuit on the practical level. It is in the interest of attaining that position that the following essays are presented.

REFERENCES

Robert Audi (1988), *Belief, Justification, and Knowledge: An Introduction to Epistemology* (Belmont, CA: Wadsworth Publishing Company).

J. Bigelow and R. Pargetter (1987), "Functions," *The Journal of Philosophy* **LXXIV**, no. 4, pp. 181-196.

Ned Block (1978), "Troubles with Functionalism" in W. Savage (ed.), *Perception and Cognition: Issues in the Foundations of Psychology*, Minnesota Studies in the Philosophy of Science, Vol. IX (Minneapolis, MN: University of Minnesota Press), pp. 261-326.

Ned Block (1980), "Introduction: What is Functionalism?" in N. Block (ed.), *Readings in Philosophy of Psychology*, Vol. 1 (Cambridge, MA: Harvard University Press), pp. 171-184.

Myles Brand (1984), *Intending and Acting* (Cambridge, MA: The MIT Press/Bradford).

Hector-Neri Castañeda (1975), *Thinking and Doing* (Boston, MA: D. Reidel Publishing Co.).

Paul M. Churchland (1984), *Matter and Consciousness: A Contemporary Introduction to the Philosophy of Mind* (Cambridge, MA: The MIT Press/Bradford).

Robert C. Cummins (1983), *The Nature of Psychological Explanation* (Cambridge, MA: The MIT Press/Bradford).

Robert D'Amico (1989), *Historicism and Knowledge* (New York, NY: Routledge, Chapman & Hall, Inc.)

Daniel C. Dennett (1978), *Brainstorms: Philosophical Essays on Mind and Psychology* (Montgomery, VT: Bradford Books Publishers, Inc.).

Daniel C. Dennett (1987), *The Intentional Stance* (Cambridge, MA: The MIT Press/Bradford).

Jerry A. Fodor (1975), *The Language of Thought* (Cambridge, MA: Harvard University Press).

Jerry A. Fodor (1980), "Methodological Solipsism Considered as a Research Strategy in Cognitive Psychology," *The Behavioral and Brain Sciences* **3**, pp. 63-73.

Jerry A. Fodor (1981), "The Present Status of the Innateness Controversy" in J. A. Fodor, *Representations: Philosophical Essays on the Foundations of Cognitive Science* (Cambridge, MA: The MIT Press/Bradford), pp. 257-316.

Jerry A. Fodor (1983), *The Modularity of Mind: An Essay on Faculty Psychology* (Cambridge, MA: The MIT Press/Bradford).

Jerry A. Fodor (1987), *Psychosemantics: The Problem of Meaning in the Philosophy of Mind* (Cambridge, MA: The MIT Press/Bradford).

J. A. Fodor and Z. W. Pylyshyn (1988), "Connectionism and Cognitive Architecture: A Critical Analysis," *Cognition* **28**, pp. 2-71.

G. Frege (1892), "On Sense and Reference" in P. Geach and M. Black (trans.), *Translations from the Philosophical Writings* (Oxford: Blackwell, 1952).

Jay L. Garfield (ed.) (1987), *Modularity in Knowledge Representation and Natural-Language Understanding* (Cambridge, MA: The MIT Press/Bradford).

Jay L. Garfield (1988), *Belief in Psychology: A Study in the Ontology of Mind* (Cambridge, MA: The MIT Press/Bradford).

C. N. Glymour (1980), *Theory and Evidence* (Princeton, NJ: Princeton University Press).

John Haugeland (1985), *Artificial Intelligence: The Very Idea* (Cambridge, MA: The MIT Press/Bradford).

Thomas Nagel (1974), "What Is It Like to Be a Bat?", *Philosophical Review* **83**, pp. 435-450; reprinted in N. Block (ed.), *Readings in Philosophy of Psychology*, Vol. One (Cambridge, MA: Harvard University Press, 1980).

R. J. Nelson (1976), "Mechanism, Functionalism, and the Identity Theory," *Journal of Philosophy* **73**, no. 13, pp. 365-385.
Robert E. Ornstein (1972), *The Psychology of Consciousness* (San Francisco, CA: W. H. Freeman and Company).
Robert E. Ornstein (ed.) (1973), *The Nature of Human Consciousness: A Book of Readings* (San Francisco, CA: W. H. Freeman and Company).
Hilary Putnam (1981), *Reason, Truth and History* (New York, NY: Cambridge University Press).
Zenon Pylyshyn (1984), *Computation and Cognition: Toward a Foundation for Cognitive Science* (Cambridge, MA: The MIT Press/Bradford).
D. E. Rumelhart, J. L. McClelland, and the PDP Research Group (1986), *Parallel Distributed Processing: Explanations in the Microstructure of Cognition*, Vols. 1 and 2 (Cambridge, MA: The MIT Press).
Kenneth M. Sayre (1969), *Consciousness: A Philosophic Study of Minds and Machines* (New York, NY: Random House, Inc.).
Stephen Schiffer (1987), *Remnants of Meaning* (Cambridge, MA: The MIT Press/Bradford).
John R. Searle (1980), "Minds, Brains, and Programs," *The Behavioral and Brain Sciences* **3**, pp. 417-424.
John R. Searle (1983), *Intentionality* (New York, NY: Cambridge University Press).
Sydney Shoemaker (1975), "Functionalism and Qualia," *Philosophical Studies* **27**, pp. 291-315.
P. Smith and O. R. Jones (1986), *The Philosophy of Mind: An Introduction* (New York, NY: Cambridge University Press).
E. E. Smith and D. L. Medin (1981), *Categories and Concepts* (Cambridge, MA: Harvard University Press).
Stephen P. Stich (1983), *From Folk Psychology to Cognitive Science: The Case against Belief* (Cambridge, MA: The MIT Press/Bradford).

HUBERT L. DREYFUS

SOCRATIC AND PLATONIC SOURCES OF COGNITIVISM

Before we can decide whether Socrates is to blame for cognitivism, we need to understand what cognitivism is and what is wrong with it. Consequently this paper has three parts: A brief account of cognitivism, supporting the claim that cognitivism treats all mental activity on the model of thinking through a problem; a phenomenological description of skill acquisition that seeks to show that expertise does not normally involve thinking at all; and a tentative foray into the history of philosophy which suggests that the source of the distorted view of expertise characteristic of cognitivism can already be found in Plato's account of Socrates's philosophizing, although its implications fully emerge only in Plato's own philosophy.

1. What Is Cognitivism?

Cognitivism is not simply a theory of cognition but, as the name, cogni*tivism*, suggests, it is the strong view prominent in cognitive science that *all* mental activity is cognitive—that perception, understanding, learning and action are all to be understood on the model of fact gathering, hypothesis formation, inference making, and problem solving. Such a view is taken for granted by current cognitive psychologists, especially those seeking to program computers to behave intelligently. They think of themselves as pioneers of a new discipline which has done for the mind what Galileo did for the universe and Darwin did for biology.

The cognitivist assumption is so self-evident to these cognitive scientists that they seldom bother to state it explicitly, let alone argue for it. For example, Jerry Fodor says in passing, as if it were obvious, mental life is essentially thought and thought involves inference.

> Conan Doyle was a far deeper psychologist—far closer to what is essential about mental life—than, say, James Joyce (or William James, for that matter).[1]

Lest my interpretation of this off-hand remark seem forced, note that Fodor's account of perception, which is least like thought, turns it even

J-C. Smith (ed.), Historical Foundations of Cognitive Science, 1–17.

into detective work: "Perception typically involves hypothesis formation and confirmation."[2] In general all cognitivist theories make two essential assumptions:

1. Our ability to deal with things intelligently is due to our capacity to think about them reasonably (including subconscious thinking).
2. Our capacity to think about things reasonably amounts to a faculty for internal "automatic" symbol manipulation.[3]

This view is the culmination of the rationalist philosophical tradition. Descartes already assumed that all understanding consisted in forming and manipulating appropriate representations, that these representations could be analyzed into primitive elements, and that all phenomena could be understood as a complex combination of these simple elements. At the same time, Hobbes implicitly assumed that the elements were formal elements related by purely syntactic operations so that reasoning could be reduced to calculation. "When a man *reasons*, he does nothing else but conceive a sum total from addition of parcels," Hobbes wrote, "for REASON...is nothing but reckoning...".[4] Finally Leibniz, working out the classical idea of *mathesis*—the formalization of everything—sought grant money to develop a universal symbol system, so that "we can assign to every object its determined characteristic number."[5] According to Leibniz, in understanding we analyze concepts into more simple elements. In order to avoid a regress of simpler and simpler elements, there must be ultimate simples in terms of which all complex concepts can be understood. Moreover, if concepts are to apply to the world, there must be logical simples in the world which the simple symbols represent. Leibniz envisaged "a kind of alphabet of human thoughts"[6] whose "characters must show, when they are used in demonstrations, some kind of connection, grouping, and order which are also found in the objects."[7]

Cognitivism is rationalism turned into a research program. It proposes to use the computer to show how on the rationalist account the mind actually works. In a seminal statement of this project, Alan Newell and Herbert Simon claim that the human brain and the digital computer, while totally different in structure and mechanism, have, at the appropriate level of abstraction, a common functional description. At this level, both the human brain and the appropriately programmed computer can be seen as two different instantiations of a single species of device—one which generates intelligent behavior by manipulating symbols according to formal rules. Newell and Simon state their view as a hypothesis:

> *The Physical Symbol System Hypothesis*: A physical symbol system has the necessary and sufficient means for general intelligent action.
> By "necessary" we mean that any system that exhibits general intelligence will prove upon analysis to be a physical symbol system. By "sufficient" we mean that any physical symbol system of sufficient size can be organized further to exhibit general intelligence.[8]

If one were looking for an easy way to challenge this hypothesis one might question its most vulnerable claim, that *perception* is thought. But I prefer to raise the stakes and take on cognitivism on the territory where it seems most plausible: its claim that *intelligent behavior* results from conscious or unconscious thinking on the model of the inference making of Sherlock Holmes. For the sake of argument, I will focus on the case which would seem most favorable to cognitivism: the ability of experts in cognitive domains such as chess.

The issue can be seen starkly by looking at the assumptions underlying the building of expert systems, since such systems involve a crude but consistent application of the cognitivist perspective. Building expert systems in isolable cognitive domains is the work of so-called *knowledge engineers*. What the knowledge engineers claim to have discovered is that in areas which are cut off from everyday common sense and social intercourse, all a machine needs in order to behave like an expert are some general rules and lots of very specific knowledge. This specialized knowledge is of two types:

> The first type is the *facts* of the domain—the widely shared knowledge—that is written in textbooks and journals of the field....Equally important to the practice of the field is the second type of knowledge called *heuristic knowledge*, which is the knowledge of good practice and good judgment in a field...that a human expert acquires over years of work.[9]

Using all these kinds of knowledge, Edward Feigenbaum, the father of expert systems, developed a program called DENDRAL. It takes the data generated by a mass spectrograph and deduces from this data the molecular structure of the compound being analyzed. Another program, MYCIN, takes the results of blood tests such as the number of red cells, white cells, sugar in the blood, etc., and comes up with a diagnosis of which blood disease is responsible for this condition. It even gives an estimate of the reliability of its own diagnosis. Such programs are impressive.

And, indeed, isn't the success of expert systems just what the tradition would lead one to expect? If we agree with Feigenbaum that: "...almost all the thinking that professionals do is done by reasoning...,"[10] we can see that once computers are used for reasoning and not just computation, they should be as good or better than we are at following rules for deducing conclusions from a host of facts. So we would expect that if the rules which an expert has acquired from years of experience could be extracted and programmed, the resulting program would exhibit expertise. Again, Feigenbaum puts the point boldly:

[T]he matters that set experts apart from beginners, are symbolic, [and] inferential....Experts build up a repertory of working rules of thumb, or "heuristics," that, combined with book knowledge, make them expert practitioners.[11]

Since each expert already has a repertory of rules in his mind, all the expert system builder need do is get the rules out and program them into the computer.

But, plausible as it seems, this project has not worked out as expected. Ever since the inception of Artificial Intelligence, researchers have been trying to produce artificial experts by programming the computer to follow the rules used by masters in various domains. Yet, although computers are faster and more accurate than people in applying rules, master level performance has remained out of reach.

Arthur Samuel's work is typical. In 1947, when electronic computers were just being developed, Samuel, then at IBM, decided to write a checker playing program. He elicited heuristic rules from checker masters and programmed a computer to follow these rules. The resulting checkers program is not only the first and one of the best experts ever built, but it is also a perfect example of the way fact turns into fiction in AI. Feigenbaum, for example, reports that "by 1961 [Samuel's program] played championship checkers, and it learned and improved with each game."[12] In fact, according to Samuel, after 35 years of effort, "the program is quite capable of beating any amateur player and can give better players a good contest." It is clearly no champion. Samuel is still bringing in expert players for help, but he "fears he may be reaching the point of diminishing returns." This does not lead him to question the view that the masters the program cannot beat are using heuristic rules; rather, Samuel thinks that the experts are poor at recollecting their compiled heuristics. "The experts do not know enough about the mental processes involved in playing the game,"[13] he says.

The same story is repeated in every area of expertise, even in areas such as medicine where expertise requires the storage of large numbers of facts, which should give an advantage to the computer. In each area where there are experts with years of experience, the computer can do better than the beginner and can even exhibit useful competence, but it cannot rival the very experts whose facts and supposed heuristics it is processing with incredible speed and unerring accuracy.

2. A Phenomenology of Skilled Behavior

In the face of this impasse, in spite of the authority of the rationalist tradition and the persuasiveness of cognitivism, we must take a fresh look at what a skill is and what the expert acquires when he achieves expertise. We must be prepared to abandon the traditional view that a beginner starts with specific cases and as he becomes more proficient, abstracts and interiorizes more and more sophisticated rules. It might turn out that skill acquisition moves in just the opposite direction: from abstract rules to particular cases. I will concentrate on chess as a cognitive skill par excellence, but I will also draw on driving to show the generality of the model which Stuart Dreyfus and I propose.[14]

Stage 1: Novice

Normally, the instruction process begins with the instructor decomposing the task environment into context-free features which the beginner can recognize without benefit of experience. The beginner is then given rules for determining actions on the basis of these features, like a computer following a program.

The student automobile driver learns to recognize such interpretation-free features as speed (indicated by his speedometer) and distance (as estimated by a previously acquired skill). Safe following distances are defined in terms of speed; conditions that allow safe entry into traffic are defined in terms of speed and distance of oncoming traffic. One learns rules such as shift to second when the speedometer needle points to 10 miles an hour.

The novice chess player learns a numerical value for each type of piece regardless of its position, and the rule: "Always exchange if the total value of pieces captured exceeds the value of pieces lost." He also learns that when no advantageous exchanges can be found, center control should be sought, and he is given a rule defining center squares and one for calculating extent of control.

Stage 2: Advanced beginner

As the novice gains experience actually coping with real situations, he begins to note, or an instructor points out, perspicuous examples of meaningful additional components of the situation. After seeing a sufficient number of examples, the student learns to recognize them. Instructional *maxims* now can refer to these new *situational aspects*.

I use the terms 'maxims' and 'aspects' here to differentiate this form of instruction from the first, where *strict rules* were given as to how to respond to *context-free features*. Since maxims are phrased in terms of aspects they already presuppose experience in the skill domain. According to Michael Polanyi, from whom I borrow the term:

> Maxims are rules [in a broad sense], the correct application of which is part of the art which they govern. The true maxims of golfing or of poetry increase our insight into golfing or poetry and may even give valuable guidance to golfers and poets; but these maxims would instantly condemn themselves to absurdity if they tried to replace the golfer's skill or the poet's art. Maxims cannot be understood, still less applied by anyone not already possessing a good practical knowledge of the art.[15]

The advanced beginner driver uses (situational) engine sounds as well as (non-situational) speed. He learns the maxim: shift up when the motor sounds like it's racing; down when it sounds like it's straining. By observing the demeanor as well as position and velocity of pedestrians or other drivers, he learns to distinguish the behavior of the distracted or drunken driver from that of the impatient but alert one. He can then apply maxims for how to proceed in various cases. No number of words can take the place of a few choice examples in learning these maxims. Engine sounds cannot be adequately captured by words, and no list of objective facts enables one to predict the behavior of a pedestrian in a crosswalk as well as can the driver who has observed many pedestrians crossing streets under a variety of conditions.[16]

Similarly, with experience the chess student begins to recognize such situational aspects of positions as a weakened king's side or a strong pawn structure, despite the lack of precise and universally applicable definitional rules. He is then given maxims to follow, such as 'attack a weakened king's side'.

Stage 3: Competence

With increasing experience, the number of features and aspects to be taken account of becomes overwhelming. To cope with this information explosion, the performer learns, or is taught, to adopt a hierarchical view of decision-making. By first choosing a plan, goal, or perspective which organizes the situation and by then examining only the small set of features and aspects that he has learned are relevant given that plan, the performer can simplify and improve his performance.

A competent driver leaving the freeway on a curved off-ramp may, after taking into account speed, surface condition, criticality of time, etc., decide he is going too fast. He then has to decide whether to let up on the accelerator, remove his foot altogether, or step on the brake. He is relieved when he gets through the curve without mishap and shaken if he begins to go into a skid.

The class A chess player, here classed as competent, may decide after studying a position that his opponent has weakened his king's defenses so that an attack against the king is a viable goal. If the attack is chosen, features involving weaknesses in his own position created by the attack are ignored as are losses of pieces inessential to the attack. Removing pieces defending the enemy king becomes most salient. Successful plans induce euphoria, and mistakes are felt in the pit of the stomach.

In both of these cases, we find a common pattern: detached planning, conscious assessment of elements that are salient with respect to the plan, and analytical rule-guided choice of action, followed by an emotionally involved experience of the outcome.

The experience is emotional because choosing a plan, a goal, or perspective is no simple matter for the competent performer. Nobody can give a useful prescription for reasoning out a perspective, so the competent performer has to make up various rules which he then adopts or discards depending on how they work out. This procedure is frustrating, however, since each rule works on some occasions and fails on others, and no set of objective features and subjective aspects correlates strongly with these successes and failures. Nonetheless the choice is unavoidable. While the advanced beginner can hold off using a particular situational aspect until a sufficient number of examples makes identification easy and sure, to perform competently requires choosing an organizing goal or perspective; hence the choice, while risky, is unavoidable. Furthermore, the choice of perspective crucially affects behavior in a way that one particular aspect rarely does.

This combination of necessity and uncertainty introduces an important new type of relationship between the performer and his environment. The novice and the advanced beginner, applying rules and maxims, feel little or no responsibility for the outcome of their acts. If they have made no mistakes, an unfortunate outcome is viewed as the result of inadequately specified elements or principles. The competent performer, on the other hand, after wrestling with the question of a choice of perspective or goal, feels responsible for, and thus emotionally involved in, the result of his choice. An outcome that is successful is deeply satisfying and leaves a vivid memory of the situation encountered as seen from the goal or perspective finally chosen. Disasters, likewise, are not easily forgotten.

Remembered whole situations, therefore, differ in one important respect from remembered aspects. The mental image of an aspect is flat; no parts stand out as salient. A whole situation, on the other hand, since it is the result of a chosen plan or perspective, has a "three-dimensional" quality. Certain elements stand out as more or less important with respect to the plan, while other irrelevant elements are forgotten. Moreover, the competent performer, gripped by the situation that his decision has produced, experiences the situation not only in terms of foreground and background elements but also in terms of ego-involving opportunity, risk, expectation, commitment, etc. These gripping holistic memories cannot guide the behavior of the competent performer since he fails to make contact with them when, conforming to a cognitivist view of himself as a computer, he reflects on problematic situations as a detached observer following better and better rules. As we shall soon see, however, if he does let these gripping, holistic experiences take over, they become the basis of the competent performer's next advance in skill.

Stage 4: Proficiency

Considerable experience at the level of competency sets the stage for yet further skill enhancement. Having experienced many situations, chosen plans in each, and having obtained vivid, involved demonstrations of the adequacy or inadequacy of the choice, the performer spontaneously associates an appropriate perspective with situations which are similar to those already experienced. Involved in the world of the skill, the performer "notices" or "is struck by" a certain plan, goal, or perspective. The connection is made by association, not inference, so the "spell" of involvement need not be broken by detached conscious choosing.

Since there are generally far fewer "ways of seeing" than "ways of acting," after associating the issue or perspective with the situation without conscious effort, the proficient performer will still have to think about what to do. During this thinking, elements that present themselves as salient are assessed and combined by rule and maxim to produce decisions about how best to deal with the situation.

On the basis of prior experience, a proficient driver approaching a curve on a rainy day may simply sense that he is traveling too fast. Then, on the basis of such salient elements as visibility, angle of road bank, criticalness of time, etc. (the elements used by the competent driver to *decide* that he is going too fast), he decides whether to take his foot off the gas or to step on the brake.

The proficient chess player, who is classed a master, can recognize a large repertoire of types of positions. Recognizing almost immediately and without conscious effort the sense of a position, he sets about calculating the move that best achieves his goal. He may, for example, know that he should attack, but he must deliberate about how best to do so.

Stage 5: Expertise

The proficient performer, immersed in the world of his skillful activity, *sees* what needs to be done but must *decide* how to do it. With enough experience with a variety of situations, all seen from the same perspective but requiring different tactical decisions, the proficient performer seems gradually to associate not only goals or perspectives with situations but also actions. Thus, for the expert not only situational understandings spring to mind but also associated appropriate responses.

The expert chess player, classed as an international master or grandmaster, in most situations experiences a compelling sense of the issue and the best move. Excellent chess players can play at the rate of 5-10 seconds a move and even faster without any serious degradation in performance. At this speed they must depend almost entirely on intuition and hardly at all on deliberate analysis and comparison of alternatives. Stuart Dreyfus has recently performed an experiment in which an International Master, Julio Kaplan, was required rapidly to add numbers presented to him as a series of beeps at the rate of about one number per second, while at the same time playing five-seconds-per-move chess against a slightly weaker but master-level opponent. Even with Kaplan's conscious analytic mind completely jammed by adding beeps, he more than held his own against the master in a series of games. Deprived of the time and resources necessary to analyze problems or construct plans, he still produced fluid and coordinated play.

Kaplan's performance seems somewhat less amazing when one realizes that a chess position is as meaningful, interesting, and important to a professional chess player as a face in a receiving line is to a professional politician. Almost anyone can add numbers and simultaneously recognize and respond to faces, even though each face will never exactly match the same face seen previously, and politicians can recognize thousands of faces, just as Julio Kaplan can recognize thousands of chess positions similar to ones previously encountered. The number of classes of discriminable chess situations, built up on the basis of experience, must be similarly immense. (It has been estimated that a master chess player can distinguish roughly 50,000 types of position.) Recognition in both cases is an immediate intuitive response.

Automobile driving also probably involves the ability to discriminate a similarly large number of typical situations. The expert driver, generally without any awareness, not only knows by feel and familiarity when an action such as slowing down is required; he knows how to perform the action without calculating and comparing alternatives. He shifts gears when appropriate and responds to a wide range of complex situations with little or no awareness of his acts. What must be done, simply is done.

It follows from this skill model that a beginner makes inferences using strict rules and features just like a computer, executing heuristic procedures, but that with talent and a great deal of involved experience the beginner develops into an expert who intuitively responds to the current situation on the basis of past experience *without* applying rules or making inferences at all. Cognitivism has given an accurate description of the beginner and of the expert facing an unfamiliar situation, but it has failed to capture the fact that normally an expert does not *solve problems*. He does not *reason*. He skillfully does what in his experience has normally worked, and naturally, it normally works.

3. The Sources of Cognitivism

We have seen that it is easy to trace cognitivism back to seventeenth century rationalism. But the roots of the cognitivist account of intelligence reach much further back into our tradition. As far as I can tell, this intellectualist account grows out of two observations made by Socrates. Both were good phenomenology, but like a good philosopher, he over-generalized them. He saw that experts can often explain why they do what they do and that these explanations reveal principles from which the behavior in question can be seen to follow rationally. Generalizing these observations, Socrates claims in the *Gorgias* that an art must have "principles of action and reason"[17] and further, in the *Laches*, that "that which we know we must surely be able to tell."[18] Let us look carefully at each of these claims.

The claim that a craft or *technê* must be based on principles which can be articulated by the practitioners leads Socrates to rule out of account all forms of intuitive expertise which do not seem to be based on any principles at all. Cooking, for example, is "unable to render any account of the nature of the methods it applies."[19] It "goes straight to its end, nor even considers or calculates anything."[20] Socrates holds that such intuitive abilities are not crafts at all but mere knacks based on trial and error. This would rule out such experts as chicken sexors, who without being able to articulate any reasons and without, as far as anyone has been able to tell, using any features (cues) or rules, are able to sort male and female one-day-old chicks at a rate of 1000 an hour with 99.5% accuracy.[21] But since I am trying to take on cognitivism where it is most plausible and in my story of skill acquisition I have not rested my argument on skills like chicken sexing and X-ray reading, which are learned purely by apprenticeship, I will take my stand on Socrates's own ground and limit my discussion to those crafts which are articulable skills learned by instruction, whose practitioners do "know many things."[22]

Socrates shares the assumptions of modern knowledge engineers that experts in a craft know principles of action and reason and that what they know they must be able to tell. In its weak form, as just stated, the claim has a basis in the experience of experts. If my account of skill acquisition is correct, some experts can at least tell the maxims of their craft—maxims they once used but have since put aside, the way a child having learned to ride a bicycle disposes of his training wheels. Polanyi has stressed that much can be said but that what can be said will never be the whole story:

> Analysis may bring subsidiary knowledge into focus and formulate it as a maxim...but such specification is in general not exhaustive. Although the expert diagnostician, taxonomist, and cotton-classer can indicate their cues and formulate their maxims, they know many more things than they can tell....[23]

According to Polanyi and my skill model, most experts can state rules of thumb they once followed—rules which can help explain how they achieve their goals—but the rules make sense only to those already familiar with the relevant skill domain.

So far so good. But Socrates seems to want much more. He seems to want to elicit rules or principles from experts in each craft domain that would enable anyone to acquire expertise in that domain. If we assume for the sake of this discussion, as Socrates assumes ironically in his dialogue *Euthyphro*, that Euthyphro, a religious prophet, is an expert at recognizing piety but Socrates is not, it looks like Socrates is asking for more than Euthyphro's piety-recognizing *maxim* when he says: "I want to know what is characteristic of piety...to use as a standard whereby to judge your actions and those of other men."[24] He seems rather to want a *strict rule* that can be used even by non-experts.

Euthyphro's first response to this demand is like that of any expert. He gives Socrates examples from his field of expertise, in this case mythical situations in which men and gods have done things which everyone considers pious. Socrates then makes his usual demand that Euthyphro tell him his rule for recognizing these cases as examples of piety. And Euthyphro, like any expert examined by Socrates, produces various maxims, which, if taken as definitions or strict rules, do not hold up under cross-examination. The problem seems to be that maxims have *ceteris paribus* conditions which do not detract from their usefulness to other experts but which make them useless to the non-expert who needs a strict rule defined over context-free features accessible to any person or machine. That Socrates demands such features seems clear in the *Laches* where he asks Laches, presumably an expert on courage, "What is that common quality, which is the same in all cases and which is called courage?"[25] But no expert can supply such features and rules. This leads Socrates to the famous conclusion that since no expert can state consistent, context-free principles that provide the rationale for his actions, no one *knows* anything at all.

This is where Plato came to the aid of Socrates. Perhaps experts were operating on principles they could not easily articulate, Plato suggested. Experts, at least in areas involving non-empirical knowledge such as morality and mathematics, had learned the principles involved in another life, Plato claimed, but they had forgotten them. The role of the philosopher was to help such moral and mathematical experts recall the principles on which they were acting. These principles would ground the knowledge of the skill. Such knowledge must be "fastened by the reasoning of cause and effect," and "this is done by 'recollection'."[26]

Knowledge engineers, generalizing the doctrine of recollection, would now say that the rules experts—even experts in empirical domains—use, have been put in a part of their mental computers where they work automatically. Thus, Feigenbaum tells us:

> When we learned how to tie our shoes, we had to think very hard about the steps involved....Now that we've tied many shoes over our lifetime, that knowledge is "compiled," to use the computing term for it; it no longer needs our conscious attention.[27]

On this modern version of Platonic recollection, the rules are there functioning in the expert's mind whether he is conscious of them or not. After all, how else could one account for the fact that the expert can perform the task?

Even Polanyi, who as far as I know was the first to see the importance of the difference between strict rules and maxims, falls for the cognitivist temptation to believe in unconscious rule-following. He tells us that:

> By watching the master...the apprentice unconsciously picks up the rules of the art, including those which are not explicitly known to the master himself.[28]

This is no lapse but an essential part of Polanyi's crypto-cognitivist view concerning the relation of rules and skilled action:

> In performing a skill we are therefore acting on certain *premisses* of which we are focally ignorant, but which we *know subsidiarily* as part of our mastery of that skill, and which we may get to know focally by analyzing the way we achieve success...in the skill in question.[28]

Polanyi is thus a phenomenologist concerning the awareness of skilled performers and a kind of cognitivist concerning the unconscious causes of skilled performance. He is not a full-fledged cognitivist, however, since he holds that the premisses from which the unconscious mind deduces actions are maxims, not strict rules.

Even a partially cognitivist view gives the knowledge engineer a venerable task. As Feigenbaum explains: "[A]n expert's knowledge is often ill-specified or incomplete because the expert himself doesn't always know exactly what it is he knows about his domain."[30] So the knowledge engineer has to help him recollect what he once knew.

> [An expert's] knowledge is currently acquired in a very painstaking way; individual computer scientists work with individual experts to explicate the expert's heuristics—to mine those jewels of knowledge out of their heads one by one....[T]he problem of knowledge acquisition is the critical bottleneck in artificial intelligence.[31]

Now, thanks to Feigenbaum and his colleagues, we have a new name for what Socrates and Plato considered the philosopher's task: *knowledge acquisition research*.[32]

But nothing has changed. When Feigenbaum suggests to an expert the rules the expert seems to be using, he gets a Euthyphro-like response: "That's true, but if you see enough patients/rocks/chip designs/instrument readings, you see that it isn't true after all."[33] And Feigenbaum comments with Socratic annoyance: "At this point, knowledge threatens to become ten thousand special cases."[34] Still, the doctrine of recollection assures the knowledge engineer that the principles must be there in the expert's mind. The expert is simply poor at remembering them.

One important question remains: Is Plato a partial cognitivist like Polanyi or a full-fledged computer-cognitivist like Feigenbaum? For Plato, are the unconscious principles that serve as premisses for deducing actions, maxims or strict rules? We have seen that Socrates seems to think that experts should be able to articulate strict rules that would enable *anyone* to share their expertise. That would make Socrates a full-fledged cognitivist, about skills at least. But the question of Socrates's intellectualism is subject to debate. Terrence Irwin, indeed, claims that Socrates is a pure intellectualist who holds that the rules experts use must fully capture their expertise. "Socrates demands an explicit understanding of the principles guiding (not necessarily explicitly) our application of 'just' or 'pious', so that we can see whether they are consistent and justifiable and can appeal to them to project our judgments to new cases."[35] Alexander Nehamas, however, has argued that Socrates, the son of a stone mason, would surely know that the rules a craftsman can articulate are not sufficiently explicit and complete to convey the craft to an outsider; rather, they would be Polanyian maxims. "Socrates, himself a statuary and a statuary's son (D.L. V.I.18), knew perfectly well that in ancient Athens the crafts were most often transmitted along with their 'secrets within the family from generation to generation'."[36]

Who is right, Irwin or Nehamas? As we have seen, Socrates does seem to hold in the *Euthyphro* that if he had Euthyphro's rules he too would be an expert on piety. Still, since this example comes from the domain of morality, where Socrates is surely to some extent already an initiate, it would be consistent with Nehamas's view that Socrates held that "only one virtuous person can recognize another."[37] The same holds for the definition of 'courage', since Socrates was acknowledged to have been a brave soldier. There is no evidence I know of that Socrates thought that an expert must be able to articulate rules that, in principle, would enable an outsider following them to become an expert.

Plato never explicitly holds this full-fledged cognitivist view either, but it seems to me that the doctrine of recollection as Plato uses it—not only as an account of why moral and mathematical experts cannot easily articulate their rules but also as an account of how they learned these rules in the first place—commits him to pure cognitivism. That is, it commits him to the view that the principles underlying all expertise must be strict rules, not merely maxims. If this is, indeed, the case, then Socrates and Plato would turn out to share the blame for cognitivism.

The argument is simple. If a soul starting from scratch is supposed to have acquired expert knowledge in a specific domain—mathematics, in Plato's example in the *Meno*—by seeing in another world the principles underlying the skill domain, then the principles learned must have been the rank beginner's strict rules applicable to features rather than the advanced beginner's maxims requiring prior familiarity with the skill domain. Since all the expert has and all the expert needs, according to Plato, are the principles he has acquired in the other world, before he had any familiarity with anything in this world, this comes to the claim that strict rules are sufficient to generate expertise, and this is precisely the pure cognitivist claim.

To sum up: What experts actually say suggests that they can at best formulate maxims that they presumably remember from when they were advanced beginners and that can be of use to others with some skill in the relevant domain. Socrates does not seem to have contradicted this observation, but he makes three moves toward cognitivism: (1) he rules out as trial and error skills such as cooking which are learned solely by apprenticeship; (2) he assumes that all true skills are based on principles; and (3) he claims that these principles must be expressible by those who exhibit the skill.

Plato, however, went further and provided an account of why experts were in fact so poor at articulating the principles underlying their performance, and in so doing he developed the theory of recollection, which committed him to the view that expertise is generated by strict rules.

Plato's account did not apply to everyday skills, like cooking. It took two thousand years before Leibniz boldly generalized the Platonic account to all forms of intelligent activity:

> [T]he most important observations and turns of skill in all sorts of trades and professions are as yet unwritten. This fact is proved by experience when, passing from theory to practice, we desire to accomplish something. *Of course, we can also write up this practice, since it is at bottom just another theory more complex and particular....*[38]

This sets things up for the last move, which makes the rules and the elements to which they apply explicitly syntactic. Thus, for modern cognitivists all mental activity is based upon rational calculations of the sort which can be implemented by a computer program.

4. Conclusion

A phenomenology of skill acquisition enables us to understand why the knowledge engineers from Socrates and Plato to Samuel and Feigenbaum have had such trouble getting the expert to articulate the rules he is using and why the rules that are dredged up, when programmed, never produce expertise. The expert is simply not following any rules! He is doing just what Socrates and Feigenbaum saw but denied he was doing, viz. discriminating thousands of special cases. The Socratic slogan, "If you understand it, you can explain it," should be reversed: Anyone who thinks he can fully explain his skill does not have expert understanding.

This explains why expert systems are never as good as experts. If one asks an expert for strict rules one will, in effect, force the expert to regress to the level of a beginner and state the rules he still remembers but no longer actually uses. If one programs these rules into a computer, one can use the speed and accuracy of the computer and its ability to store and access millions of facts to outdo a human beginner using the same rules. But no amount of rules and facts can capture the knowledge an expert has when he has assimilated his experiences of the actual outcomes of tens of thousands of situations.

The knowledge engineer will no doubt respond that in spite of appearances the mind and brain *must* be reasoning—making millions of rapid and accurate inferences like a computer. After all, the brain is not "wonder tissue," and how else could it work? But there *are* other models for what might be going on in the hardware. Researchers who call themselves "new connectionists" are building devices that operate somewhat like neural networks. In a connectionist machine what are stored are not representations at all but connection strengths between simulated neurons. Moreover, the states of the machine cannot always be interpreted as symbols representing invariant features of the skill domain.[39] Such "parallel distributed processing" systems can recognize patterns and detect similarity and regularity without using rules and features at all.[40]

Once one gives up the cognitivist assumption that skills must be produced by detached intellects making inferences and admits instead the role of involvement and intuition in the acquisition and exercise of expertise, one will have no reason to cling to the cognitivist program as a model of human intellectual operations. Those, like Feigenbaum, who boast that expert systems will soon have "access to machine intelligence—faster, deeper, better than human intelligence"[41]—can be seen as false prophets blinded by Socratic and Platonic assumptions, while Euthyphro, the alleged expert on piety who kept giving Socrates examples instead of rules, may turn out to have been a true prophet after all.

University of California, Berkeley

NOTES

1 J. A. Fodor, "Fodor's Guide to Mental Representation: The Intelligent Auntie's Vade-Mecum," *Mind* **94**, 1985, p. 91.

2 J. A. Fodor, *The Language of Thought* (New York: Thomas Y. Crowell Company, 1975), p. 44.

3 John Haugeland, *Artificial Intelligence, The Very Idea* (Cambridge: MIT Press/Bradford Books, 1985), p. 113.

4 Hobbes, *Leviathan* (Library of Liberal Arts, 1958), p. 45.

5 Leibniz, *Selections*, ed. Philip Wiener (New York: Scribner, 1951), p. 18).

6 Ibid., p. 20.

7 Ibid., p. 10.

8 Allen Newell and Herbert Simon, "Computer Science as Empirical Inquiry: Symbols and Search" in John Haugeland (ed.), *Mind Design* (Cambridge: MIT Press/Bradford Books, 1981), p. 41.

9 Edward Feigenbaum and Pamela McCorduck, *The Fifth Generation* (New York: Addison-Wesley Publishing Co., 1983), pp. 76-77.

10 Ibid., p. 18.

11 Ibid., p. 64.

12 Ibid., p. 179.

13 Interview with Arthur Samuel released by the Stanford University News Office, April 28, 1983.

[14] For a detailed account of this model and its defense, see Hubert Dreyfus and Stuart Dreyfus, *Mind over Machine* (Free Press, 1986).

[15] Michael Polanyi, *Personal Knowledge*, 2nd ed. (London: Routledge & Kegan Paul, 1962), p. 31.

[16] Of course, this description does not *prove* that unconscious rules and features are not involved, but the argument that they are is shifted to the cognitivist. Moreover, a what-else-could-it-be response is no longer possible since connectionist architectures have shown that a system can learn to recognize prototypical cases without feature extraction and rule-based inferences. Similarity recognition may well be basic.

[17] *Gorgias*, 501a.

[18] *Laches*, 190.

[19] *Gorgias*, 465a.

[20] Ibid., 501a. Aristotle, on the contrary, stays close to the normal everyday phenomenon and so sees the immediate, intuitive response precisely as characteristic of an expert craftsman. "Art (*technè*) does not deliberate," he says in *Physics*, Bk. II, Ch. 8.

[21] John H. Lunn, "Chick Sexing," *American Scientist* 36, 1948.

[22] *Apology*, 22d.

[23] Polanyi, op. cit., p. 88.

[24] *Euthyphro*, 6e3-6.

[25] *Laches*, 191e.

[26] *Meno*, 98a.

[27] Feigenbaum, op. cit., p. 55.

[28] Polanyi, op. cit., p. 53.

[29] Polanyi, op. cit., p. 162 (my italics).

[30] Feigenbaum, op. cit., p. 85.

[31] Ibid., pp. 79-80.

[32] Ibid., p. 79.

[33] Ibid., p. 82.

[34] Ibid., p. 82.

[35] Terrence Irwin, *Plato's Moral Theory* (Oxford: Oxford University Press, 1977), p. 65.

[36] Alexander Nehamas, "Socratic Intellectualism" in John Cleary (ed.), *Proceeding of the Boston Area Colloquium in Ancient Philosophy*, Vol. II (1986), p. 299.

[37] Ibid., p. 308.

[38] Leibniz, *Selections*, ed. Philip Wiener (New York: Scribner, 1951), p. 48 (my italics).

[39] Hubert and Stuart Dreyfus, "Making a Mind vs. Modeling the Brain: AI Back at a Branchpoint," *Daedalus*, Winter, 1988.

[40] See David Rumelhart and James McClelland, *Parallel Distributed Processing* (Cambridge: MIT Press/Bradford Books, 1986).

[41] Feigenbaum, op. cit., p. 236.

CHRISTOPHER SHIELDS

THE FIRST FUNCTIONALIST

Aristotle sternly rejects Platonic dualism, but nevertheless resists an ancient version of the identity theory of mind and body.[1] Equally dissatisfied with each of these alternatives, he endeavors to analyze the relationship between soul and body by employing the notions of form and matter introduced in the *Physics* and articulated and expanded in the *Metaphysics* (Meta.). Thus, in the *De Anima* (DA) we find the familiar claim that "the soul is a substance as form of a natural body having life in potentiality" (DS 412a19-22),[2] which Aristotle proceeds to illustrate by suggesting that the soul bears the same relation to the body as the shape of a candle does to its wax (DA 412b6-7). Both the claim and its illustration are in some ways obscure. But it is reasonably clear that these remarks represent Aristotle's attempt to provide a workable alternative to both Platonism and an austere form of the identity theory. Aristotle is perhaps the first philosopher to seek an account of mind-body relations which captures the insights of these theories while avoiding their individual shortcomings. Indeed, Aristotle self-consciously views his position in the philosophy of mind in much the same way contemporary functionalists view their own. Like the contemporary functionalist, Aristotle seeks a theory of the mental which avoids what he regards as the excesses of his predecessors: his preferred account would capture the supervenience of the mental on the physical without identifying mental state types with physical state types. Moreover, beyond having analogous historical vantage points, Aristotle and contemporary functionalists share deep theoretical commitments. So deep are these commitments that it is fair to regard Aristotle as the first functionalist.

1. Contemporary Functionalism

It is not at all fantastic that Aristotle should be a functionalist in the philosophy of mind, since functionalism as such need not be regarded as peculiarly contemporary.[3] This can be appreciated by considering a plausible textbook version of the history of the philosophy of mind in the latter twentieth century. In *Matter and Consciousness*, Paul Churchland

J-C. Smith (ed.), Historical Foundations of Cognitive Science, 19–33.

reasonably treats functionalism as a descendant of logical behaviorism which is sensitive to the failures of both behaviorism and the identity theory.[4] Recognizing that any identification of mental state types with neural state types is parochial in the extreme, the functionalist embraces a central behaviorist strategy: mental states are to be defined relationally, but with reference to other mental states as well as to causal inputs and behavioral outputs. Mental state types are to be identified with functional state types, which in turn are best thought of in humans as properties of neural states: neural state *n* has the property of being a belief just because it has a certain functional role.[5] States of non-human systems—computers, Martians, angels, or whatever—can in principle have that same functional role, and so, strictly speaking, can equally be belief states. On the functionalist analysis, then, whatever can be in a state that has the functional role associated with, e.g., pain, will be in pain. Thus, functionalism is neutral in terms of its ontological commitments: as far as functionalism *per se* is concerned, any system whatsoever can realize a given mental state type—so long as it can be in a state which has the functional role associated with that type.[6] Even so, nearly all modern functionalists, as a matter of fact, are materialists,[7] and expect the neurological sciences to specify which physiological states in humans realize the functional roles characteristic of pains, beliefs, and so on. Unlike identity theorists, however, their definitions of mental states do not by themselves require that mental states be physical states.

This history rightly regards the functionalist as reacting to two distinct pressures. First, the functionalist has recognized the need to structure an account of the mental compatible with a commitment to ontological materialism in such a way that its theoretical underpinnings will dovetail with the advances in the physical sciences. At the same time, the functionalist has sought to avoid the overly constricted view of the manner in which mental states are realized which sometimes accompanies these advances: mental states must be regarded as multiply realizable. It is important to recognize that these pressures are conceptual as opposed to empirical, in the sense that they are theoretical considerations not immediately informed by any ongoing empirical research. Although compatibility with the results attained in the physical sciences is a *desideratum*, the emergence of functionalism has not depended essentially upon advances in the neurological sciences. Indeed, as I will argue, there were analogous pressures and theoretical constraints already in place in antiquity, and these gave rise to an Aristotelian theory with striking parallels to the variety of functionalism espoused today.[8] In particular, we find in Aristotle the suggestions that: (1) mental states are multiply realizable; i.e., that entities which realize mental states are compositionally plastic; and more importantly, (2) mental states are definable in terms of their causal relations to inputs, outputs, and other mental states. (2) constitutes a minimal functionalist theory and so establishes that Aristotle accepts functionalism in the philosophy of mind. (1) is a consequence of (2), taken together with some plausible assumptions, and has been one of the governing insights of functionalism. That Aristotle embraces (1) further demonstrates his acceptance of the implications of this theory.

2. Functional Determination and Compositional Plasticity

Aristotle argues in many passages that in order to determine whether a given thing *x* belongs to a class or kind *F*, one must determine the function (*ergon*) of that class or kind, and then determine whether *x* has the capacity to fulfill that function. For example, in the *Meteorologica* Aristotle claims:

> All things are defined by their function: for [in those cases where][9] things are able to perform their function, each thing truly is [*F*], e.g., an eye, when it can see. But when something cannot [perform that function] it is homonymously [*F*], like a dead eye or one made of stone, just as a wooden saw is no more a saw than one in a picture. The same, then, [holds true] of flesh....(*Meteorologica*, 390a10-15)[10]

Here Aristotle directly asserts a thesis of Functional Determination:

> **(FD)** An individual *x* will belong to a kind or class *F* iff *x* can perform the function of that kind or class.[11]

According to FD, if the function of a knife is to cut, *x* will be a knife just in case *x* can cut. Aristotle sometimes makes this point by claiming that after a living organism has died, its parts retain their names only by courtesy, "only homonymously" in his words.[12] The import would seem to be, for example, that whatever—and only whatever—has the functional role of seeing will be an eye.

Not surprisingly, then, Aristotle suggests that "if an old man were to receive an eye of the right sort, he would see just as a young man" (DA 408b21-23), and then goes on to observe that the afflictions of old age result solely from the decrepitude of the body. The suggestion is an interesting one for Aristotle to make: he allows in principle that one could gradually replace bodily parts at will with others of the right sort, viz. ones capable of performing the functional role assigned to those parts, and still end up with a functioning human being.

Now, Aristotle does not specify what "the right sort" (*toiondi*) of material for an eye will be. But he thinks that the suitability of matter for realizing an F is determined by its ability to perform the functional role characteristic of F's.[13] Hence, what matters for Aristotle in determining whether something belongs to a class of kind F will depend, as a general principle, not so much on its specific material composition as on whether it has the functional role assigned to F's. This allows him to hold that certain types of things are, within certain specifiable limits, multiply realizable. That is, a commitment to compositional plasticity allows Aristotle to hold that entities or states can be realized in distinct types of materials, so long as those materials are functionally suitable. It is striking that Aristotle accepts this consequence not only for relatively innocuous cases, like tables and chairs, but, as I will argue, extends it to include human beings and mental states.

This is most readily appreciated by considering one important but often neglected feature of Aristotle's hylemorphic analysis of body and soul. Since, according to Aristotle, the soul is the form of the body, and the body its matter, one would expect the relationship between soul and body to be nothing more than a special case of the relationship between form and matter, and so to be explicable in terms of this more general relationship. And indeed it is, but, unfortunately, it is rather difficult to specify with any precision the relationship between form and matter. Aristotle's own reticence in addressing this question is surely part of the cause of the difficulty here. But there are passages where Aristotle endeavors to address this issue.

Especially relevant in this regard are *Meta*. VII 10 and 11, where Aristotle sets himself the problem of determining "Whether it is necessary that the account (*logos*) of the parts be present in the *logos* of the whole" (*Meta*. 1034b22-24). This may seem an oblique way of attempting to assess the relationship between matter and form, but it is made clear in the course of these chapters that among Aristotle's concerns is the question of whether forms must be regarded as *essentially* material. While investigating this topic, Aristotle draws a striking analogy between humans and circles:

> In those cases where things appear realized in [materials] differing in kind, e.g., a circle in bronze or stone or wood, it seems to be clear that none of these, the bronze or the stone, belongs to the essence of the circle, because it is separated from them; but in those cases which are not seen separated, nothing hinders them from being similar to these, [viz. circles]; just as if circles were always seen to be brazen, nonetheless bronze would not belong to the form [of circle], but it would be difficult to abstract this in thought. For example, the form of man always appears in flesh and bones and these sorts of parts; are these, therefore, parts of the form and account [of man]? No, they are but matter, but

because [man is not seen] coming to be in other [materials], we are not able to separate them.[14] (1036a-31-b7)

In this passage Aristotle argues that despite the fact that we always see human beings realized in flesh and blood, "nothing hinders" their being realized in other ways. The comparison with brazen circles is particularly helpful. It would be a completely contingent fact if all circles happened to be composed of bronze; it is similarly contingent that all human beings have been composed of flesh and bones. Human beings are compositionally plastic. Anything which is able to realize the states and activities characteristic of the human *ergon* will suffice: if an appropriate configuration of silicon chips could be in the complex state with the functional roles of beliefs, desires, and other thoughts, it would qualify as a human being. Thus, the only restriction Aristotle (sensibly) places on matter is that it be functionally suitable (DA 414a25-27). That Aristotle should have this insight is not surprising, given his acceptance of FD, the principle that a given *x* will be an *F* so long as it has the characteristic function associated with *F*'s. His view that the states and activities of human beings are functional states follows plausibly from his commitment to FD.[15]

3. The Multiple Realizability of the Mental

In a general way, then, Aristotle suggests that human beings can be realized in any functionally suitable matter. This is already sufficient to show that Aristotle has functionalist inclinations in the philosophy of mind. But his commitment is more precise and runs deeper than just this vague orientation would seem to indicate: Aristotle, not surprisingly, endorses the multiple realizability of the mental. This endorsement is not surprising since FD entails that whatever plays the functional role of a given mental state type will count as an instance of that state. Together with the plausible assumption that entities with physiological structures different from humans can have states with the same functional roles as the states of humans, this entails a rejection of any type-type identification of the mental with the physiological. But it is nevertheless noteworthy that Aristotle demonstrates his recognition of this consequence in two ways, by suggesting that both sensory and intellective psychic states are multiply realizable.

In his biological writings, Aristotle repeatedly notes that a given sensory based mental state can be realized in different physical systems. Different species of animals have radically different physiological systems, but can nevertheless equally be in the state of smelling a strong odor. For example, in *De Partibus Animalium*, after describing differences in the sense organs of different animals, Aristotle plausibly remarks that "in animals other [than man] these sense organs are well [arranged] in accordance with the peculiar nature of each" (657a10-12).[16] Dogs and humans will realize the property of smelling orchids differently. Consequently, for Aristotle, two tokens of the same mental type, such as smelling an orchid, will have different physiological realizations. Therefore, he rejects any straightforward (type-type) identification of such psychic states and (non-disjunctive) physiological properties and is committed to the multiple realizability of mental states that are sensory based.

Aristotle's commitment to the multiple realizability of the mental includes mental states beyond these, however. In *DA* III 4, he argues that intellective states narrowly construed cannot be realized in material systems at all; and in the *Meta*. XII 9 he describes the activity of an immaterial God as a rarefied sort of thinking. Just as we can think, so God can think; but, according to Aristotle, our thinking necessarily involves the use and manipulations of images, while God's evidently does not (DA 432a8-9; cf. *De Memoria* 450b30-451a3). Consequently, our thinking, for example, that Socrates is mortal will differ in its realization from God's thinking this same thought. More precisely, the state in us which has the property of being the thought that Socrates is mortal will differ in salient ways from the corresponding state in God, since in one case it will involve the use of images and in the other it will not. That this thought can be betokened in different entities in different ways reinforces the suggestion that Aristotle recognizes and accepts the thesis that mental states are multiply realizable. Hence, Aristotle rightly rejects one central tenet of a type-type identity theory and accepts an intuition which motivated functionalism, viz. the multiple realizability of the mental.

4. Functional Definability

Aristotle promotes the Functional Determination thesis and recognizes as a consequence of that thesis that mental states are multiply realizable. It follows directly from FD that a certain physiological state in humans will count as a token of the mental state type "being a pain" just in case it can perform the function associated with that type; i.e., just in case it realizes the functional role of that type. Aristotle embraces this consequence in an important and striking passage from an early portion of the *De Anima*. There Aristotle suggests that anger is best defined as "a certain sort of motion of such and such a body—or part or faculty [of a body]—by this on account of that" (DA 403a26-27). In other words, anger is best defined as a movement of a certain sort of body with a particular cause and for the sake of a certain end.[17] He not only specifies that mental states are to be defined relationally in terms of their inputs and outputs but is also willing to allow that such states may be construed as second-order properties, viz. properties of physical properties of the agent. Thus, Coriscus is angry when the blood around his heart boils *and* this state has an appropriate set of causal relations.[18] In so speaking, Aristotle explicitly advocates the use of functional definitions. Consequently, this passage is particularly important. It represents one of Aristotle's few explicit remarks about the appropriate form an adequate definition of a mental state must take: a token of a physical state type will count as a pain when that token has the functional role definitive of being a pain. There is evidently nothing peculiar about *anger* in this regard; rather, anger simply serves as an example of one sort of mental state. Consequently, Aristotle's point is really quite general. Mental states are to be defined relationally, with reference to causal inputs, outputs, and other mental states. By advancing this view, Aristotle satisfies the conditions for a minimal functionalist theory.

Aristotle's acceptance of FD, together with his acknowledgment of the multiple realizability of the mental, entails a rejection of any identification of physical state types with mental state types; it further entails that a token of a physical state type will count as a token of a mental state type when and only when that token has the functional role associated with that mental state type. Aristotle embraces these consequences and explicitly avows that mental states should be *defined* in terms of their functional roles. To this extent Aristotle has a clear appreciation of the motivations and consequences of functionalism in the philosophy of mind and self-consciously commits himself to that view.

5. Constrained Realizability

Although some Aristotelian commentators have been amenable to the suggestion that Aristotle is best represented as a functionalist in philosophy of mind,[19] some have resisted this interpretation, and one author has been positively hostile toward it.[20] Doubtful commentators have reasonably expressed two main concerns: (1) Aristotle seems to think that certain psychic states are *essentially* material states, so he cannot be a functionalist since this sort of materialism is incompatible with functionalism; and (2) Aristotle often invokes one mental state when defining another and so may not accept the commitment to topic neutrality required by functionalism; i.e., Aristotle may not suppose that every mental state is in principle definable in functional terms wholly devoid of mentalistic vocabulary.[21] Neither of these objections is defeating to my thesis, but an assessment of each in turn will serve to buttress the points I have made.

Deborah Modrak, for example, articulates the first of these concerns and concludes that "the differences between Aristotelian hylomorphism and modern functionalism tell against the assimilation of the one to the other,"[22] though she concedes that "Aristotle might be a functionalist in a broad sense."[23] The problem with any such assimilation, according to Modrak, is that functionalism identifies mental state types with functional state types, and this allows that anything which can be in a state which has the functional role associated with a given mental state will be in that mental state. Aristotle, however, "includes a physiological description in the full definition of a psychological state."[24] The suggestion is that the contemporary functionalist, Aristotle restricts the range of potential subjects for mental states.

Now the problem here cannot simply be, as Modrak suggests in one place, that "psychological states are regarded as functional states of certain bodily organs."[25] Clearly the contemporary functionalist believes this as well: *S* is in pain whenever she is in a neural state which has the functional role (together will the requisite beliefs and desires) of causing *S* to grimace in discomfort, pursue avoidance behavior, etc. So the concern cannot be just that Aristotle *in fact* thinks that being in this or that physiological state (e.g., being in the state of having the blood boil around one's heart) can realize a pain.

Perhaps the worry is deeper. The concern may be that Aristotle is himself guilty of precisely the sort of parochialism functionalism seeks to overcome. That is, functionalism is supposed to supercede the identity theory by allowing the possibility that mental states can be realized in physical systems radically unlike human beings and, as a matter of metaphysical or logical possibility, even in non-material systems. If this is the sort of concern Modrak means to express, the argument will be:

(1) given its commitment to multiple realizability, functionalism presupposes an ontological neutrality as regards the sorts of systems which can in principle realize mental states;

(2) if Aristotle incorporates physiological descriptions into his definitions of mental states, he tacitly rejects this presupposition;

(3) Aristotle does frequently incorporate physiological descriptions into his definitions of mental states; and

(4) therefore, Aristotle rejects a central tenet of functionalism and consequently cannot be characterized as a functionalist in the philosophy of mind.

Perhaps (3) is true; indeed in the canonical definition of anger adduced as evidence of his functionalism, Aristotle holds that there must be some bodily movement involved.[26] But this by itself provides no reason to suppose that (2) is true. Aristotle, like the contemporary functionalist, can maintain that mental states are *in fact* realized in humans in certain specifiable ways without regarding this as a consequence of his functionalism *per se*.[27] It is perfectly compatible to hold: (a) mental state types are identifiable with functional state types; (b) it is logically possible for any sort of system whatsoever to be in a particular state with the functional role determinative of a given mental state; and even so, (c) in fact only physical systems of specifiable sorts do come to be in states with such functional roles. Indeed, this would seem to be precisely Aristotle's position as regards all non-intellective psychic states. To this extent, Aristotle seems all the closer to the contemporary functionalist.[28] In any case, Aristotle's references to physiological states do not entail a rejection of functionalism.

Still, there is an important conceptual connection between a thing's function and its matter for Aristotle. Though many functional states are multiply realizable, the range of materials appropriate for realizing a given function is constrained or circumscribed by that function. Even when articulating FD in the *Meteorologica*, Aristotle notes that not just any matter can realize the functional property of being able to cut wood: "a wooden saw is no more a saw than one in a picture" (*Meteorologica* 390a14-15). Similarly, if being an eye involves seeing, and seeing involves receiving light waves, then only something so constituted as to be able to receive light waves, can be an eye. But these sorts of constraints do not entail a rejection of functionalism. Rather, they represent the plausible independent judgment that only structures of suitably sophisticated sorts can realize mental states.[29]

6. Reductionism and Aristotle's Functionalism

It may seem implausible to suppose, even given the parallels we have seen, that Aristotle has a fully articulated functionalist theory. There may be concern that the textual evidence underdetermines the issue and that consequently Aristotle is made to conform to a functionalist theory, even though he is non-committal about some crucial issues. One such crucial issue concerns whether Aristotle's functionalism is appropriately regarded as reductive. The concern, perhaps, is that Aristotle is not sufficiently self-conscious about the aims of his theory and so has not systematically considered all relevant alternatives. Although in some ways unfair to Aristotle, this sort of concern is not entirely unwarranted. Indeed, it should be noted that the minimal conditions for a functionalist theory specified above are conditions for what has been called "weak functionalism."[30] Weak functionalism is simply the view that mental states are definable in terms of their causal relations to inputs, outputs, and other mental states.[31] Weak functionalism is weak because it does not affirm that when mental state *b* is referred to in the functional definition of mental state *a*, there must likewise be a functional definition of *b* which contains no mental predicates at all, in order to avoid circularity and to provide definitions of mental states which are topic neutral in the sense that they contain no ineliminable mention of the mental. Strong functionalism is weak functionalism plus this explicit affirmation. (So, strong functionalism entails weak functionalism, but not conversely.) As one commentator has noted, Aristotle sometimes employs mental vocabulary in explicating mental states (e.g., *Rhetoric* 1378a31), and so it is not immediately clear whether he accepts strong as well as weak functionalism and so not evident that his view is reductive in the relevant sense.[32]

Even so, there is reason to suppose that Aristotle accepts strong functionalism. It is true that he nowhere explicitly insists that mental states must in principle be definable without reference to other mental states ("in principle" because it is open to the strong functionalist in practice to avail herself of mental states when defining mental states, so long as *some* of those states can be given functional definitions expunged of all mental vocabulary and the rest defined in terms of those). But to return to his canonical definition of anger at DA 403a26-27, it is instructive to note that here at any rate Aristotle provides a definition in wholly non-mentalistic terms. This, of course, does not by itself demonstrate that Aristotle's considered view will be that every mental state admits to such a definition or even that every mental state is definable in terms of other states which admit of such a definition. Nonetheless, it shows that in his most self-conscious account of the appropriate form for definitions of the mental, Aristotle avoids the circularity problem in the manner of a strong functionalist. Thus, the textual evidence, although inconclusive, suggests that Aristotle inclines toward strong functionalism. His being a strong functionalist is compatible with his sometimes invoking mentalistic vocabulary in defining mental states, and in his most explicit account of the proper form of definitions of mental states, he takes care to avoid mentalistic vocabulary. This practice constitutes evidence, albeit inconclusive, that Aristotle would be willing to embrace strong as well as weak functionalism.

7. Conclusion

Aristotle's functionalism in the philosophy of mind is not articulated with the same sort of precision we have come to expect from contemporary theorists. Even so, it develops out of a highly systematic analysis of kinds and kind membership and shares both substantive and methodological principles in common with contemporary functionalism in the philosophy of mind. Even though the contours of Aristotle's functionalism are teleological in character, one need not embrace teleology as such to accept his brand of functionalism. But there are theoretical advantages to Aristotle's expository strategy. Aristotle's methodology in the philosophy of mind carries the suggestion that only by working to situate functionalist theories within broader, more systematic analyses of property and kind individuation will philosophers provide the deep theoretical support for functionalism required to establish its plausibility as a workable analysis of the mental.[33]

University of Colorado at Boulder
Seminar für Klassische Philologie,
Universität Mainz

NOTES

1 A great deal of work has been done on the vexed matter of Aristotle's ontological commitments in the philosophy of mind. See my "Body and Soul in Aristotle," *Oxford Studies in Ancient Philosophy*, Vol. VI (forthcoming), for a discussion of this literature.

2 All translations from the Greek are my own.

3 See, e.g., Ned Block, "Introduction: What is Functionalism?" in N. Block (ed.), *Readings in Philosophy of Psychology*, Vol I (Cambridge: Harvard University Press, 1980), p. 172: "The functionalisms of philosophy of psychology are, however, a closely knit group; indeed, they appear to have a common origin in the works of Aristotle."

4 Paul Churchland, *Matter and Consciousness* (Cambridge: The MIT Press, 1984), Chapter Two. Cf. Block, Readings, Vol. I, pp. 268-275; Jerry Fodor, *Psychosemantics* (Cambridge: The MIT Press, 1987), pp. 67-68; and Georges Rey, "Functionalism and the Emotions" in A. O. Rorty (ed.), *Explaining Emotions* (Berkeley: University of California Press, 1980), pp. 164-165.

[5] This analysis of functional properties derives from Stephen Schiffer, *Remnants of Meaning* (Cambridge: The MIT Press, 1987), Chapter Two. Of course, not all functionalists agree on precisely this analysis, but such differences as there are will not affect the present discussion.

[6] See, e.g., Hilary Putnam, "Philosophy and Our Mental Life," in *Mind, Language, and Reality* (Cambridge: Cambridge University Press, 1975), pp. 291-303. Putnam rightly suggests that "two systems can have quite different constitutions and be functionally isomorphic" (p. 292) and cites Aristotle favorably in claiming:

> ...what we are really interested in, as Aristotle saw [citing DA 412a6-6b], is form and not matter. *What is our intellectual form?* is the question, not what the matter is. And whatever our substance may be, soul-stuff, or matter or Swiss cheese, it is not going to place any interesting restrictions on the answer to this question. (p. 302; his emphases)

[7] See, e.g., Sydney Shoemaker, *Personal Identity* (Oxford: Blackwell, 1984), p. 139:

> It is because I believe that there is an acceptable account of the mental and personal identity that is compatible with materialism, and because I believe there are independent reasons (roughly, the explanatory success of the physical sciences) for believing materialism to be true, that I am a materialist.

[8] This is not to suggest that all who call themselves functionalists are in agreement about the commitments of their theories. One main source of disagreement concerns whether (with "analytical functionalism") functional definitions are formulated *a priori* or (with "psychofunctionalism") are products of an empirical investigation. See Sydney Shoemaker, "Some Varieties of Functionalism," *Identity, Cause, and Mind* (Cambridge: Cambridge University Press, 1984), p. 272. Also cf. Block, *Readings*, Vol. I, pp. 171-184 and 268-306. The account of functional ism I have just provided is intended only to sketch the minimal commitments any theory must satisfy to be regarded as functionalist. Although the theory I attribute to Aristotle does not move significantly beyond satisfying these minimal conditions, he would seem inclined toward what Shoemaker calls analytical functionalism.

[9] Translations from the Greek incorporate brackets where I have expanded Aristotle's characteristically terse prose for clarity of meaning.

[10] Several points must be made about this passage from the *Meteorologica*: (1) it comes from a work whose authenticity has been doubted; (2) some translators regard it as locally qualified (it occurs in a passage concerning homoeomerous bodies, and some, e.g., Lee, suppose it concerns these entities only); and (3) it states in an unrestricted form what elsewhere Aristotle tacitly restricts to classes with clear functions (*erga*), e.g., natural kinds. But these observations do not undermine its utility here, since: (1) the *Meteorologica* is not spurious, or at any rate Book IV is genuine (cf. Konrad Gaiser, *Theophrast in Assos: Zur Entwicklung der Naturalwissenschaft zwischen Akademie und Peripatos* (Heidelberg: Carl Winter, 1985), pp. 64-65, with review by Troels Engberg-Penderson, *Classical Review*, N. S. XXXVII, 1987, pp. 53-57); (2) the language Aristotle uses suggests that he intends a general rather than qualified principle, and he elsewhere in similar contexts extends it beyond this restricted class, e.g., *De Partibut Animalium* (PA), 640b18-23 (and even within this passage it is unclear why Aristotle would mention the eye if he were concerned only with homoeomerous bodies); and (3) human beings and mental states will certainly be included in any restricted formulation of the claim that function determines kind membership. Cf. *Meta.* 1029b23-1030a17 and *Nicomachean Ethics*, 1098a7-8.

[11] FD is a thesis about class or kind membership. But it can be translated without violence into a thesis about state types or properties:

> (**FDS**) *x*'s being *F* is a token of a state type being-*F* iff *x* is in a state which has the functional role associated with being-*F*.

[12] This is a common claim of Aristotle's. Cf. DA 412b17-22, Meta. 1036b31-33, and esp. PA 640b34-641a5: "And yet a dead body has exactly the same configuration as a living one; but for all that it is not a man. So also no hand of bronze or wood constituted in any but the appropriate way can possibly be a hand except homonymously. For like a physician in a painting, or like a flute in a sculpture, it will be unable to perform its function. Precisely in the same way, no part of a dead body, e.g., as its eye or its hand, is really an eye or a hand." T. H. Irwin, "Homonymy in Aristotle," *Review of Metaphysics* **34** (March, 1981), pp. 523-544.

[13] See DA 403b3 (with Hicks's note, *Aristotle: De Anima* (Cambridge: Cambridge University Press, 1908), p. 202), 412b11, and esp. 414a25-27: "...the actuality of anything naturally comes to be in what is present potentially [as that thing], i.e., in suitable manner."

[14] See W. D. Ross, *Metaphysics*, Vol. II, p. 207, n. 14, for the motivation for translating this difficult passage as I have. Shortly after this passage Aristotle rejects the position of one Socrates the Younger (*Meta.* 1036b25ff.) and insists that some matter must figure into the definition of man. I take him to mean only that since in humans non-intellective states are realized in functionally suitable matter, such matter must figure into the definition of man. See esp. 1036b30; cf. PA 645b15-22.

[15] In this sense, Aristotle's functionalism derives from a broadly based analysis of properties and kind membership. It is striking that on his account, mental states will not be peculiar in being susceptible to functional definitions. At least one contemporary functionalist, Shoemaker, has adopted this more comprehensive approach advocated by Aristotle. Shoemaker suggests that his commitment to a causal theory of properties (CTP; roughly the view that a property is identified by its causal potentialities) entails weak functionalism, where weak functionalism is the view that mental properties, like all other proper ties, are individuated in terms of their causal features. Though CTP neither entails nor is entailed by FD, the positions are analogous in that both Aristotle and Shoemaker suggest that functionalism in the philosophy of mind tumbles out of a broader analysis of properties and state types. The principal difference is that we find in Aristotle no clear attempt to differentiate strong from weak functionalism. (See section VI below.) Cf. Shoemaker, "Causality and Properties," and "Some Varieties of Functionalism," sections V-VIII together with McGuinn's review of Shoemaker's *Identity, Cause, and Mind, Journal of Philosophy* **LXXXIV** (April, 1987), p.230.

[16] See also PA 656a35-37, 658b27-659b19; DA 421b9-422a6.

[17] The Oxford translation of J. A. Smith has: "Consequently, their definitions ought to correspond, e.g., anger should be defined as a certain mode of movement of such and such a body (or part or faculty of a body) by this or that cause and for this or that end." Hicks has: "Hence, they must be defined accordingly: anger, for instance, as a certain movement in a body of a given kind, or some part or faculty of it, produced by such and such a cause and for such and such an end." Both Smith and Hicks are right to expand their translations to make explicit that Aristotle intends the *causal* relations of the bodily state to be essential to the definition. If we generalize Aristotle's definition here, we end up with the claim that a bodily state will realize a (non-intellective) psychic state only when that state has certain specifiable relations to inputs, outputs, and other mental states. Cf. *Rhetoric* 1378a31 for a parallel definition where Aristotle, significantly, suppresses the

reference to the bodily state and makes explicit the role of other mental states in the definition of anger.

18 Indeed, Artistotle suggests at DA 403a22-24 that simply being in a token of a certain physiological state type normally associated with a given mental state type is not sufficient for being in that mental state. Thus, having one's blood boil is not sufficient for anger. This suggests: (a) that Aristotle rightly resists any type-type identification of the mental with the physical, and more importantly, (b) that a given physiological state will have the property of being a certain mental state only when that state has the appropriate causal inputs and outputs, as when it plays the functional role constitutive of anger (cf. Meta. 1036b25-27); and (c) that Aristotle marks a rudimentary distinction between what Shoemaker calls core and total realizations of mental states. See "Some Varieties of Functionalism," section II. Cf. also Edwin Hartman, *Substance, body, and Soul* (Princeton: Princeton University Press, 1977), p. 147.

19 See, e.g., R. Manning, "Materialism, Dualism, and Functionalism in Aristotle's Philosophy of Mind," *Apeiron* **IXX** (1985), pp. 11-23. In an unpublished manuscript, "Changing Aristotle's Mind," M. Nussbaum and H. Putnam establish some similarities. An earlier discussion of some of these issues occurs in A. Kosman, "Perceiving That We Perceive: *On the Soul* III, 2," *Philosophical Review* **84** (1975), pp. 499-519. Two philosophically sophisticated discussions of this topic are Edwin Hartman, *Substance, Body, and Soul*, Chapter Four, section V, and S. Sparkman, "Aristotle's Philosophy of Mind: Does He Give a Functionalist Account?," M. A. Thesis, University of Arizona, 1987.

20 H. M. Robinson, "Mind and Body in Aristotle," *Classical Quarterly* **28** (1978), pp. 105-124. A more concessive rejection of my thesis is to be found in D. K. W. Modrak, *Aristotle: The Power of Perception* (Chicago: University of Chicago Press, 1987), Chapters One and Two.

21 Robinson launches a series of other objections, the most plausible of which concerns the role of *qualia* in functional definitions. Robinson duly notes that Aristotle offers what appear to be functional definitions of mental states but argues that although Aristotle believes that "every biological process has a function which explains why it exists," he has no interest in statements which provide a "reductive account of the experience, feeling, or sensation of [e.g.] pain." Robinson's objection here is obscure, but presumably he understands functionalist definitions to be reductive in the sense that they make no essential reference to their qualitative character, viz. to *qualia*, and supposes that Aristotle would resist any such approach. If so, his characterization of functionalism is apt, but his presentation of Aristotle is unconvincing. First, he does not present a single passage which supports the contention that Aristotle would resist such an approach; and second, we have seen passages (e.g., DA 403a26-27) in which Aristotle proceeds by offering definitions of mental state types which make no reference at all to their qualitative character. Consequently, Robinson's objection is not at all compelling.

He also offers the following objections: (1) there is no positive evidence that Aristotle accepts functionalism; (2) Aristotle has an alternative theory incompatible with functionalism; (3) Aristotle has worries about his theory which no functionalist could have and so "is enmeshed in a more convoluted theory than functionalism;" and (4) Aristotle's distinction between potentiality and actuality preempts a functionalist interpretation. My responses, in short, are: (1) is false, and evidence has been presented to show that this is so; as I argue below (2) and (3) involve the mistaken assumption that a functionalist cannot *independently* articulate the physiology of perception in humans; and (4) is asserted without discernible argument.

22 Modrak, p. 28.

23 Modrak, p. 184, n. 18; cf. also p. 187, n. 53, and p. 190, n. 26.

24 Modrak, p. 28.

25 Modrak, p. 28.
26 But cf. *Rhetoric* 1378a31, where Aristotle makes no reference to such movement.
27 That most functionalists are materialists suggests they have, for independent reasons, come to believe that only physical systems can be in states which realize the functional roles associated with mental states. Cf. note 7 above. But a functionalist *qua* functionalist is not a materialist. As Block succinctly observes in his "Introduction: What Is Functionalism?": "Functionalists can be physicalists in allowing that all the entities (things, states, events, and so on) that exist are physical entities, denying only that what binds certain types of things together is a physical property" (p. 172). Cf. Fodor, *Psychosemantics*, p. 68.
28 Aristotle himself provides a good example of an independent motivation for supposing that certain sorts of states, viz. material states, are incapable of realizing thought. See DA III 4. Of course, the soundness of his arguments for this conclusion is not at issue here. The point is that Aristotle, like the contemporary functionalist, may have independent reasons for restricting the range of actual systems which can realize a state with this or that functional role and may do so without undermining a commitment to functionalism. In any case, it should be clear that for Aristotle any *general* definition of mental states cannot include essential references to bodily states.
29 The point may also be put as follows. No functionalist commits herself to the perverse claim that just anything can be in pain. No functionalist is constrained to say, e.g., that a carving knife can feel pain when made to cut an especially tough piece of beef. She will endorse the conditional, that *if* it can be in a state with the functional role definitive of pain, the knife will be in pain; but she will rightly point out what is obvious, that the knife is not sufficiently sophisticated in terms of its structure to be in a state with that functional role. Similarly, Aristotle is free to maintain, despite his functionalism, that only certain sorts of material systems can realize mental states.
30 Shoemaker, "Some Varieties of Functionalism," section I. Shoemaker is clear that he takes it as a task of functionalism to provide non-circular definitions of the mental; but not all functionalists have agreed with him.
31 The account of functionalism one often sees advanced is, then, properly weak functionalism. See, e.g., Ned Block, "Are Absent Qualia Impossible?", *Philosophical Review* **89** (1980), p. 257; cf. Georges Rey, "Functionalism and the Emotions," esp. p. 165: "Suffice it to say that philosophers [viz. functionalists] gradually came to find mental states inextricably intertwined, in such a way that no one state could properly be identified without mention of the others, no one of them, there fore, being 'reducible' to nonmental terms." Consequently, even if Aristotle's functionalism is non-reductive, this will not distinguish him from many practicing contemporary functionalists.
32 Cf. Modrak, p. 28. Note, however, that Modrak regards this as a reason for supposing that Aristotle is not a functionalist at all. I would prefer to say that without a clear commitment to topic neutrality, Aristotle is a weak functionalist (see note n. 31). In any case, there may be some evidence to suppose that Aristotle accepts strong as well as weak functionalism.
33 I gratefully acknowledge the Alexander von Humboldt-Stiftung for its generous support. I have profited from conversations with S. Sparkman, J. Anderson, and P. Mitsis in preparing this essay. I am also pleased to thank J-C. Smith and J. Whiting for instructive written comments.

MARTIN M. TWEEDALE

MENTAL REPRESENTATIONS IN LATER MEDIEVAL SCHOLASTICISM

Starting from Aristotle's psychological theories as interpreted by the immensely influential Arab thinkers, Avicenna and Averroes, the scholastics of the high middle ages developed a program for understanding cognition in ways that, I believe, are philosophically interesting. But the whole topic is an area in the history of philosophy and science that languishes for lack of attention by scholars who are sufficiently sensitive to the significance of the problems the scholastics faced. The present mentalist turn in cognitive science, however, may help to awaken an interest in this side of scholastic philosophy. Both programs, the current computational mentalism and the scholastic, are very much concerned with mental representation and think more deeply about this subject than other approaches to cognition in Western philosophy and psychology. Of course, this concern with representations is embedded within two very different programs, and it will be part of my purpose here to state their differences explicitly. But in addition I shall try to delimit the area of common concerns.

1

First, let us get a picture of the general program more or less shared by all the scholastics from the mid-thirteenth century to the early decades of the fourteenth. The following will serve as a very brief summary. Cognition manifests itself first and foremost in acts of apprehension (*cognitiones*) carried out by each of the various cognitive faculties of the soul: the five external senses, several internal senses,[1] including the imagination, and the intellect. These acts have objects, viz. whatever it is that is directly apprehended, and these objects are in varying degrees partial causes of the acts. Corresponding to any act there is a readiness on the part of the agent to perform this act, and that readiness must exist prior to the act and may very well continue to exist after the act is finished. This is the Aristotelian distinction of first actuality (the readiness) from second actuality (the performance).[2] Behind the readiness lies the power of the faculty itself.

J-C. Smith (ed.), Historical Foundations of Cognitive Science, 35–51.

How does apprehension of objects take place? Aristotle had been influenced by a view at least as old as Empedocles, that if we are to apprehend things in the outside world then those things must produce something like themselves in the cognitive faculty. Aristotle himself modified this so that the first actuality of the object *qua* object was *identical* with the first actuality of the faculty; i.e., with its very act of apprehending the object.[3] But this theory, with its quasi-idealist tendencies, does not seem to have been recognized in the middle ages. Instead we have the view that the object external to the soul produces in the cognitive faculty a "species," a sort of likeness of itself.[4] The color of the leaf I am looking at, for example, creates in my eye something like itself, and this immediately brings about my seeing the object, the color of the leaf. It is important to note here that everyone agreed it was not the likeness itself which was seen but what that likeness was a likeness of.[5] Direct realism in epistemology co-existed happily for the most part with the belief that cognition occurs through a causally mediating entity.

The external senses are basically passive faculties in that what we call their acts are really their being actualized, or moved from a state of readiness to a state of act by something acting on them. On the other hand, imagination, the faculty of apprehending in a sensory fashion what may very well not be present in external reality, is not so passive as that. It can have an act that is clearly not just an automatic response to the external presence of an object. Yet some object must be present to it in some way; i.e., there must be a likeness in that faculty of what is imagined. The scholastics were led then to propose that species of sensed objects exist in the imagination, as well, and in fact are produced there by the species in the sense organs. But unlike the species in the sense organs, which produce acts of sensory apprehension whenever and as long as they are present, those in the imagination only produce a readiness on the part of the imagination to engage in its apprehensive act. These "phantasms," as they were called, are stored for use whenever needed.

The Aristotelian view tied the sense faculties very closely to their organs, to their physiology, and this made possible a fairly literal interpretation of the view that the species are likenesses of external objects. For example, the visual species can be viewed as a little colored image that is propagated through the air and comes to exist in the eye. In general what happens is that the property of external things that is perceived is recreated in the sense organ without its actually being a property of that organ. What we have in the organ is a species which has a color, or a sound, or a taste, etc., in the way that a mirror image has a color. Even the imagination supposedly involves some organ which has this capacity to receive an image—just which organ this is was a matter for speculation.

It is when the same sort of theory is applied to the intellect that a literal similarity becomes difficult to understand, for all the scholastics accepted that the intellect has no material organ.[6] In fact, how intellectual cognition can in any way be viewed as analogous to the sensory is something of a question, but the Aristotelian program insisted on just such an analogy. The intellect has apprehensive acts whose objects are universals or "quiddities," what we would describe as properties considered on their own in isolation from the particular subjects they are properties of. Besides apprehending such entities, the intellect can combine them to form propositions (*complexa*) and then go on to affirm or deny such propositions. Everyone assumed that intellectual acts of apprehension must precede any acts of affirmation and negation, as well as the more dispositional intellectual states of knowing and believing and in general all those acts and states which we now call propositional attitudes.

The distinction of intellectual from sensory apprehension was crucial to all the scholastics. The senses apprehend properties too, but always in some completely determinate form (e.g., color is apprehended but only as some determinate shade or other) and as belonging to a particular subject with all that subject's other sensible properties. The senses and even the imagination are incapable of abstractly considering a property; i.e., considering it just as it is in itself and apart from the contingencies of some particular instantiation. More than any other thinker, Avicenna was responsible for this view of the intellect as the apprehender of "quiddities" without any of the trappings found with them in their actual existence,[7] and the theme was repeated in one way or another by all the scholastics from the mid-thirteenth century onward.

But then, how can an abstract "quiddity" be an object for any act if all that exists are quiddities enmeshed in the particulars of sensory experience? Here proposals differed quite a bit, but they all attributed a role to each of three entities: (1) the phantasm in the imagination, (2) the agent or active intellect, and (3) the possible or passive intellect. Everyone agreed that in contrast to the more or less passive way in which the senses and the imagination have objects (the sensory species and the phantasm are "impressed" from outside the faculties), the intellect is relatively active in this process. The phantasm in the imagination cannot on its own serve either to present the intellect with its object directly or to create some species in the intellect which will present to the intellect its object in the way the phantasm presents to the imagination its object. Something is needed to abstract the quiddity from all that is accidental to it but included in the phantasm of whatever instantiates that quiddity. The something proposed was the rather mysterious "agent intellect," which for Avicenna and Averroes was a superhuman entity but which most scholastics thought of as a creative power of each human soul.

Although the influence of external objects via the imagination is not sufficient for intellectual apprehension, neither is this apprehension an act the intellect can perform while not being in any way affected from the outside. If we had no contact through the senses with the outside world, our intellect, at least in this life (the scholastic theologians always had to keep in mind the activities of disembodied souls), would have precious little in the way of objects to apprehend. To account for this passive side to our intellectual life, the scholastics posited the passive or possible intellect, which is affected by the phantasms as well as by the active or agent intellect in such a way that apprehensions of quiddities could occur.[8]

It is at this point that a controversy developed in later scholastic philosophy, one which I want to sketch in some detail since it is revealing of how the main scholastic thinkers conceived of their program. Most thinkers in the latter part of the thirteenth century accepted the view that the agent intellect and the phantasm work together in some way to produce a species in the possible intellect and that this "intelligible species" enables the intellect to have an object for its apprehensive acts (*intellectiones*) analogous to the way a phantasm enables the imagination to have an object for its acts of imagining. The talk of similarity remained to some extent, but on account of the immateriality of the intellect this tends to give way to talk of the species representing the quiddity. In fact it is hard to see at this point how being similar to the object could mean anything more than representing it.

With such talk of representation also came the vocabulary of intentional or represented existence (*esse intentionale vel repraesentatum*). As long as an object is represented it has this sort of existence, and it can continue to have this sort of existence even when it has ceased to have real existence (*esse reale*). It is then a mere objective or represented being (*ens objectivum vel repraesentatum*), as opposed to a real or "subjective" being (*ens reale vel subjectivum*). (An objective being is like a character in a story. It may be that this character is some real person, but even so, the character may continue to exist as a character in the story after that person has ceased to exist in reality.) As with the imagination, then, the intellect can have objects which do not have real existence but only represented existence. Such existence, however, is always dependent on the real existence of a representation, viz. the species. In this case the intentional or represented existence was often spoken of as "shining forth" (*relucet*) in the species.

One of the first to dissent from this way of positing intelligible species was Henry of Ghent, a very influential figure in theology and philosophy at Paris in the 1280's. Henry's view[9] was that the phantasm, once "illumined" by the agent intellect, could serve to present the intellect with its object, so that there was no need for a specifically "intelligible" species prior to acts of intellectual apprehension. Henry, in fact, was doubtful that it made any sense to talk of a phantasm, a physical phenomenon, being even a partial cause of an immaterial or non-physical species in the intellect. If there were intelligible species, they were created by the intellectual acts of apprehension themselves (which would account perhaps for the intellect's not needing after a while to call up phantasms in order to consider quiddities already considered before). According to Henry, the whole business of receiving species from outside was something appropriate only to cognitive faculties that have a physiology which can be affected by an external physical object operating through some medium between it and the organ. The intellect is obviously not this sort of faculty.

John Duns Scotus, who had listened to Henry's lectures at Paris and clearly thought of him as the foremost thinker of the day, took up an opposing view on this question, as on so many others.[10] Scotus's objections were basically the following four.

(1) The phantasm, since it certainly represents what it represents as a singular, cannot represent it as a universal as well, no matter what "light" we may care to say the agent intellect casts on it. This is simply because one and the same representation cannot represent something under opposed aspects. But everyone agreed that the intellect has universals *qua* universals as at least some of its objects. Consequently, some other representation or species is required to represent the universal *qua* universal.[11]

(2) The agent intellect performs some real operation here, not just an intentional operation, and every real operation ends up producing something real. But it will be of no use for it to produce something real in the imagination, for that will only be some physical state that will no more be capable of affecting the intellect than the original phantasm. Neither will it suffice for the agent intellect to produce the "actual universal" merely by making the phantasm represent it, for such a represented universal would only have intentional existence, not real existence. The only alternative seems to be production of something real in the possible intellect, something by which the possible intellect has its object present to it. This is the intelligible species.[12]

(3) If we depended on the phantasm to present the intellect with its object, it would not be possible to apprehend a highly determinable

universal in its full "indifference" without also apprehending some determinate universal under the determinable one, so in the end apprehending some singular thing. But we can in fact apprehend a genus without apprehending any of the species under it, let alone any of the singular things contained under those species.[13]

(4) If the intellect could not in principle have its object present to it without depending on the operation of some other cognitive faculty, it would be a particularly imperfect faculty as compared with the sensory powers. But the intellect generally seems to be a more perfect power than these.[14]

This summary shows that the basic problem for Henry of Ghent is that the end output of all the sensory cognitive powers is a representation that is not the sort of representation required for intellectual operations. Sensory output is incapable of representing an abstraction, or at least an abstraction of the sort which we take to be referred to by words with a highly generic or determinable meaning. Scotus's objections only get off the ground if we grant that the intellect is not just another way of apprehending the same things the senses apprehend, but has a set of objects of its own, viz. the abstractions which he calls the "actual universals."

Scotus's own position was, then, that the phantasm in the imagination and the agent intellect work together in some way to produce in the possible intellect a representation of an abstract quiddity.[15] This process occurs prior to the intellect's engaging in an act of apprehending that quiddity and also, of course, prior to any acts with regard to propositions, since propositions are complexes of quiddities. What we usually call knowledge is a habit that results from acts of apprehending the truth of propositions, but in a potential sense that knowledge is already present as soon as the quiddities are represented in the intellect. The basic cognitive work is done by subconscious operations on the output of the sensory faculties; i.e., by the agent intellect, whose job it is to take what is represented by the senses only as a particular enmeshed in many extraneous features and get it represented as a universal abstracted away from everything accidental to itself. This takes it into the form that is suitable to intellectual knowledge. The intellect's conscious activity consists in apprehending these represented objects, then combining them into propositions and making judgments as to the truth of those propositions.

This whole theory was attacked in the early fourteenth century by William of Ockham.[16] While admitting that little can be strictly proved in this field, he claimed that a theory in which species were entirely eliminated was the most plausible proposal, in that it explained all the facts while positing the fewest types of entities.[17] Ockham constructed an alternative to Scotus's theory by adopting another side of it, specifically, Scotus's distinction between intuitive and abstractive apprehension of objects. The former occurs when the object existing outside the faculty affects it in such a way that the knower realizes that the object is present and existent; the latter occurs when the external object does not itself bring about the apprehension and the knower is not aware in that apprehension of the external existence of the object.[18] For intuitive apprehension Scotus allowed that a species in the faculty is not necessary to make the object be present, but it is for the abstractive since without the species the object would not be present in any way at all. Scotus also held that the nature of the intellect allows it to have intuitive apprehensions, just as the sensory faculties have, even if in this life it is prevented from exercising this capacity. Finally, he thought that the nature of the intellect allows it to apprehend singulars directly, but once again this is not a capacity it can exercise in this life.[19]

Ockham reconstructed the theory of cognition using these notions.[20] The intellect, he said, primarily apprehends singulars and can do this either intuitively or abstractively.[21] The intellectual intuitions of sensible singulars are carried out through the offerings of the senses, but they are nonetheless intellectual. These intuitive apprehensions immediately and simultaneously cause abstractive apprehensions of the same singulars, and these in turn cause a "habit" which remains in the intellect after the acts of apprehension have ceased. It is this habit which enables the intellect to renew at will the abstractive apprehensions in the absence both of intuitive apprehension and of its object.

The intellect also apprehends universals, but universals have only a mental existence.[22] Earlier in his career Ockham thought that the universals might be images or "fictive" entities the mind creates for itself, but his mature view was that the acts of intellectual apprehension are themselves the universals. Some acts of abstractive apprehension are equally of each of many distinct singulars, all of which are apprehended by the one act. When the intellect reflects on a universal, it is only reflecting on the way a certain act apprehends its many singular objects. The objects of science (in the Aristotelian conception) are universals and are consequently mental entities, but they are mental entities which themselves are connected to extra-mental particulars.[23] These acts or concepts are then thought of by Ockham as natural *signs* of the entities they apprehend, and thus they provide the terms for a mental language. Spoken language derives its ability to signify conventionally by the association of its terms with terms in the mental language.[24] Out of the simple mental terms, propositional complexes which are the objects of judgments, assent, dissent, belief, etc., can then be formed.

As should be clear, species have entirely disappeared from this account. Instead, singular entities outside the mind are the objects for the basic intuitive and abstractive apprehensions, and the rest require no more than the habit established by the former. Also, Ockham substantially dismantled the distinction of the active and passive intellects, for once we drop the idea that universals are the intellect's primary objects, the need for an active intellect to mediate between the sensory faculties and the passive intellect disappears.[25]

The dispute which Ockham's theory engendered was long and intricate and has not been well charted by scholars. I shall mention here only some of the criticisms developed by John of Reading, a Scotist writing in the 1330's.[26] John sensed that Ockham's heavy reliance on habit as the basis for abstractive apprehensions once the object is no longer present may well be a weak point in the theory. He said that a habit must serve two functions in Ockham's theory: (1) it must cause later acts of abstractive apprehension, and (2) it must represent the object of such acts.[27] Now a habit is something that is caused by particular examples of the sort of act it inclines the agent toward, and it is strengthened by repetition of such acts. But a cause is not itself strengthened by its effects. It follows, then, that the habit should not be treated as a cause of the acts it inclines the agent toward; i.e., of the acts of abstractive apprehension which Ockham's proposal says it does cause.

Further, a habit caused by acts of apprehension is a likeness of them if it is a likeness of anything. That is, if the habit represents anything it represents those acts that cause it. But the acts of abstractive apprehension that come after the formation of the habit do not have as their objects other acts of apprehension like themselves. Rather, their objects of apprehension are the same as the objects of the original acts that preceded and caused the habit.

This doubt could be expressed as follows. After we have had an act of apprehension, we can apprehend later both the object of the original act and also the original act itself. Now are both these later apprehensions due to the same habit left by the original act or to different habits? If Ockham says it is due to the same habit, then it would seem that our knowledge of, say, a rose is the same intellectual disposition as our knowledge of the act of apprehending the rose, and thus knowledge of the external world and knowledge of our own acts would be conflated. But it is equally absurd to claim that two different habits are left over by the single original act. Nor will it suffice to say that along with the original act of apprehension went a reflexive act of apprehending that act and that then two habits resulted, one from each act. For it is clear from experience that I can intensely apprehend some object without also apprehending my own act of apprehending, and nevertheless afterwards I can recall that act. In such a case any habits that were left over resulted from the single original act.

Reading made it clear that he did not take arguments such as these to be demonstrative but only to establish that Ockham's theory was less plausible than Scotus's.[28] A determined Ockhamist is not likely to have been deterred, however. A full development of Ockham's line would lead to the conclusion that nothing at all is needed to represent an object to an act of apprehension. Whatever it is that makes such an act be an apprehension of a certain object is internal to the act itself. Or, in other words, only the act itself represents or signifies anything. As long as some sort of disposition to revive past acts can be retained by the mind, we can account for the phenomena of recall of objects previously apprehended without supposing that they are retained in the form of actual representations.

The fundamental idea of species theory as it developed in the attempts to account for sensory cognition was that there are real physical entities emanating from physical objects that act as carriers of intentional or represented existence. These carriers can exist in the various external and internal sense organs, can to some degree be preserved there, and can cause the sensory faculties' apprehensions. Aristotle had encouraged the idea that the intellect was to be treated as analogous to a sensory faculty, so naturally it was suggested that some carrier of the intentional was possible there, too, despite the fact that the intellect has no organ: hence the idea of an "intelligible" species. Henry of Ghent's position should be seen as an attempt to keep species theory within the realm of the material and corporeal. Perhaps it was in part motivated by a fear that extension of the theory to the intellect would be the beginning of a process that would eventually lead to a materialist conception of the intellect. Ockham's alternative is much more sweeping. Both sensible and intelligible species are rejected in favor of a view where the only carriers of intentional existence, the only actual representations, are acts of apprehension themselves.

2

For contemporary cognitive scientists the scholastic reliance on mental representations points in the direction of current computational theories of mind such as those espoused by Zenon Pylyshyn and Jerry Fodor.[29] But the foregoing quick survey of scholastic theorizing should have left no doubt that the two programs are far apart on some basic issues.

First of all, the computational theories are embedded within a functionalist theory of mental entities, while the scholastics seem never to consider such a philosophy of mind. Functionalism is the claim that things belong to mental types (i.e., are desires, apprehensions, willings, etc.) and have mental properties by extrinsic denomination; i.e., by their relationships, often causal relationships, to other things including behavior.[30] It differs from behaviorism by endorsing the causation of behavior by mental events internal to the organism, and it differs from reductive "central state" materialism by not holding that mental types and properties are even in principle eliminable from a full account of an organism's life.

It is the very first claim that all scholastics, I believe, would have rejected in its unqualified form. This point deserves some clarification. It is not that the scholastics were unaware of definition by extrinsic denomination or even refused to use such definitions in their philosophical psychology. For example, the definitions given by Scotus and Ockham of intuitive apprehension rely on relations of whatever it is that is an intuitive apprehension to the object which causes it, as well as relations to the subsequent judgment of the existence of the object. This is all agreeable to functionalist practice. But when it was time to say what it is that bears these relations, in most cases the scholastic held that it is something in itself which we can all experience and which we experience as a mental phenomenon.

But what are we experiencing it *as* when we experience it as mental? Here we ask a question the scholastics never posed, because they did not really have a category of the mental that exactly corresponds to ours. There were acts and dispositions of the soul, but these extended to all operations indicative of life, including breathing, reproducing, and other activities we would not at all call mental. What came closer to our mental category were just all events, acts, and dispositions constitutive of or presupposing cognition. But I think perhaps the most useful scholastic way of characterizing what we mean by the mental is whatever it is in a cognitive organism that carries intentional or represented existence. Now the scholastic divergence from functionalism can be put more precisely: Entities are mental, or carry intentional existence, in virtue of what they are intrinsically, not in virtue of relations they bear to things outside themselves. In other words, the intentional content of a mental state or act can be read off of that state or act itself without considering its relations to other things.

This it seems to me is the import of the scholastic insistence that species, acts of cognition, and even habits are likenesses of their objects. The idea is, crudely put, that you can tell just by looking at them what they are likenesses of. This was maintained by all the thinkers of the period we have considered, even when it is obviously quite difficult to see how the entity can be like what it is supposed to be a likeness of. Ockham, for example, insisted that intellectual acts of apprehension are likenesses of what they apprehend.[31] An apprehension of a bulldog is, then, like a bulldog! A functional account, on the other hand, will insist that the intentional content of a state is determined by the network of relations, often causal relations, which that state fits into. Indeed, its being a carrier of intentional content at all must be seen as a matter of its fitting into some network or other that meets some very general requirement. We have here a fundamental philosophical divide between the scholastics and computational mentalists.

Another difference which strikes one immediately is that when they deal with intellectual cognition the computationalists basically concern themselves with propositional attitudes, whereas the scholastics took attitudes toward non-propositional objects to be fundamental. This difference, I believe, is not nearly so serious. The scholastics held that to know or believe that P, I must first apprehend P or know the content of the belief I might come to have. So to apprehend P I must apprehend the terms that are combined to produce P, for every proposition is a complex of intentions related by the so-called relations of second intention. Ultimately, then, everything rests on apprehensions of simple, non-complex, non-propositional items.

A caveat is in order here. The scholastics were not in the business of giving analyses of what the various propositional attitude verbs ('know', 'believe', 'hope', etc.) mean, and they would have freely allowed that not all cases of knowing and believing fit this pattern, even where the grammatical object is some sentence or *that*-clause. The scholastic theory was meant to apply only to intellectual cognition. Yet the senses also know, judge, and believe, and there the object is not in reality a proposition. The scholastics would have admitted, I believe, that much of what passes for knowledge is only implicitly propositional; i.e., is something which would normally result in the intellect's consciously accepting some proposition only if called upon to do so.

Now, how does this differ from current mentalist thinking? First of all, because the computational mode of explanation of cognition rests on there being logical relations among the states of the organism[32] and because our logical theory is primarily a logic of relations among propositions, there is naturally a tendency to find propositions where the unsophisticated would least expect to find them. Since the mentalist holds that much of cognition rests on computational processes of which we are unconscious, he is led to posit propositions existing in the mind other than the ones articulated in the public language of sounds and inscriptions the cognizer consciously uses. These propositions are symbols in the mind's own language, a language which is perhaps not even translatable into a spoken language such as English. Certainly there need be no ability on the part of the cognizer to express publicly the content of all the propositions of his mind's language. The scholastic, on the other hand, restricted mental propositions pretty much to the ones that correlate with the sentences of natural human language. To have a mental proposition in one's head is then tantamount to being able consciously to consider that proposition and, given that we do not face a *penuria verborum*, to articulate it.

This difference is really part and parcel of a difference in approach to the positing of mental representations generally. In the sensory faculties, too, the scholastics never posited representations that cannot under normal conditions easily be brought to consciousness. The only use they saw for subconscious knowledge is for purposes of storing what can be made conscious on demand.[33] Consequently, explanation by logical relations among the contents of representative mental states was mainly limited by the scholastics to where the unsophisticated would expect to find it, viz. in conscious reasoning and deliberation. On the other hand, a non-conscious process such as abstraction of the universal from sensible particulars was explained by positing fairly mechanical causal interaction between the entities involved.

What we see, then, is that the scholastic emphasis on apprehensions of non-propositional items is not *in itself* a very fundamental difference from the modern mentalist program. One can very well imagine that Jerry Fodor might be driven by empirical evidence of some sort to allow that "cognizings" of terms of propositions are more fundamental than "cognizings" of whole propositions. But the scholastic emphasis is symptomatic of a wider difference over the plausible range of explanation via *logical* relations among cognized entities. In this respect the scholastic cognitive science was, like every other cognitive science that has preceded the modern computational approach, much more restrictive of that range.

A final obvious difference is that the scholastics were all convinced that some mental states, events, etc., are possible in a non-material medium. In fact, most held that intellectual processes are possible *only* in a non-material medium. This divergence I want to claim, however, is genuinely superficial and actually masks an important area of agreement.

In the first place, the scholastic notion of the material was more restrictive in extension than our notion of the physical, and consequently the immateriality of the intellect does not imply that it is non-physical in our sense. Electromagnetic fields and such other matterless entities common in our physics would have been immaterial beings to a scholastic despite their inhabiting the domain of physics.

In the second place, the scholastics clearly situated psychology, and hence the science of cognition, within the domain of natural science, just as do modern computational mentalists.[34] If anything, the scholastics went further than the moderns in integrating theories of cognition with physics. The whole theory of species was grounded in proposals about how material entities emanate rays from themselves and was connected to the intense interest in light during this period. Species were not just invented for the purposes of cognitive theory.

In the third place, the scholastics, following the lead of Aristotle, clearly held to something very much like the psycho-physical monism current among functionalist philosophers of mind. What I mean is that, although they would not have thought that mental features of organisms were somehow eliminable in favor of non-mental physical features, they would have allowed that every mental entity also has non-mental features and is part of a nexus of efficient causes and effects; i.e., that it is also a physical entity in our sense of physical. This point can be put in another way: The intentional existences only exist in being carried by ("shining forth in") "real" existences, so the mental character of portions of reality belongs not to entities that lie outside the realm of causal principles that apply to all of nature but to entities that are squarely situated within that realm. Representations cannot just have content, they must have form as well, and that form is in the "real" realm. One might say that in this regard modern functionalist mentalism and scholastic psychology share a deep anti-Cartesianism when it comes to questions of ontology, for although Descartes used the distinction of "objective" from "formal" being, the formal being of his mental items was outside the realm that natural science investigates.

3.

In my opinion the most startling point of convergence between the scholastic and mentalist programs is the need felt by both to run simultaneously two quite different sorts of explanation of at least some psychological phenomena. On the one hand, we have connections at the level of intentional existences, including the panoply of logical relationships which can be used along with some view of human rationality to explain why we come to believe the things we do and choose to do what we choose to do and which in the computational program extend to processes at a less than conscious level. Here we operate on the meanings of our //

mental representations. But at the same time intentional existences, or meanings, cannot exist on their own. They have to be carried by something that is a real thing in the world and that interacts with other real things according to the principles that natural science attempts to formulate. The soul, its faculties, its states, its acts are all real entities, and the explanation of how the soul changes from one state to another or comes to act must involve a causal story that depends on the real rather than intentional character of these things. In modern terms the actual mechanics of the mind depends on the form of the representations rather than their content.[35]

Given all this, it is fairly easy to see where the computational mentalist's sympathy would lie in the dispute between Ockham and the Scotists. The species theory held out the possibility of extending the explanation at the level of represented contents to mental processes of which we are not conscious. The positing of an agent intellect was at least a beginning toward understanding the interface between sense and intellect by way of an "input system," as Fodor calls it.[36] Ockham's proposal essentially limited this level of explanation to that of conscious acts; the rest of mental processes are not mental at all but purely physical, since no bearers of intentional existence are involved. For this reason the modern mentalist must look on Ockham's proposals as a step toward a superficial mentalism in which what is mental is limited pretty much to the introspectible and away from the hidden-structure mentalism which computational theorists find so promising.

University of Alberta

NOTES

1 Avicenna lists five inner sensory faculties: "common" sense, imagination, a cogitative power, an estimative power, and memory. See [6], cap. 1, and also pp. 49*-52* in the introduction.

2 *De Anima* II, 1, 412a22-29; II, 5, 417a9-14.

3 Aristotle's main general discussions of sensation are found in *De Anima* II, 5 and III, 2. The two accounts there are different in important ways.

4 Avicenna and Averroes do not use the term 'species' in this way but speak instead of "intentions." Avicenna also had a doctrine of forms radiating out from sensible objects. In Roger Bacon's work the whole subject becomes tied in with optics and the study of light generally. He uses the term 'species' to refer to the likenesses that emanate from sensible bodies, come to exist in the sense organs, and account for sensation. See [12], introduction.

5 For example, Aquinas, *Summa Theologica* I, qu.85, art. 2.

[6] An extended defense of the immateriality of the intellective soul can be found in Duns Scotus's *Ordinatio* (*Opus Oxoniense*) IV, d. 43, q. 2. The latin text with an English translation is in [14], pp. 134-162.

[7] See [7], vol. I, p. 237.28-238.42.

[8] There was considerable disagreement at this point as to whether the phantasm or the agent intellect or both act on the possible intellect. Aquinas let the phantasm be the immediate causal agent here; the agent intellect transforms the phantasm into something intelligible; i.e., something which can affect the possible intellect. See *Summa Theologica* I, qu. 84, art. 6. Duns Scotus claimed it is the agent intellect which operates on the possible intellect; the phantasm only determines which quiddity the agent intellect will impress. See [9], vol. III., 381. John of Reading seems to have attributed causal agency to both working together. See [5], par. 172.

[9] Not having Henry's works at hand, I have relied on the accounts of it by Scotus in his *Ordinatio* I, d. 3, p. 3, q. 1 (see [9], vol. III, 333-338) and by John of Reading in his *Super Sententias* I, d. 3, q. 3 (see [5], par. 9-31).

[10] Scotus's position is most fully expressed in his *Ordinatio* I, d. 3, p. 3, q. 1 ([9], vol. III, pp. 201-244), and in his *Lectura* I, d. 3, p. 3, q. 1 ([9], vol. XVI, pp. 325-348). Also John of Reading recounts Scotus's arguments in op. cit. (see [5], par. 76-168).

[11] [9], vol. III. pp. 348-352, 357-358.

[12] Ibid., pp. 359-360.

[13] Ibid., pp. 364-365.

[14] Ibid., pp. 368-369.

[15] Ibid., p. 370.

[16] See Ockham's *Quaestiones in Librum Secundum Sententiarum*, q. 13 ([13e], p. 269.20-276.11). Also, John of Reading gives a verbatim rendition of Ockham's position and arguments in op. cit. (see [5], par. 38-74).

[17] [13e], p. 256.5-9.

[18] Scotus's formulation of the distinction can be found in his *Quaestiones Quodlibetales*, q. VI (see [8], vol. 12, p. 145, art. 1) and q.XIII (ibid., pp. 309-310). (English translation in [1], pp. 135-7, 291-2.) It can also be found in his *Ordinatio* (*Opus Oxoniense*) II, d. 3, p. 2, q. 2 ([9], vol. VII, pp. 318-323).

[19] See Scotus's *Ordinatio* (*Opus Oxoniense*) IV, d. 45, q. 3 ([8], vol. 10, p. 217, (21)); II, d. 3, q. 11, ([8], vol. 6, p. 491, (9)).

[20] Ockham defines intuitive apprehension somewhat differently, but the differences are irrelevant to the concerns of this paper.

[21] See [13a], *Prologus*, q. 1, pp. 15-38 (English translation in [2], pp. 18-25). Also see Ockham's *Quodlibeta Septem* I, q. 13 (relevant portions in [2], pp. 27-32). Also, his position is set out in the references cited in note 16.

[22] See [13b], d. II, q. 8, pp. 266-292 (English translation of part of this in [2], pp. 41-43); also, [13d], pp. 351-358 (partly translated in [2], pp. 43-45); and in [13c], *Pars Prima*, cap. 14 & 15, pp. 47-54.

[23] See Ockham's *Expositio Super Octo Libros Physicorum, Prologus* (relevant portion with English translation in [2], pp. 2-16).

[24] But spoken terms do not primarily signify the mental terms; rather they signify just what the mental terms signify. See ch. 11 of [13c], *Pars Prima*, pp. 38-41.

[25] See [10], pp. 40-42.

[26] See [5].

27 My summary greatly abbreviates Reading's polemic. See [5], par. 283-292.
28 [5], par. 263.
29 I claim no first-hand expertise in cognitive science but rely mainly on [3], [4], and [11]. When I refer to "computational mentalism," I am referring to theories operating within the general philosophical frame work and methodological principles espoused by Jerry Fodor and Zenon Pylyshyn.
30 See [3], pp. 8-11.
31 [2], p. 45.
32 [4], p. 5: "Now, the notion of computation is intrinsically connected to such semantical concepts as implication, confirmation, and logical consequence. Specifically, a computation is a transformation of representations which respects these sorts of semantic relations....It is, however, a point of definition that such semantic relations hold only among the sorts of things which can be said to *mean that* P" (Fodor's emphasis).
33 The sole exception is the active intellect, which according to some views innately contains in a virtual way all intellectual representations. But the active intellect generally tends to become the dumping ground for whatever it is that is needed to explain intellectual cognition in addition to what is proposed by the current theories. No one pretends to understand it very well. One of the best things to be said in favor of Ockham's proposals is that they dispense with the need for such an entity.
34 The intellect is sometimes an exception, as with Avicenna and even more with Averroes. But in this respect the Christians rejected the Arab view on grounds drawn mainly from natural science and metaphysics, although clearly a theological motivation was present, too.
35 Cf. Pylyshyn, [11], p. 39: "...the semantics of representations cannot literally cause a system to behave the way it does; only the material form of the representation is causally efficacious....What actually do the causing are certain physical properties of the representational state—but in a way that reflects the representational state's content."
36 [4], p. 41.

REFERENCES

[1] F. Alluntis and A. Wolter (1975), *John Duns Scotus, God and Creatures, The Quodlibetal Questions* (Princeton, N.J.: Princeton University Press).

[2] P. Boehner (1957), *Ockham, Philosophical Writings* (Edinburgh: Nelson).

[3] Jerry A. Fodor (1983), *Representations* (Cambridge: The MIT Press).

[4] Jerry A. Fodor (1983), *The Modularity of Mind* (Cambridge: The MIT Press).

[5] Gedeon Gal (1969), "Quaestio Ioannis de Reading de Necessitate Specierum Intelligibilium Defensio Doctrinae Scoti," *Franciscan Studies* **29**, pp. 66-156.

[6] Abu 'Ali al-Husayn Ibn Sina (1968), *Avicenna Latinus, Liber de Anima seu Sextua de Naturalibus*, IV-V, S. Van Riet (ed.), with an introduction by G. Verbeke (Leiden: E. J. Brill).

[7] Abu 'Ali al-Husayn Ibn Sina (1977, 1980), *Avicenna Latinus, Liber de Philosophia Prima sive Scientia Divina*, S. Van Riet and E. Peeters (eds.), 2 vols. (Leiden: E. J. Brill).

[8] John Duns Scotus (1639), *Ioannis Duns Scoti Opera Omnia*, L. Wadding (ed.), 12 vols. (Lyons).

[9] John Duns Scotus (1950-?), *Ioannis Duns Scoti Opera Omnia*, P. Balic (ed.) (Vatican; incomplete at present).

[10] Gordon Leff (1975), *William of Ockham, The Metamorphosis of Scholastic Discourse* (Totowa, N.J.: Rowman and Littlefield).
[11] Zenon W. Pylyshyn (1984), *Computation and Cognition* (Cambridge, MA: The MIT Press).
[12] Peter G. Sobol (1984), *John Buridan on the Soul and Sensation* (Ph.D. dissertation for Indiana University).
[13] William of Ockham (1967--), *Guillelmi de Ockham Opera Philosophica et Theologica* (St. Bonaventure, N.Y.: Franciscan Institute of St. Bonaventure University; incomplete at present).
[14] William of Ockham (1967), *Scriptum in Librum Primum Sententiarum Ordinatio, Prologus et Distinctio Prima* in G. Gal and S. Brown (eds.), *Opera Theologica* I.
[15] William of Ockham (1970), *Scriptum in Librum Primum Sententiarum Ordinatio, Distinctiones* II-III in S. Brown and G. Gal (eds.), *Opera Theologica* II.
[16] William of Ockham (1974), *Summa Totius Logicae* in P. Boehner, G. Gal, and S. Brown (eds.), *Opera Philosophica* I.
[17] William of Ockham (1978), *Expositio in Librum Perihermenias Aristotelis* in A. Gambatese and S. Brown (eds.), *Opera Philosophica* II.
[18] William of Ockham (1981), *Quaestiones in Librum Secundum Sententiarum* (Reportatio) in G. Gal and R. Wood (eds.), *Opera Theologica* V.
[19] Alan Wolter (1962), *Duns Scotus, Philosophical Writings* (Edinburgh: Nelson).

CALVIN G. NORMORE

OCKHAM ON MENTAL LANGUAGE

Thanks largely to the work of Noam Chomsky, we have witnessed over the last thirty years a revival of interest in two closely related ideas: that there is a universal grammar, a set of structural features common to every human language, and that the exploration of this grammar is, in part, an exploration of the structure of thought.

Fourteenth century grammarians and philosophers were also interested in this complex of questions, and debate about them raged as vigorously then as now. One tradition in this debate grew out of thirteenth century terminist logic and seems to have been given a distinctive shape by William Ockham. This tradition posited a fully-fledged language of thought common to all rational beings and prior to al linguistic convention. In this essay I will attempt to outline Ockham's account of this mental language, to consider some fourteenth century objections which lead to the refinement of the account by others in the fourteenth century, and finally to suggest that Ockham's approach has something to contribute to current debate about the relationship between the theory of meaning and any language of thought.

At the very beginning of his *Summa Logicae* Ockham claims that there are three distinct types of language: written, spoken, and mental. He insists that written and spoken language are distinct in kind and that there is a type of language whose terms are concepts and which exists only in the mind.[1]

Ockham's mental language plays several distinct roles within his philosophy. On the one hand, mental language figures crucially in the semantics of spoken and written language. On the other hand, mental language is a fully articulated language which is suited to be spoken by natural telepaths and is spoken by the angels. These two kinds of role require very different features of mental language, features which, as we shall see, sometimes pull its structure in opposite directions.

1. Mental Language as Semantics

The key terms in Ockham's employment of mental language as a semantics are 'signification' and 'subordination'. 'Signification' is a

J-C. Smith (ed.), Historical Foundations of Cognitive Science, 53–70.

primitive term within Ockham's semantics, though he explains its use by saying that a term signifies an object if it can be truly predicated of a pronoun picking out that object.[2] 'Human' signifies you because "That is a human" (pointing at you) is true. Ockham is one of the people later called "nominalistae," and his "nominalism" is reflected in his account of signification. On Ockham's view all signification is of particular concrete objects, concepts, and words. A singular term like 'Socrates' signifies a single thing, Socrates; a common noun like 'animal' signifies all the animals.

In the passage at the beginning of the *Summa Logicae* in which mental language is introduced Ockham claims that while terms of spoken language do not (usually) serve as signs of terms in mental language but rather serve as signs of things in the world ('human' signifies human beings, not the concept *human being*), nevertheless spoken terms signify what they do because they are subordinated to mental terms. If a mental term were to change its signification, the subordinated spoken term would as well. I will speak of the mental expression to which a spoken expression is subordinated as its *corresponding* mental expression.

The correspondence between spoken language and mental language induced by subordination is not simply a one-one mapping of term onto term. Ockham makes this clear in a crucial passage in his fifth *Quodlibet*:

> I say that just as among spoken and written terms some are nouns, some verbs, some pronouns, some participles, some conjunctions, and some prepositions, so it is in the mind as is evident from this consideration that to every true or false spoken expression (oratio) there corresponds some mental expression made up of concepts. Therefore just as those parts of spoken sentences which are imposed to signify things because they are needed for signification or expression...are distinct parts, so those parts of mental sentences corresponding to words are distinct [if needed] to make distinct sentences true or false.[3]

Here Ockham asserts that mental language has distinct terms corresponding to those terms of spoken language which cannot be substituted one for another *salva veritate*. This is amplified by his claim in *Summa Logicae* I, c.13, that strictly speaking no term or concept of mental language can be equivocal. This, as Ockham explains, is because "a word is equivocal if, in signifying different things, it is a sign subordinated to several rather than to one concept or intention of the soul."[4] On Ockham's account, then, to be equivocal requires being subordinated and so is confined to spoken and written speech. A mental term has, we might say, a single sense. Ockham amplifies this by remarking that "a term can be equivocal by the very fact that it can function as different parts of speech, i.e., as both a noun and a participle, a noun and an adverb, etc."[5]

These constraints on the structure of mental terms taken in conjunction with the subordination relation between spoken and mental terms suggest that mental language functions very much like a theory of sense for spoken language. Indeed, we can summarize many of the semantic roles mental language plays in four theses:

1. The signification of a spoken term is a function of that of the corresponding mental term.

2. The truth-value of a sentence is a function of the truth-value of the corresponding mental sentence.

3. If two expressions have the same corresponding mental expression then they have the same signification.

4. Expressions with the same corresponding mental expression are everywhere interchangeable *salva veritate*.

If we replace 'mental expression' by 'customary sense' and 'signification' by 'customary reference' in these claims, we obtain the theses of Frege's semantics often thought of as crucial for determining the relationship between sense and reference. Indeed, we can assert that the significations of spoken terms and the truth-values of spoken sentences are completely fixed by the corresponding mental terms and the subordination relation.

2. Mental Language and Epistemology

Mental language is fitted for its semantic roles not only by the subordination relation which connects it to spoken language but by its structure and, in particular, by the fact that it contains neither synonyms nor equivocal expressions. As John Trentman pointed out in his pioneering work on mental language, these structural features also make mental language look like the ideal languages of early twentieth century analytic philosophy.[6] Both contain just those distinctions relevant to the truth-values of sentences. Moreover, just as the structure of an ideal language was thought to be the proper study of the logician, so for Ockham, as we shall see, the proper study of the logician is the structure of mental language.

It must be emphasized, however, that mental language (Mental) was understood by Ockham to be a medium of communication. If we were telepaths we would communicate in mental language, and the angels, who are telepaths, do so.[7] To see how it could function as a medium of communication, we must take a closer look both at its structure and at its relations to the world.[8]

Spoken and written terms signify things in the world, but they do so in subordination to mental terms. Ockham distinguishes the signification of mental terms from that of spoken and written terms. He calls the former *natural signification* and the latter *conventional signification*. There is a clear sense in which, for Ockham, natural signification is primary. Indeed, as Paul Spade has pointed out, we can think of conventional signification as,

> ...the relative product of the subordination relation into the natural signification relation. A term conventionally signifies a thing *x* if there is some concept to which the term is subordinated and which in tern naturally signifies *x*.[9]

It is because the relation of conventional signification can be so defined that mental language can serve as a semantics for spoken language. Spoken language is connected to the world by its relations to the connections which mental language has to the world.

Just what are those connections? The problem of the nature of the natural signification relation is very nearly the analogue within Ockham's philosophy of the general problem of the nature of representation. It should come as no surprise, then, that it is a controversial subject whose full treatment would take us very far afield. What does seem clear is that for Ockham natural signification involves two natural relations between the mental term and what it signifies. One is a relation of similarity, while the other is a causal relation: mental terms are the natural effects in the soul of encounter with the objects which they signify. There is no doubt that Ockham thinks that both causality and similarity are involved in determining what a mental term signifies, but there are two reasons for thinking that causality is the more fundamental. First, the process of concept formation, that is, the process whereby simple mental terms are acquired, is a causal process. As Ockham explains:

> This is the process. First a human is apprehended by some particular sense, then that same human is apprehended by the intellect, and from this having been apprehended there is had a general notion common to all humans. This apprehension (*cognitio*) is called a concept, intention, or passion, and from it existing in the intellect, the intellect at once knows that a human is something without any process of reasoning (*sine discursu*). Then another animal or other animals distinct from a human having been apprehended, there is. elicited a general notion of all animals, and that general notion of all animals is called a passion or intention of the soul or concept common to all animals.[10]

Second, Ockham considers the possibility that a singular concept may be as similar to one thing as to another and explicitly claims that it is the causal connection which determines which thing the concept signifies:

> I say that we have an apprehension proper to one singular thing not on account of a greater likeness to one than to another but because this intuitive apprehension is naturally caused only by the one and not by the other, and cannot be caused by the other.[11]

So causality plays a crucial role in the acquisition of concepts, and it figures in the tying of singular concepts to the singulars of which they are concepts. Nevertheless, as Marilyn Adams has emphasized, similarity does figure prominently in Ockham's account of why concepts signify as they do. Indeed, it seems clear that a sufficient degree of natural similarity between the mental term and what it signifies is a necessary condition for signification on Ockham's view.[12]

The process of concept acquisition which Ockham mentions is central to his epistemology. Following a tradition which goes back to Scotus, Ockham posits two kinds of intellectual apprehension: *intuitive cognition* and *abstractive cognition*. An intuitive cognition is an apprehension which supports an evident judgment that what is apprehended exists. An abstractive cognition is an apprehension which does not support such a judgment.[13] Intuitive cognition can only be of singulars, and it and the associated process of "eliciting" concepts give rise to what he calls *absolute concepts*. Absolute concepts are those which constitute the basic significative elements of mental language.

The fundamental distinction among the terms of any language, for Ockham, is that between categorematic and syncategorematic terms. Syncategorematic terms are those like 'and' and 'all' which do not by themselves signify; categorematic terms are those which do. Syncategorematic terms create special epistemological problems for Ockham, and I shall return to them below. First, however, there is a distinction among categorematic terms, namely their division into the *absolute* and the *connotative*, which must be clarified. Ockham distinguishes absolute and connotative terms by the ways they signify:

> Purely absolute names are those which do not signify one thing principally and another or even the same thing secondarily. Rather, everything which is signified by the same absolute name is signified primarily [and]...properly speaking, such names have no definition expressing the meaning of the term (*quid nominis*). For, strictly speaking, a name that has a definition expressing the meaning of the term has only one such definition, and consequently no two sentences which express the meaning of such terms are so different in their parts that some part of the first sentence signifies something that is not signified by any corresponding part in the second. The meaning of an absolute term may, however, be explained in some manner by several sentences, whose respective parts do not signify the same things. Therefore, properly speaking none of these is a definition giving the meaning of the term.

....A connotative name, however, is one which signifies something primarily and something else secondarily. Such a name has, properly speaking, a definition expressing its meaning. In such a definition expressing the meaning of the term it is often necessary to put one of its terms in the nominative case and something else in an oblique case.[14]

In this passage Ockham distinguishes primary and secondary signification. Paul Spade has suggested that we can understand this distinction if we take the basic relation of signification to be given by the claim that for a simple term signification is as explained above while a complex expression signifies exactly what is signified by its constituent non-complex categorematic terms.[15] Here 'signifies' is taken more broadly than in my earlier discussion, and a complex expression will, in this broader sense, signify objects of which it cannot be truly predicated. We can distinguish primary and secondary signification within this broader sense by combining two criteria each of which has initial plausibility but each of which is insufficient by itself. The resulting distinction is that a term primarily signifies that of which it can be correctly predicated, and if it is complex then it secondarily signifies whatever is signified by those of its non-complex constituents which are in an oblique case *and* whatever it signifies but does not primarily signify.

According to the text quoted above, absolute terms have only primary signification. Ockham claims that from this it follows that they have no nominal definition. His argument seems to be that any term which has a nominal definition is synonymous with its definition. A definition, being a complex expression, typically signifies some things which it does not primarily signify. (For instance a definition by genus and differentia of some species always signifies the members of the genus outside the species but does not primarily signify them.) Since an absolute term does not signify anything which it does not primarily signify, then, if it had a nominal definition, the signification of the defining expression would not be exactly that of the expression defined. But, Ockham claims, two expressions are synonyms only if they signify the same thing in every way.[16] Hence the term and its putative definition would not be synonyms after all.

Absolute terms do not have nominal definitions; connotative terms do. If mental language has no synonyms, then since a term and its nominal definition must be synonyms, there is no place in mental language for both. The terms which appear in the definition of a connotative term typically have other uses and so will usually be required for the expressive adequacy of mental language. It seems, then, though Ockham appears never to draw the conclusion explicitly, that if mental language has no synonyms it has no simple connotative terms. Each spoken connotative term and its defining expression will be subordinated to the same complex expression of the mental, and that expression will have at least as much structure as the definition. For example, in discussions of analyticity one sometimes finds the claim that 'bachelor' and 'unmarried man' are synonyms. Suppose this to be so. Then 'bachelor' and 'unmarried man' will be subordinated to the same mental terms. An unmarried man is something which is both a man and unmarried; hence if there is a mental term corresponding to 'man', one corresponding to 'unmarried', and one doing the work of conjunction, we could form a mental term corresponding to the written expression 'unmarried man'. Since mental language is a language with a grammar, it will have such a term if it has all of its components. Of course, 'unmarried' is itself logically complex; hence if there is a mental term for 'married' and one doing the work of term negation, there will be a complex term corresponding to 'unmarried'. Thus, the process of *definition* in spoken language has as analogue the process of *composition* in mental language.[17]

Absolute terms are the effect in the mind of causal contact with objects.[18] Connotative terms can be built out of absolute terms and syncategoremata. But where do the syncategorematic terms come from? It is certainly not plausible to suppose that they are obtained by intuitive and abstractive cognition in the way absolute terms are. The world doesn't contain 'ifs' and 'cans' in the way it does pots and pans.

This is a subject on which Ockham seems to have changed his mind as he changed his mind about the nature of concepts themselves. Ockham's earlier view of concepts as we find it reported in his *Reportatio*, for example, had it that concepts were intentional objects, *ficta* in Ockham's terminology. During the period in which he held this theory of concepts in general, Ockham seems to have emphasized a distinction between *two* kinds of mental language. One was the mental language with which this paper has so far been concerned; the other was a language whose terms were the mental representations of the spoken expressions of natural language itself. Terms of the first kind of mental language (what other writers sometimes called 'proper' mental language) were, as we have seen, prior to spoken language. Terms of the second kind of mental language ('improper' in some terminologies) were posterior to spoken language. While he held the 'fictum' theory, Ockham seems to have held the astonishing view that all syncategorematic concepts were terms of improper mental language. He writes that:

> ...matters are similar with this word 'not' and with these 'per se', 'insofar as', 'if', and syncategoremata of this sort. From these words, so signifying, the intellect abstracts common concepts predicable of them and imposes these concepts to signify those same things which those words signify outside [the intellect], and in the same way and from such as these it forms similar sentences having properties like those which the spoken sentences have.[19]

This would make hash of the claim put forward in the *Summa Logicae* that spoken language depends on mental language for its sense. Indeed since spoken language is arbitrary (*ad placitum*), it would make hash of the claim that mental language is natural. Hester Gelber has pointed out that Ockham's contemporary Walter Chatton criticized Ockham's early approach along just these lines.[20] Perhaps in response to Chatton's criticisms of it, Ockham abandoned the 'fictum' theory in favor of an 'adverbial' view which identified the concept with the act of conceiving itself or with a quality in the mind and seems also to have abandoned his early view of the origin of syncategoremata. But it is less clear to what view of the origin of syncategoremata he then turned.

One suggestion is that they are innate. According to Ockham the intellect prior to all experience is a *tabula nuda*. Thus it has no intellectual habits and no concepts. It does, however, have the power to perform certain mental acts. It can, for example, assent to the proposition that a whole is greater than a part if this proposition is presented to it. But syncategorematic terms just are certain operations which the mind performs on categorematic terms. They are not so much acts of thought as acts of will. Hence the mind has innately the power to perform them.

This innatist position may well have been Ockham's. But there are other possibilities. Although syncategoremata could not be acquired by intuitive cognition and abstraction in the way categorematic terms are (because there are no syncategorematic *objects*), this does not establish that they cannot be acquired at all during intuitive cognition. If the objects of intuitive cognition are sufficiently complex, if, for example, one intuits not just Socrates's paleness or Socrates himself but, say, pale Socrates sitting on a fence, then it may be that the syncategorematic elements of the complex are presented in intuition.

But it seems unlikely that Ockham would allow that we have such complex intuitive cognitions. T. K. Scott has argued that he does at least sometimes hold that "what is intuited is a substance as subject of an accident, not merely an accident."[21] Since Ockham holds that certain accidents, such as color, are real things which could exist apart from any substances, the claim that one could intuit Socrates as white would be the claim that one can intuit the *inherence* of whiteness *in* Socrates. Such an 'inherence in' would be a syncategorematic element of the kind in question. Ockham does sometimes speak in this vein. For example in II Sentences, q. 12-13, Ockham writes that "if someone experiences something to be white he sees the whiteness to be in it." But it is not obvious how we are to take such claims. What is clear is that Ockham thinks that our intuitive cognitions ground evident judgments that contingent objects are related as they contingently are—that Socrates is white, for example. What is not clear is whether we, strictly speaking, ever *intuit* that Socrates is white. Indeed, as Marilyn Adams has argued, it is not even clear that Ockham thinks we intuit substances at all.[22]

Even if we cannot determine the source of syncategorematic mental terms within Ockham's philosophy, the picture which emerges so far is a remarkably simple and attractive one. On this picture causal contact with particular objects gives rise to concepts which serve simply as signs of things sufficiently like the original cause. These are the absolute mental terms. Which absolute terms there are in someone's mental vocabulary will depend on the person's causal history, and so there are differences among mental languages, but only in the lexicon. Mental expressions consist of absolute terms combined by mental syncategoremata. If the combination is itself something which could be the subject or predicate of a sentence, then it is a mental connotative term. All simple terms in mental language are absolute. All complex terms are connotative. Mental sentences, we may suppose, are composed of mental terms joined by a mental copula. Mental language has no equivocal expressions and no synonyms. Spoken expressions are subordinated to mental expressions and, though they signify directly, signify as they do because of the expressions to which they are subordinated. Since subordination is conventional, so is spoken language. Mental expressions function in many ways like Fregean senses for the expressions subordinated to them. In particular, the process of definition corresponds to the process of decomposition in the mental.

This picture can be made to do work elsewhere in Ockham's philosophy, as well. One might suggest, for example, that only those items should be admitted into ontology which are signified by some absolute term, a familiar theme for Ockham.

3. Mental Language as a Medium of Communication

The picture painted above also unifies much of Ockham's epistemology and semantics, but it is almost certainly too good to be true. Some of the difficulties begin to arise when we remember that Ockham intends mental language not only as a semantics for spoken language but further as a medium of communication in its own right. If we were telepaths we would communicate in Mental, and the angels, who supposedly are telepaths, are held in fact to do so. What must the mental be like in order to play such a role?

One thing it must be is in some sense concrete. Ockham thinks that the mind is an immaterial substantial form and that we must be prepared for the possibility that an implementation of a language in such a medium will be rather different from implementations of spoken or written language. Yet there will be questions about Mental as a language which would not arise about Mental as a semantics. One of these is a question raised by Gregory of Rimini and revived recently by Paul Spade. It is the question of whether mental sentences are made up of their parts.[23]

This is the problem. Subordination is a relation between terms of spoken language and terms of mental language. Hence to each significant element of a spoken sentence there corresponds a significant mental element. But some spoken sentences contain exactly the same significant elements and yet have different truth-values. That is how it is with:

> "Every human is an animal" and
> "Every animal is a human"

for example. How do the mental sentences corresponding to these differ?

The written sentences above and the corresponding spoken sentences differ in *word order*. In a written sentence spatial configuration is significant, and in a spoken sentence temporal configuration is significant. But neither spatial nor temporal arrangement is essential to a mental sentence: not spatial arrangement simply because the mind is not a spatial object, and not temporal arrangement because Ockham believes that mental sentences can be produced all at once. Indeed, Ockham believes that the practiced logician grasps long chains of argument at once.[24]

One solution to this problem would be to suppose mental language to be so highly inflected that no two mental sentences ever would contain exactly the same significant elements. Ockham is already committed to markers for case and tense, so why not commit him to markers for subject, predicate, etc.? This is, after all, a popular solution in the tradition which follows him.[25]

There are two reasons why we should not. The first is that it would not solve a deeper and related problem. What, on such an account, would be the difference between simply conceiving the parts of a mental sentence and perceiving the sentence itself? How are the parts of a mental sentence bound into a whole?

Ockham is well aware of this problem. He discusses it in several places, notably in Book II of his *Commentary on the Sentences* and in the Introduction to his *Commentary on the Perihermenias*. The first discussion sets the problem within Ockham's earlier theory of concepts as *ficta* or intentional objects; the discussion in the *Commentary on the Perihermenias* sets it within his mature view that concepts are *intellectiones* or acts of understanding, themselves.

In II *Sent.*, q.1 and q. 12-13, Ockham tells us how the mind understands a mental proposition composed of *ficta* and how such comprehension is distinguished from the understanding of the mental terms and the copula. The immediate problem under discussion is whether there has to be a separate item of knowledge (*ratio cognoscendi*) for each part of the sentence known. Ockham claims not, suggesting that when we have habitual knowledge of a sentence we know the whole sentence and its terms by the same habit. Moreover, even when our grasping of the sentence is for the first time we do not need to grasp the subject, predicate, and copula in separate acts.[26]

Within the 'fictum' theory we can distinguish between the *ficta*, which are the subject, predicate, and copula, and the acts of 'uttering' and 'grasping' the sentence, whose objects are those *ficta*. So within this theory the problem of distinguishing the mere collection of *ficta* (the list of 'words' in the sentence) from the sentence as an ordered whole is particularly acute. Ockham's solution is to treat the act of uttering or grasping the copula as a very special act. As he says:

> ...the act terminating at the copula of at the concept of the copula does not terminate there absolutely but at the same time [terminates] at the subject and the predicate. Therefore the acts which terminate only at the subject and the predicate are incomplex, but the act which terminates at the copula is complex inasmuch as it terminates immediately at the whole complex; and this is called composition and division.[27]

Within Ockham's later 'intellectio' theory of the nature of concepts there no longer is a distinction between the act of grasping something by a concept and the concept itself. There are just the acts themselves. On Ockham's version of the 'intellectio' theory the subject and predicate of a sentence are acts, presumably the acts analogous to those which on the 'fictum' theory grasped the subject and predicate. If this is so, then what on the 'fictum' theory is the act terminating at the copula *is*, on the 'intellectio' theory, the copula of the mental proposition. Thus the copula itself is a complex act of compounding or dividing, one which somehow makes essential use of the two "incomplex" acts which are the extremes. This makes it possible to think of a sentence (or even an argument) as a single complex mental act. As Ockham says:

> ...a sentence could be an act of understanding equivalent to the one whole sentence made up of distinct parts [which there would be] if [the parts] had the order they have in speech, and then there will be distinct sentences whenever there are distinct corresponding sentences in speech.[28]

The problem with this approach is that it threatens the recursive character of mental language and so also the possibility that it could be a medium of communication. If each sentence is a distinct primitive mental act, then decoding each sentence will be a completely novel experience in which my knowledge of other acts will be of no help. Ockham tries to evade this problem by treating some acts, such as the copula on the 'intellectio' view, as complex acts in which the simpler parts are somehow represented, but exactly how this works remains a mystery.

A second difficulty with attributing to Ockham the view that mental language is so highly inflected that problems like word order never appear is that there are cases in which he could appeal to such a device but does not do so, even though the cost of not doing so is making Mental in one way equivocal. For instance, Ockham's account of supposition requires that a term as it appears in a sentence have personal supposition when it is used to stand for what it signifies, simple supposition when it is used to stand for the concept to which it is subordinated, and material supposition when it is used to stand for itself. The sentences,

1) Humans have been on the moon.
2) *Humans* is not a genus.
3) 'Humans' has six letters.

illustrate the three types of supposition for the term 'humans'.

In his discussion of equivocation Ockham expressly claims that, while mental terms cannot be equivocal in the sense of having two different ranges of signification, they are equivocal in the sense that they can have personal supposition in one sentence and material supposition in another.

For spoken sentences this raises no special problem, because one can presume that the intention of the speaker determines the supposition of the subject term even if it is difficult for a hearer to discover it. But for mental sentences there is a real problem because, on Ockham's view, once the concepts which make up the sentence have been specified the supposition of the subject is still underdetermined. But what could determine it?

Ockham seems never to have faced this problem. By insisting that the concepts which make up a mental sentence do underdetermine the supposition of the subject term, he rules out the solution taken by Pierre d'Ailly which maintains that the subject term of a mental sentence is always in personal supposition. So it cannot be assumed that he would rely on inflection to solve problems about the structure of Mental. And Ockham's willingness to leave equivocity about supposition in Mental raises a question about his commitment to the view that it is a thoroughly univocal language.

4. Logic and Mental Language

Besides its role as semantics for spoken language and as a medium of angelic communication, mental language serves Ockham as a medium of logical analysis. It is here, I think, that we find some of the most striking successes and the most disturbing difficulties of Ockham's approach. I will consider one of each.

One of the tasks any logician must face, and most especially a logician who works as much with natural languages as medieval logicians did, is that of providing analyses of the logical structure of arguments and their component sentences. If one holds that spoken expressions are subordinated to mental expressions and that mental language is the appropriate medium for logic, then one can formulate a rather powerful constraint on acceptable analysis: an analysis will be acceptable only if the replacement of the analysandum by the analysans preserves grammaticality in mental language. This need not imply that such replacement will preserve grammaticality in the spoken language, however.

The idea can be illustrated by seeing how an analysis along these lines treats a problem which Norman Kretzmann has discussed.[29] Kretzmann draws attention to the first sophism in Richard Kilvington's *Sophismata*:

> Socrates is whiter than Plato begins to be white.

Kilvington analyzes 'Plato begins to be white' as "Plato is not now white and immediately after now he will be white." As Kretzmann points out, Albert of Saxony in his own *Sophismata* objects that either the analysis is wrong or the original sophism is ill-formed. He supports this claim by pointing out that if we replace the embedded occurrence of 'Plato begins to be white' by its analysis, we get:

> Socrates is whiter than Plato is now white and immediately after now he will be white.

And this is simply ungrammatical. Kretzmann suggests that Albert has the wrong side of this dispute but that it is not easy to see exactly why. My proposal is that Albert is mistaken in thinking that the correct analysis of an expression has to preserve the grammatical category of the expression. This amounts, I think, to the claim that grammatical form mirrors logical form in natural language and is refuted by theories, like Russell's theory of descriptions, in which expressions of one grammatical category are analyzed as being of another or even, as with Russellian descriptions, of no grammatical category at all. Of course, an analysis has to preserve meaning, but that, within a framework like Ockham's mental language, just amounts to the claim that analysandum and analysis be subordinated to the same mental expression. If we insist that analysis preserve grammaticality (indeed, *identity*) in Mental but not in spoken language, Albert's argument fails to be sound.

The suggestion just made is that the mental sentence to which a spoken sentence is subordinated wears its logical form on its sleeve, so to speak, whereas spoken sentences themselves may conceal their logical form to some degree. In mental language all logical relations are to be explicit. Mental just is the language of logic.

This view is tested by Ockham's account of the fallacies. As Hester Gelber has argued, Ockham refurbished the classification of fallacies, treating the traditional Aristotelian distinction between fallacies 'in speech' and 'outside speech' as a distinction between fallacies which occur because someone has mistaken how spoken or written expressions are related to mental expressions and those which occur because someone has mistaken the structure of the mental argument itself.[30] But if the logical structure of mental expressions is explicit, how could a speaker be mistaken about it?

Ockham's paradigm of a fallacy *extra dictione* is the fallacy of accident. As Gelber has pointed out, Ockham rejects the diagnosis of this fallacy as arising from variation in the middle term and argues that it arises from mistaking the structure of the mental argument itself. How is this possible?

Ockham argues in several places that we cannot be mistaken about certain kinds of sentences whose subject and predicate are both absolute terms.[31] It is tempting to generalize from this to the conclusion that any mental sentence whose only categorematic terms are absolute terms would, since its logical structure is explicit, be transparent to the thinker. If, as I have argued above, *all* mental sentences contain as simple components only absolute terms and syncategoremata, then on this view all mental sentences would be transparent and there would be no fallacies *extra dictione*.

The key to this puzzle is the assumption that the structure which is explicit in a sentence is manifest to the thinker. Ockham claims evidence only for certain categorical sentences involving only absolute terms, and there is no reason to suppose that he thought bolder claims would be true. Moreover, Ockham does not in general hold that our mental acts are evident to us. It is perfectly plausible, then, to suppose that fallacies arise because thinkers fail to notice such things as the scope of negation operators in their thoughts and so make inferences which are not warranted: what is happening in such cases is that one mental expression is being treated as if it were of a different structure than it is. Complicated mental acts are *complicated*, and the human mind, being the finite instrument that it is, may not react correctly to this complexity. Ockham does not explicitly endorse such a view, but it is, I think, very much in the spirit of his work.

5. Conclusion

It is always tempting and always dangerous to suggest that a long dead thinker's views have a close contemporary analogue. Nevertheless, this is precisely what I wish to suggest for Ockham on the basis of the foregoing. It seems to me that Ockham's approach is being rediscovered at this time, and that many of the problems and theoretical devices of the fourteenth century should be and are being considered again.

Perhaps the contemporary thinker whose work seems in many ways closest to Ockham's is Jerry Fodor. In a series of works going back at least to *The Language of Thought*, Fodor argues that the explanation of an organism's behavior requires determination of how external stimuli are represented internally.[32] Representation requires a medium of representation, and such a medium must be symbolic in character. Thus there is no internal representation without an internal language. Moreover, this internal language cannot be a natural language. Natural languages cannot be the medium of thought because non-humans and pre-verbal children exhibit the behavior we need to postulate internal representation in order to explain. In particular, they learn. Fodor argues that learning, especially language learning, presupposes a language within which hypotheses about the material to be learned can be represented. Such a language, he argues, must be expressively as powerful as the natural language to be learned.

Thus Fodor believes that he has shown that no language can be learned unless one already has a language within which to formulate hypotheses about the meaning or use of predicates in the new language. But, as he readily admits, this does not mean that no language could be *acquired* without another language. On the contrary, a first language could be acquired *ex nihilo* provided its acquisition did not involve hypothesis formation.

This is exactly how Ockham thinks mental language is in fact acquired. As we have seen, Ockham holds that absolute categorematic concepts are either the effects in the mind of acquaintance with the world or the result of performing an apparently innate operation of comparison and abstraction on such effects. In any case they involve no hypothesis formation.

Although the situation with syncategoremata is less clear, it is, as we have also seen, plausible to attribute to Ockham the view that these are operations of the mind which arise simply from the exercise of its innate powers. If this is so then a mental language of the sort Ockham describes seems very attractive. It offers the avid psychosemanticist all the internal representational and computational character that could be wanted, while preserving for empiricism the spirit and most of the letter of the dictum that there is nothing in the mind which was not previously in the senses.

Nor is this all. Ockham stands near the beginning of a long tradition.[33] He and his successors make a number of substantive hypotheses about the structure of mental language. The mature Ockham, for example, claims that it contains neither synonyms nor equivocal terms and that it will be inflected only to the point needed for expressive adequacy. Buridan claims that it does contain simple connotative terms. Albert of Saxony claims that every categorical mental sentence contains a subject term, a predicate term, and a copula. The tradition contains discussions about the very nature of mental representation, as well. Chomsky has traced the history of some elements in his approach to language back to Descartes. I would like to suggest that it is in Ockham and his successors that we find other and perhaps more significant elements, embedded in a research program which spans much of the later Middle Ages and which may well turn out to have more than merely historical significance.

University of Toronto

NOTES

[1] Cf. W. Ockham, *Summa Logicae* (hereafter S.L.) I, c.1, in P. Boehner, G. Gal, S. Brown (eds.), *Opera Philosophica* (hereafter O.P.) I (St. Bonaventure, NY: The Franciscan Institute, 1974).

[2] Cf. S.L. I, c.33, in O.P.I, p. 95 ff.

[3] W. Ockham, Quodlibeta V, Q.8, Art. 1, in Joseph C. Wey (ed.), *Opera Theologica* (hereafter O.T.) IX (St. Bonaventure, NY: The Franciscan Institute, 1980), p. 509.

[4] W. Ockham, S.L. I, c.13, in O.P. I, p. 45.

[5] Ibid., p. 46.

[6] John Trentman, "Ockham on Mental," *Mind* **LXXIX** (1970), pp. 586-590.

[7] That they do creates some interesting problems. Is it possible to lie in mental language, for example? For discussion of this question, cf. Hester Gelber, "I Cannot Tell A Lie: Hugh of Lawton's Critique of William of Ockham on Mental Language," forthcoming in *Franciscan Studies* 1984.

8 A full study of mental language as a medium of communication would require some account of what it is to 'speak' and 'hear' it. There will be some indirect remarks about 'speaking' it below, but a full treatment of the subject will not be attempted here. For a beginning consideration, cf. Hester Gelber, op cit., and Joan Gibson, *The Role of Mental Language in the Philosophy of William of Ockham*, Ph.D. dissertation, University of Toronto, 1976.

9 Paul Spade, "Ockham's Distinctions between Absolute and Connotative Terms," *Vivarium* **13** (1975), pp. 55-76.

10 Ockham, *S.L.* III-2, cap.29, in O.P. I, p. 557, 1.14-22.

11 Ockham, *Quodlibeta* I, q.13, in O.T. IX, p. 74. Ockham holds the view that the causes of each thing are naturally proper to it, so that a particular thing, in this case an apprehension, could be produced naturally by only one set of causes.

12 Cf. Marilyn M. Adams, *William Ockham* (Notre Dame Press, 1987), ch. 4.

13 Cf. Marilyn M. Adams, op. cit., ch. 13.2, for a full discussion of this and for texts. There are interesting similarities between Ockham's distinction between intuitive and abstractive cognition and B. Russell's distinction between knowledge by acquaintance and knowledge by description.

14 Ockham, *S.L.* I, cap.10 (O.P. I, pp. 35-36).

15 Paul Spade, loc. cit.

16 For example, Ockham writes in *S.L.* I, c.6 (O.P. I, p. 19, 1.8-16):

> Loosely speaking those are called synonyms which simply signify the same in every way with the result that nothing is in any way signified through the one but that it is in the same way signified through the others even if not those using it believe them to signify the same....It is in this second way that I intend to use this noun 'synonym' in this chapter and in many others.

17 Some of Ockham's successors do seem to have admitted simple connotative terms into Mental. For evidence that Buridan does, cf. the introduction to P. King, *John Buridan's Logic: The Treatise on Supposition; The Treatise on Consequences*, Synthese Historical Library, Vol. 27 (Boston: D. Reidel Publishing Co., 1985).

18 Cf. Ockham, *S.L.* III-II, c.29, in O.P. I, p. 559.

19 Ockham, I *Sent.*, d.2, q.8, in O.T. II, p. 285.

20 Cf. Chatton, *Reportatio* I, d.3, q.2, quoted in Gelber, op, cit.

21 T. K. Scott, "Ockham on Evidence, Necessity, and Intuition," *Journal of the History of Philosophy*, Vol. 7 (1969), pp. 27-49, note 22.

22 Ockham, Sentences, Prol. q.1, in O.T. I, pp. 6, 31.

23 Paul Spade has drawn attention to medieval discussions of this problem. Cf., for example, P. V. Spade, *Peter of Ailly: Concepts and Insolubles*, an annotated translation (Dordrecht: D. Reidel Publishing Co., 1980). Attend especially to his note 313.

24 Ockham II *Sent.*, q. 12-13, in O.T. V, G. Gal and R. Wood (eds.) (St. Bonaventure, NY: The Franciscan Institute, 1981), p. 279.

25 For information on this tradition, cf. P. Spade, *Peter of Ailly*, note 217 and the references therein.

26 Cf. the citation in note 24.

27 Ockham, II *Sent.*, q.12-13, in O.T. V, p. 280.

28 Ockham, Proemium to *Expositio in Librum Perihermenias Aristotelis* in O.P. II, E. A. Moody, G. Gal, A. Gambatese, S. Brown, P. Boehner (eds.) (St. Bonaventure, NY: The Franciscan Institute, 1978), p. 356.

29 Norman Kretzmann, "Socrates is Whiter than Plato begins to be White," *Nous* **XI** (1977), pp. 3-15.

30 H. Gelber, "The Fallacy of Accident and the *Dictum de Omni*: Late Medieval Controversy over a Reciprocal Pair," *Vivarium* **25** (1987).

31 Cf., for example, S.L. III-II, c.29, in O.P. I, p. 558.

32 J. A. Fodor, *The Language of Thought* (New York: T.Y. Crowell, 1975). Cf. also *Psychosemantics* (Cambridge: The M.I.T. Press, 1987).

33 Among the significant figures in this tradition one may mention Jean Buridan, Buridan's student Pierre d'Ailly, and John Maior. For some discussion of Buridan's view, cf. P. King, *John Buridan's Logic*. For work on Pierre d'Ailly, cf. P. Spade, *Peter of Ailly*. This tradition was controversial, and Hester Gelber has discovered at least one opposing voice in Hugh of Lawton. Cf. H. Gelber, "I Cannot Tell A Lie." For an introduction to the later tradition, cf. E. J. Ashworth, "The Structure of Mental Language: Some Problems Discussed by Early Sixteenth Century Logicians," *Vivarium* **20**, 1 (1982).

34 This paper has benefited from many kind people over many years. Paul V. Spade helped enormously when we were both graduate students at the University of Toronto. The earliest version of the paper itself was read at Princeton University in 1977. The most recent version has benefited from ongoing argument with Hester Gelber and Peter King about whether there are simple mental connotative terms and from the very concrete help of Amelie Rorty.

NOAM CHOMSKY

LINGUISTICS AND DESCARTES

Although Descartes makes only scant reference to language in his writings, certain observations about the nature of language play a significant role in the formulation of his general point of view. In the course of his careful and intensive study of the limits of mechanical explanation, which carried him beyond physics to physiology and psychology, Descartes was able to convince himself that all aspects of animal behavior can be explained on the assumption that an animal is an automaton.[1] In the course of this investigation, he developed an important and influential system of speculative physiology. But he arrived at the conclusion that man has unique abilities that cannot be accounted for on purely mechanistic grounds, although, to a very large extent, a mechanistic explanation can be provided for human bodily function and behavior. The essential difference between man and animal is exhibited most clearly by human language, in particular, by man's ability to form new statements which express new thoughts and which are appropriate to new situations. It is quite easy, in his view, to

> understand a machine's being constituted so that it can utter words, and even emit some responses to action on it of a corporeal kind, which brings about a change in its organs; for instance, if it is touched in a particular part it may ask what we wish to say to it; if in another part it may exclaim that it is being hurt, and so on. But it never happens that it arranges its speech in various ways, in order to reply appropriately to everything that may be said in its presence, as even the lowest type of man can do.[2]

This ability to use language must not be confused with "natural movements which betray passions and may be imitated by machines as well as manifested by animals." The crucial difference is that automata "could never use speech or other signs as we do when placing our thoughts on record for the benefit of others." This is a specific human ability, independent of intelligence. Thus,

> it is a very remarkable fact that there are none so depraved and stupid, without even excepting idiots, that they cannot arrange different words together, forming of them a statement by which they make known their thoughts; while on the other hand there is no other animal, however perfect and fortunately circumstanced it may be, which can do the same.

J-C. Smith (ed.), Historical Foundations of Cognitive Science, 71–79.
Reprinted from *Cartesian Linguistics: A Chapter in the History of Rationalist Thought* (pp. 3-13), with permission of Harper & Row, Publishers, Inc.

Nor can this distinction between man and animal be based on peripheral physiological differences. Thus Descartes goes on to point out that

> ...it is not the want of organs that brings this to pass, for it is evident that magpies and parrots are able to utter words just like ourselves, and yet they cannot speak as we do, that is, so as to give evidence that they think of what they say. On the other hand, men who, being born deaf and dumb, are in the same degree, or even more than brutes, destitute of the organs which serve the others for talking, are in the habit of themselves inventing certain signs by which they make themselves understood.

In short, then, man has a species-specific capacity, a unique type of intellectual organization which cannot be attributed to peripheral organs or related to general intelligence[3] and which manifests itself in what we may refer to as the "creative aspect" of ordinary language use—its property of being both unbounded in scope and stimulus-free. Thus Descartes maintains that language is available for the free expression of thought or for appropriate response in any new context and is undetermined by any fixed association of utterances to external stimuli or physiological states (identifiable in any noncircular fashion).[4]

Arguing from the presumed impossibility of a mechanistic explanation for the creative aspect of normal use of language, Descartes concludes that in addition to body it is necessary to attribute mind—a substance whose essence is thought—to other humans. From the arguments that he offers for the association of mind to bodies that "bear a resemblance" to his, it seems clear that the postulated substance plays the role of a "creative principle" alongside of the "mechanical principle" that accounts for bodily function. Human reason, in fact, "is a universal instrument which can serve for all contingencies," whereas the organs of an animal or machine "have need of some special adaptation for any particular action."[5]

The crucial role of language in Descartes's argument is brought out still more clearly in his subsequent correspondence. In his letter to the Marquis of Newcastle (1646), he asserts that "there is no one of our external actions which can assure those who examine them that our body is anything more than a machine which moves of itself, but which also has in it a mind which thinks—excepting words, or other signs made in regard to whatever subjects present themselves, without reference to any passion."[6] The final condition is added to exclude "cries of joy or pain and the like" as well as "all that can be taught to any animal by art."[7] He goes on, then, to repeat the arguments in the *Discourse on Method*, emphasizing once again that there is no man so imperfect as not to use language for the expression of his thoughts and no "brute so perfect that it has made use of a sign to inform other animals of something which had no relation to their passions;" and, once again, pointing to the very perfection of animal instinct as an indication of lack of thought and as a proof that animals are mere automata. In a letter of 1647 to Henry More, he expresses himself in the following terms:

But the principal argument, to my mind, which may convince us that the brutes are devoid of reason, is that, although among those of the same species, some are more perfect than others, as among men, which is particularly noticeable in horses and dogs, some of which have more capacity than others to retain what is taught them, and although all of them make us clearly understand their natural movements of anger, of fear, of hunger, and others of like kind, either by the voice or by other bodily motions, it has never yet been observed that any animal has arrived at such a degree of perfection as to make use of a true language; that is to say, as to be able to indicate to us by the voice, or by other signs, anything which could be referred to thought alone, rather than to a movement of mere nature; for the word is the sole sign and the only certain mark of the presence of thought hidden and wrapped up in the body; now all men, the most stupid and the most foolish, those even who are deprived of the organs of speech, make use of signs, whereas the brutes never do anything of the kind; which may be taken for the true distinction between man and brute.[8,9]

In summary, it is the diversity of human behavior, its appropriateness to new situations, and man's capacity to innovate—the creative aspect of language use providing the principle indication of this—that leads Descartes to attribute possession of mind to other humans, since he regards this capacity as beyond the limitations of any imaginable mechanism. Thus a fully adequate psychology requires the postulation of a "creative principle" alongside of the "mechanical principle" that suffices to account for all other aspects of the inanimate and animate world and for a significant range of human actions and "passions" as well.

Descartes's observations on language in relation to the problem of mechanistic explanation were elaborated in an interesting study by Cordemoy.[10] His problem in this study is to determine whether it is necessary to assume the existence of other minds.[11] A great deal of the complexity of human behavior is irrelevant to demonstrating that other persons are not mere automata, since it can be explained on hypothetical physiological terms, in terms of reflex and tropism....

In short, Cordemoy is arguing that there can be no mechanistic explanation for the novelty, coherence, and relevance of normal speech. He emphasizes, however, that care must be exercised in using ability to speak as evidence for the inadequacy of mechanistic explanation. The fact that articulate sounds are produced or that utterances can be imitated in itself proves nothing, as this can be explained in mechanical terms. Nor is it of any relevance that "*signes naturels*" may be produced that express internal states or that specific signs may be produced that are contingent on the presence of external stimuli. It is only the ability to innovate, and to do so in a way which is appropriate to novel situations and which yields coherent discourse, that provides crucial evidence....

Cordemoy consistently maintains that the "experiments" that reveal the limitations of mechanical explanation are those which involve the use of language, in particular, what we have called its creative aspect. In this, as in his discussion of the acoustic and articulatory basis for language use and the methods of conditioning, association, and reinforcement that may facilitate acquisition of true language by humans and nonlinguistic functional communication systems by animals, Cordemoy is working completely within the framework of Cartesian assumptions.

For our purposes what is important in this is the emphasis on the creative aspect of language use and on the fundamental distinction between human language and the purely functional and stimulus-bound animal communication systems, rather than the Cartesian attempts to account for human abilities.

It is noteworthy that subsequent discussion rarely attempts to meet the Cartesian arguments regarding the limitations of mechanical explanation. Descartes argued that a "thinking substance" must be postulated to account for the facts that he cites. This proposal is generally countered by the claim that a more complex organization of the body is sufficient to account for human abilities, but no serious attempt is made to show how this might be possible (as Descartes, Cordemoy, and others tried to show how animal behavior and human bodily functions of many kinds can be explained on the basis of assumptions about physical organization). La Mettrie, for example, holds that man is simply the most complex of machines. "He is to the ape, and to the most intelligent animals as the planetary pendulum of Huyghens is to a watch of Julien Leroy" (p. 140).[12] There is, in his opinion, no difficulty in accounting for thought on mechanical principles. "I believe that thought is so little incompatible with organized matter, that it seems to be one of its properties on a par with electricity, the faculty of motion, impenetrability, extension, etc." (pp. 143-144). There should, furthermore, be no obstacle in principle to teaching an ape to speak. It is only "some defect in the organs of speech" that stands in the way, and this can be overcome by proper training (p. 100). "I have very little doubt that if this animal were properly trained he might at least be taught to pronounce, and consequently to know a language. Then he would no longer be a wild man, nor a defective man, but he would be a perfect man, a little gentleman" (p. 103). Similarly, a talking machine is not beyond imagination. "Vaucanson, who needed more skill for making his flute player than for making his duck, would have needed still more to make a talking man, a mechanism no longer to be regarded as impossible..." (pp. 140-141). In short, animal "language" remains completely within the bounds of mechanical explanation as this was conceived by Descartes and Cordemoy.

Evidently, neither La Mettrie nor Bougeant comes to grips with the problem raised by Descartes—the problem posed by the creative aspect of language use, by the fact that human language, being free from control by identifiable external stimuli of internal physiological states, can serve as a general instrument of thought and self-expression rather than merely as a communicative device of report, request, or command.[13] Modern attempts to deal with the problem of intelligent behavior are hardly more satisfactory. Ryle, for example, in his critique of "Descartes's myth"[14] simply avoids the issue entirely. He claims that the Cartesians should have been "asking by what criteria intelligent behavior is actually distinguished from non-intelligent behavior" (p. 21), rather than seeking an explanation for the former. Properly understood, these are not mutually exclusive alternatives. The criteria that Ryle discusses differ little, in principle, from Cordemoy's proposed "experiments"; but whereas Ryle is content simply to cite the fact that "intelligent behavior" has certain properties,[15] the Cartesians were concerned with the problem of accounting for such behavior in the face of their inability to provide an explanation in mechanical terms. It can hardly be claimed that we have advanced significantly beyond the seventeenth century in determining the characteristics of intelligent behavior, the means by which it is acquired, the principles that govern it, or the nature of the structures that underlie it. One may choose to ignore these problems, but no coherent argument has been offered that suggests that they are either unreal or beyond investigation.

Modern linguistics has also failed to deal with the Cartesian observations regarding human language in any serious way. Bloomfield, for example, observes that in a natural language "the possibilities of combination are practically infinite," so that there is no hope of accounting for language use on the basis of repetition or listing, but he has nothing further to say about the problem beyond the remark that the speaker utters new forms "*on the analogy* of similar forms which he has heard."[16] Similarly, Hockett attributes innovation completely to "analogy."[17] Similar remarks can be found in Paul, Saussure, Jespersen, and many others. To attribute the creative aspect of language use to "analogy" or "grammatical patterns" is to use these terms in a completely metaphorical way, with no clear sense and with no relation to the technical usage of linguistic theory. It is no less empty than Ryle's description of intelligent behavior as an exercise of "powers" and "dispositions" of some mysterious sort, or the attempt to account for the normal, creative use of language in terms of "generalization" or "habit" or "conditioning." A description in these terms is incorrect if the terms have anything like their technical meanings, and highly misleading otherwise, insofar as it suggests that the capacities in question can somehow be accounted for as just a "more complicated case" of something reasonably well understood.

Massachusetts Institute of Technology

NOTES

[1] He leaves open, as beyond the limitations of human reason, the question whether the explanatory hypotheses that he proposes are the "correct" ones in any absolute sense, limiting himself to the claim that they are adequate, though obviously not uniquely so. Cf. *Principles of Philosophy*, Principle CCIV.

The context of this discussion of the limits of mechanical explanation must be kept clearly in mind. The issue is not the existence of mind, as a substance whose essence is thinking. To Descartes, this is obvious from introspection—more easily demonstrated, in fact, than the existence of body. What is at stake is the existence of other minds. This can be established only through indirect evidence of the sort that Descartes and his followers cite. These attempts to prove the existence of other minds were not too convincing to contemporary opinion. Pierre Bayle, for example, characterizes the presumed inability of the Cartesians to prove the existence of other minds "as perhaps the weakest side of Cartesianism" ("Rorarius," in the *Historical and Critical Dictionary*, 1697, selections translated by Richard H. Popkin, Bobbs-Merrill, Indianapolis, 1965, p. 231).

[2] *Discourse on Method*, part V. In *The Philosophical Works of Descartes*, translated by E. S. Haldane and G. R. T. Ross, vol. I, p. 116. The other quotations here are from pp. 116-117 of this edition.

In general, I will use English translations where these and the original are readily available and will cite the original otherwise, if available to me. In citing original sources, I will occasionally regularize spelling and punctuation slightly.

[3] For some recent views and evidence on this question, see E. H. Lenneberg, "A Biological Perspective of Language" in E. H. Lenneberg (ed.), *New Directions in the Study of Language* (Cambridge: MIT Press, 1964).

[4] Obviously, the properties of being unbounded and being stimulus-free are independent. An automaton may have only two responses that are produced randomly. A tape recorder or a person whose knowledge of a language extends only to the ability to take dictation has an unbounded output that is not stimulus-free in the intended sense. Animal behavior is typically regarded by the Cartesians as unbounded, but not stimulus-free, and hence not "creative" in the sense of human speech. Cf., for example, François Bayle, *The General System of the Cartesian Philosophy*, 1669 (English translation dated 1670, p. 63): "And because there may be an infinite variety in the impressions made by the objects upon the sense, there may also be an innumerable variety in the determination of the Spirits to flow into the Muscles, and by consequence, an infinite variety in the Motions of Animals; and that the more, because there is a greater variety of parts, and more contrivance and art in the structure." The unboundedness of human speech, as an expression of limitless thought, is an entirely different matter, because of the freedom from stimulus control and the appropriateness to new situations.

It is important to distinguish "appropriateness of behavior to situations" from "control of behavior by stimuli." The latter is characteristic of automata; it is the former that is held to be beyond the bounds of mechanical explanation, in its full human variety.

Modern studies of animal communication so far offer no counterevidence to the Cartesian assumption that human language is based on an entirely distinct principle. Each known animal communication system either consists of a fixed number of signals, each associated with a specific range of eliciting conditions or internal states, or a fixed number of "linguistic dimensions," each associated with a nonlinguistic dimension in the sense that selection of a point along one indicates a corresponding point along the other. In neither case is there any significant similarity to human language. Human and animal communication fall together only at a level of generality that includes almost all other behavior as well.

5 In general, then, "although machines can perform certain things as well as or perhaps better than any of us can do, they infallibly fall short in others, by the which means we may discover that they did not act from knowledge, but only from the disposition of their organs." There are, then, two "very certain tests" by which we can determine whether a device is really human, the one provided by the creative aspect of language use, the other, by the diversity of human action. "It is virtually impossible [in the Haldane-Ross translation, "morally impossible"] that there should be sufficient diversity in any machine to allow it to act in all the events of life in the same way as our reason causes us to act." In taking this position, Descartes expands on his conception of the "cognitive power" as a faculty which is not purely passive and which "is properly called mind when it either forms new ideas in the fancy or attends to those already formed," acting in a way that is not completely under the control of sense or imagination or memory (*Rules for the Direction of the Mind*, 1628; Haldane and Ross, p. 39). Still earlier, Descartes remarks that "from the very perfection of animal actions we suspect that they do not have free will" (c. 1620; cited by L. C. Rosenfield, *From Beast-Machine to Man-Machine* (New York: Oxford University Press, 1941), p. 3, as the first reference by Descartes to the problem of animal soul).

The idea that the "cognitive power" is properly called "mind" only when it is in some sense creative has earlier origins. One source that might well have been familiar to Descartes is Juan Huarte's *Examen de Ingenios* (1575), which was widely translated and circulated (I quote from the English translation by Bellamy, 1698). Huarte understands the word '*Ingenio*' to have the root meaning "engender," "generate"—he relates it to '*gigno*', '*genero*', '*ingenero*' (p. 2). Thus "one may discover two generative Powers in Man, one common with the Beasts and Plants, and the other Participating of Spiritual Substances, God and the Angels" (p. 3). "With [*Ingenio*] is a generative power....the Understanding is a Generative Faculty" (p. 3). As distinct from divine "Genius," the human "rational soul" and "spiritual substances" do not have "sufficient Force and Power in their Generation to give real being to what they Ingender" but only "to produce an accident in the Memory," "an Idea and Image of what we know and understand" that must be given concrete existence by work and art (pp. 4-5). Similarly the arts and sciences are "a sort of Images, and Figures, begotten by [men's] Minds in their Memory, which represent to the Life the Posture and natural Composition of the Subject relating to the intended Science" (p. 6). One who learns some subject must "Engender within himself an entire and true Figure" that represents its principles and structure (p. 6). Truly active minds will be "such, that assisted by the subject only, [they will] without the help of any Body, produce a thousand Conceits they never heard spoke of" (p. 7). The empiricist maxim, "That there is nothing in the Understanding, but what has past through the Sense," attributed to Aristotle, applies only to "docile wits" that lack this capacity. Although the "perfect wit" is only an ideal case, "yet it must be granted, we have observ'd many Persons approach very near it, inventing and saying such things as they never heard from their Masters, nor any Mouth" (p. 16). There is even a third kind of wit "by means of which, some have without Art or

Study spoke such subtle and surprising things, and yet true, that were never before seen, heard, or writ, no nor ever so much as thought of" and which may involve "a mixture of Madness" (p. 17); these three types of wit involve the memory, understanding, and imagination, respectively. In general,

> ...all [man's] Honour and Nobility, as *Cicero* observed, consists in his being favour'd with, and having an Eloquent Tongue: *As Wit is the Ornament of a Man, so Eloquence is the Light and Beauty of Wit.* In this alone he distinguishes himself from the Brutes, and approaches near to God, as being the greatest Glory which is possible to be obtained in Nature. (p. 22)

The most severe "disability of wit," under which men "differ not at all from Brute Beasts," is the disability, which "very much resembles that of Eunuchs...unable for Generation," that prevents the rational faculty from arriving at

> ...the first Principles of all Arts implanted in the Scholar's Mind, before he begin to learn, for which the Wit can give no other proofs of itself, than to receive them as things already known; and if he be not able to form an Idea of them in his Mind, we may strongly conclude him wholly incapable of the Sciences....[In this case,] neither the Lash of the Rod, nor his Cries, nor Method, nor Examples, nor Time, nor Experience, nor anything in Nature can sufficiently Excite him to bring forth anything. (pp. 27-28)

See K. Gunderson, "Descartes, La Mettrie, Language and Machines," *Philosophy*, vol. 39 (1964), pp. 193-222, for an interesting discussion of Descartes's arguments as related to contemporary discussions of "intelligence" of automata. For general background on the development and critique of Descartes's theory of the extent and limits of mechanical explanation, see Rosenfield, op. cit., and H. Kirkinen, "*Les origines de la conception moderne de l'homme-machine*," *Annales Academiae Scientiarum Fennicae*, ser. B, vol. 122 (1961), Helsinki.

[6] Translated (in part) in H. A. P. Torrey, *The Philosophy of Descartes*, (New York: Holt, 1892), pp. 281-284.

[7] That is, by conditioning. When animals are taught "by art," their actions are produced with reference to a passion, in the sense that this behavior is associated with the "stir of expectation of something to eat" or the "motions of their fear, their hope, or their joy" that constitute the original contingency for the teaching. Descartes is therefore pointing out that, just as in its normal use "verbal behavior" is free of identifiable external stimuli or internal physiological states, so it is evidently not developed in the individual by conditioning. He does not elaborate on this, regarding it perhaps as too obvious to merit discussion. It is noteworthy that modern behaviorist speculation about human learning denies these truisms. For some discussion, see Chomsky, "Review of Skinner, 'Verbal Behavior,'" Language, vol. 35, pp. 26-58; *Aspects of the Theory of Syntax* (Cambridge: MIT Press, 1965), chap. 1, section 8; J. Katz, *Philosophy of Language* (New York: Harper & Row, 1966); J. Fodor, "Could Meaning be an 'r_m'," *Journal of Verbal Learning and Verbal Behavior*, vol. 4 (1965), pp. 73-81.

[8] Torrey, *The Philosophy of Descartes*, pp. 284-287. The Descartes-More correspondence, insofar as it relates to animal automatism, is translated in full by L. C. Rosenfield (L. Cohen) in the *Annals of Science*, vol. I (1936), no. 1.

[9] Descartes goes on to explain that he does not deny to animals life or sensation or even feeling, insofar as it depends only on the bodily organs.

10 *Discours Physique de la Parole*, 1666. Page references are to the second edition, 1677. There is an English translation, dated 1668. Rosenfield remarks that Cordemoy develops Descartes's argument involving lack of true speech among animals so fully that after him "the point was given very little attention, as if subsequent authors considered this the last word on the subject" (*From Beast-Machine to Man-Machine*, p. 40).

11 There is no problem, for Cordemoy (as for Descartes), in determining whether he himself possesses a soul, since it is evident to him, by introspection, "*que certaines pensèes accompagnoient toujours en moy la pluspart des mouvemens de mes organes*" (p. 3).

12 La Mettrie, *L'Homme Machine* (1747). A critical edition with notes and background material was published by A. Vartanian (Princeton University Press, 1960). The translations given here are from *Man a Machine* (La Salle, IL: Open Court, 1912); page references are to the 1961 edition, which contains a French text with English translation.

13 This is not to deny that the method of explanation suggested by La Mettrie may be in principle correct. What concerns me here is not the adequacy of the proposed explanations of Descartes and others, but the observations on human language that elicited these attempts.

14 G. Ryle, *The Concept of Mind* (London: Hutchinson's University Library, 1949). See J. Fodor, "Is Psychology Possible?", chap. 1 of *Psychological Explanation* (New York: Random House, 1968), for a critique of the views of Ryle and others regarding psychological explanation.

15 These are described in terms of "powers," "propensities," and "dispositions," which are characterized only through scattered examples. These constitute a new "myth" as mysterious and poorly understood as Descartes's "mental substance."

16 L. Bloomfield, *Language* (New York: Holt, 1933), p. 275. When a speaker produces speech forms that he has not heard, "we say that he utters them *on the analogy* of similar forms which he has heard." For Bloomfield, human language differs from animal communication systems in no fundamental way, but only by its "great differentiation." Other wise, its function is similar. "Man utters many kinds of vocal noise and makes use of the variety: under certain types of stimuli he pro duces certain vocal sounds, and his fellows, hearing these same sounds, make the appropriate response" (p. 27). He holds that "language is a matter of training and habit" (p. 34) and that with careful statistical investigation "we should doubtless be able to foretell how many times any given utterance...would be spoken within a fixed number of days" (p. 37) (a conclusion that is certainly correct, since for almost all normal utterances the predicted number would be zero).

17 C. F. Hockett, *A Course in Modern Linguistics* (New York: Macmillan, 1958), sections 36, 50. He remarks that "it has been said that when ever a person speaks, he is either mimicking or analogizing," and he accepts this view, stating that "when we hear a fairly long and involved utterance which is evidently not a direct quotation, we can be reasonably certain that analogy is at work" (p. 425). Among modern linguists, Hockett is unusual in that he has at least noticed that a problem exists.

In discussing innovation, Hockett seems to imply that novel expressions can be understood only through reference to context (p. 303). In fact, failure to consider the linguistic mechanisms that determine the meaning of the ordinary, generally quite novel sentences of every day life is typical of modern linguistics.

J. THOMAS COOK

SPINOZA'S SCIENCE OF THE "IDEA OF THE BODY"

In her recent book entitled *Neurophilosophy* (1986), Patricia Churchland provides a brief overview of the ideas of selected philosophers whose views have been historically influential in the development of epistemology and philosophy of mind. She credits Kant with the insight that introspective awareness may not yield immediate truths about the way the mind really works, since the mind's observations of itself are mediated by the categories and the form of inner sense.

Kant did of course hold such a view, and in doing so he contributed much to the overthrow of the Cartesian and early empiricist programs of gaining knowledge of the mind and its workings via straightforward introspection. But to suggest that Kant was the first proponent of such a view is to ignore important earlier philosophers. Hobbes and other materialists, for example, had voiced doubts about Descartes's view that introspective awareness provides a reliable source of knowledge regarding the nature and workings of the mind. Of yet greater interest to contemporary cognitive scientists, I think, is the position of Spinoza. Though in many ways a Cartesian, Spinoza explicitly held (in conscious opposition to Descartes) that our introspective knowledge of the mind and its workings is woefully deficient, and that the human mind can gain knowledge of itself only rarely and with great difficulty. According to Spinoza, the mind initially knows itself only by means of the ideas of the ways in which the body is affected by external things. These ideas are highly truncated and fragmentary, and as a result the mind "does not have an adequate, but only a confused and mutilated knowledge of itself...".[1] Overcoming this confusion and mutilation requires first a functional psycho-physical account of ideas and affects, and ultimately the adoption of an entirely different set of concepts and categories for the understanding of one's mind—concepts and categories derived from the study of the body and its interactions with other physical things.

In this view—that understanding the mind requires the adoption of a different set of categories from those of introspection and of everyday "folk psychology"—Spinoza can be seen as a forerunner of influential schools of current cognitive science. Indeed, as I will argue in the following pages, in many ways Spinoza anticipated specific issues and

J-C. Smith (ed.), Historical Foundations of Cognitive Science, 81–97.

controversies in contemporary cognitive science. In one important respect, however, he differed radically from current theorists—a respect which deserves mention at the outset. Spinoza's motive in seeking to gain understanding of the mind was not that of a disinterested philosophical or scientific inquirer. On the contrary, his efforts to understand our cognitive powers were rooted in his conviction that full knowledge of one's own mind is identical with the attainment of "beatitude." At the beginning of Part II of the *Ethics*, we are told that of the infinitely many things which he might undertake to explain, he will focus only on "...those that can lead us, by the hand, as it were, to the knowledge of the human Mind and its highest blessedness."[2] To know one's own mind, he suggests elsewhere, is to know of the "union that the mind has with the whole of Nature,"[3] and in such knowledge lies "salvation."

Given these extravagant claims about the soteriological consequences of "knowledge of the mind," one might suspect that what Spinoza meant by this phrase bears little relation to the issues and problems which interest contemporary cognitive scientists.[4] I shall argue, to the contrary, that Spinoza's approach to the understanding of the mind is remarkably pertinent to present-day concerns, combining elements from what we would call "functionalist," "reductionist," and "replacement" programs. The greater part of his account of our cognitive and affective lives is couched in functional/intentional terms, with little direct attention to the physical states and processes in which the functionally-described states are realized. But Spinoza would have had little sympathy for the notion that such a functionally based cognitive psychology could or should maintain autonomy as a research program. He was profoundly influenced by a belief that there must be "...[one] way of understanding the nature of anything of whatever kind...,viz. through the universal laws and rules of nature" (*Ethics* 3, Preface). Ultimately, the workings of the mind must be understood in terms which reflect the basic laws of motion of the "simplest bodies." Spinoza tried his hand at reductionist identifications of specific types of ideas and affects (emotive states) with specific physically described states, but as I hope to show, his efforts to understand the mind "through the universal laws and rules of nature" brought him into conceptual difficulties quite familiar to the contemporary cognitive scientist. His view that the mind (the "idea of the body") is both identical with and epistemically related to the body, raised for him difficulties much like the problems of intentionality which beset recent efforts to achieve a naturalistic understanding (or elimination) of the mind. And the tensions in Spinoza's final view anticipate, to a remarkable extent, contemporary controversies between those who urge an intentional/propositional approach to understanding the mind and those who would abandon "folk psychology" in favor of a directly neurophysiological account (or elimination) of the mental.

Spinoza deserves our historical attention as a thinker in the early phase of the new "natural philosophy" who sought to bring that natural philosophy to bear upon the understanding of the mind. More than that, he deserves to be reckoned with as a powerful thinker who dealt with problems very similar to those which bedevil us now. We might even hope to learn from him something about the possibly salutary affective and characterological consequences of coming to know one's own mind in terms of the concepts and categories of natural science.

1. The Mind as the "Idea of the Body"

Notoriously, Spinoza was a monist. He held that there is a single substance (indifferently referred to as "substance," "God," or "Nature"), which we know by means of two of its infinite attributes—thought and extension. Substance or God expresses itself as infinitely many modes, each of which can be understood through the attribute of thought (as an idea) or through the attribute of extension (as a physical thing). Whether viewed under the attribute of extension or under the attribute of thought, the "order and connection" of modes is the same. For each and every (simple or complex) mode of extension (physical thing), there is a (simple or complex) mode of thought (an idea) which is identical with the former (since these are in fact one mode of substance understood through two different attributes). The mode of thought thus identified with a certain physical thing Spinoza calls "the idea of that physical thing."

This odd (and oddly attractive) account of the relationship between the physical and the mental has been interpreted in various ways through the centuries. Spinoza has been classified as a panpsychist based on his claim that all things are, in varying degrees, "animata."[5] More recently he has been seen as an early proponent of the view that mental and physical vocabularies are but two ways of talking about one singular thing (usually "the person"[6]). Without entering into the details of these controversies, we need only note, for our purposes, the uses to which Spinoza puts this doctrine in his account of the mind and its workings.

The human mind, we are told, is "the idea of the human body." Thus the mind and the body are one thing (a mode of substance) understood through the attributes of thought and extension, respectively. But there is also a cognitive relation between the two in that the mind is the "idea of the body" in an *epistemic* sense. The mind is cognition of the body—it is an idea which has the body as its intentional object. This view of the identity and epistemic relatedness of mind and body bears a lot of weight in Spinoza's system. For example, having laboriously explained the individuation of bodies (early in Part 2 of the *Ethics*), he is not obliged to provide a further account of the individuation of minds. Since the mind is substantially identical with the body, his account of the individuation of bodies is *eo ipso* an account of the individuation of minds. And since the mind is epistemically related to the body, mental processes can be understood and discussed by reference to the physical processes of which they are cognitive reflections.[7] In more contemporary terminology, the mind can be seen as a monitoring of various physical processes in the body, and thus mental changes can be discussed as reflections of physical changes.

This account of the relationship of mind and body, coupled with the view that all ideas "as they are in God" are true ideas, gives rise to epistemological problems peculiar to Spinoza and to other philosophers of a loosely Neoplatonic orientation. To put it crudely, the problem is not one of understanding how we can come to know about ourselves, but one of understanding how we could *fail* to know all that there is to know about ourselves. If all of God's ideas are true ideas, and if our minds are in fact constituted by various of God's ideas (specifically the ideas of our bodies), how can it be that we seem to be so ignorant and to be so ridden with false ideas about ourselves? Spinoza's answer is that our cognition initially begins with ideas of perception which are in fact ideas of the ways in which our bodies (our sense organs) are affected by external things around us. These ideas are referred to as "images," and this kind of perceptual cognition is called "imagination."[8] There is serious confusion at this level of cognition, for in these ideas there is no distinction made between the nature of the external objects affecting us and the peculiar nature of our own sense organs. We confusedly think that the ideas of the ways in which our bodies are affected by things around us are ideas of those things themselves. We take a complex environmental monitoring device (the body) to be a smooth "mirror of nature," and the confusion thereby engendered is the source of ignorance and error regarding the world and our own bodies and minds.[9]

This remarkably modern view sets the task and indicates the direction for the remainder of Spinoza's philosophical program. The mind and body must be understood in order that we might be freed from the confusion and error to which we are subject in the course of normal sense-perception (imagination). Understanding of the mind requires a greater understanding of the body (whose idea the mind is) and of the ways in which that body is affected by things in the natural environment.

2. Confused and Inadequate Ideas of Imagination

Profoundly influenced by the Stoics,[10] Spinoza holds that among the most debilitating consequences of our ignorance and confusion are certain affects (passions) which disturb our peace of mind, undermine our ability to think in an orderly way, and generally keep us in their thrall. Thus, his fullest discussion of the mind's ability to achieve knowledge of itself is pursued under the rubric of "overcoming the passions." Since these destructive emotions just are confused and inadequate ideas of certain kinds, an account of their etiology and the means whereby they can be overcome provides an introduction into the details of Spinoza's science of the mind.

According to Spinoza, the human body is a physical organism whose identity consists of a certain relatively constant ratio of motion and rest among the parts which make it up.[11] The body is characterized by a homeostatic tendency to maintain its individual physical integrity while being affected in various ways by things in the surrounding environment. This tendency toward self-maintenance Spinoza calls an individual's endeavor to "persevere in its being."[12] In the normal course of things, the body is sometimes affected in ways which increase its power to preserve itself intact, and sometimes in ways which decrease that power. The mind of an individual, as the idea of the body, reflects all of this, though in a way which is subject to the confusion characteristic of all ideas of the "imagination." The individual's endeavor toward self-preservation, when understood through the attribute of thought, is called "will" or "desire."[13] The idea of the increase in the body's power for self-maintenance is identified as joy, and the idea of the body's decrease in such power is sadness. From these three basic affects (desire, joy, and sadness), Spinoza constructs an elaborate and systematic taxonomy and etiological theory of the emotions—a theory whose objects play a role in his account of behavior, as well as of cognition and affect.

Further passions are understood in terms of these basic three plus the mind's (confused) ideas of the causes of these. For example, love is seen to be "joy, with the accompanying idea of an external cause." Our desire or will to bring about states of affairs which we love is but the mental reflection of the body's natural homeostatic endeavor to realize that which increases its power of self-preservation. The possibilities for understanding other emotions in terms of these (plus the ideas of the causes of these) are nearly endless, and Spinoza exploits these possibilities at great length in Part III of the *Ethics*. We need not follow his exposition of these matters, however, for we already have enough before us to sustain an investigation of his program of overcoming the detrimental passions by understanding the workings of the body and its mind.

3. Functional and Fine-Structure Accounts

On Spinoza's account, the mind is constituted by ideas of the body and of the "affections" of the body (ways in which the body is affected by external things). These affections of the body are generally understood in functional terms. The relevant states of the body are identified as those which indicate the presence of external bodies or which indicate the body's increased or decreased power to preserve itself in the environment. These states are further identified as the causes of certain behaviors of the body (shunning destructive things, seeking the presence of loved things). Two questions naturally suggest themselves to us: (1) Can these functionally described states be identified with specific physiologically identified states or events at the micro-level? (2) Can the quasi-teleological and intentional remnant preserved in the phrase "the body's endeavor to persevere in being" be eliminated at a deeper level of description? Both of these are questions hotly disputed among Spinoza scholars. I will briefly defend an affirmative answer to both.

Two factors help to explain Spinoza's emphasis upon functional descriptions of the bodily states whose ideas constitute the mind. First, and most obviously, his knowledge of the fine structure of physiology was primitive at best (see below). Secondly, his primary purpose was therapeutic, and for this purpose he needed states which corresponded neatly with the passions as commonly referred to in "folk-psychology."[14] The explanatory and predictive success of folk psychology is best accounted for on the assumption that its referents are functionally defined states.[15] As we will see, his first-stage remedy for overcoming the passions does not require a detailed structural account of every minute mechanism of the workings of the body.

Nonetheless, I think that he was committed to the view that there is at least a token-token identity between any functionally described bodily state and a state described in purely fine-structure physical terms. As evidence, we might cite the passage in which he is concerned to explain how we sometimes regard things as present to us which are in fact not present at all. He might simply have spoken of the "affection of the body which indicates the presence of the thing," but he goes into much more detail:

> While external bodies so determine the parts of the human body that they often thrust against the softer parts, they change their surfaces with the result that they are reflected from it in another way than they used to be before; and still later, when the fluid parts, by their spontaneous motion, encounter those new surfaces, they are reflected in the same way as when they were driven against those surfaces by the external bodies. Consequently while, thus reflected, they continue to move, they will affect the human body with the same mode, concerning which the mind will think again; i.e., the mind will again regard the external body as present. (*Ethics*, 2, P. 17, Cor., Dem.)

Aside from revealing the primitive character of the kind of physiological theory available to Spinoza, this passage illustrates the fact that for a typical "affection of the body" there is a fine-structure description of the physiological state or process in question. Spinoza admits his ignorance and uncertainty regarding the details of the story, but in discussing this passage says that "...it is sufficient for me here to have shown one [cause] through which I can explain it as if I had shown it through its true cause." There is never any question about there being a true causal story at the fine-structure level. There certainly is such a story, but it is unknown and is in any case unnecessary for Spinoza's philosophical-therapeutic program.

Spinoza holds that every action undertaken by the human body can be (and must be) thoroughly explained in terms of the body and its (corporeally described) interactions with the things and bodies around it. He is resolutely opposed to the (Cartesian) claim that explanation of the body's actions requires, at some points, the introduction of mental acts as such.[16] The desire or will to eat, for example, is the mental reflection (a better phrase would be 'mental version') of the body's endeavor to be affected in ways which increase its powers of self-maintenance. The will or desire cannot explain that endeavor nor explain the event of the body's eating. Explanation of the body's endeavor and of its eating must ultimately be in purely corporeal terms.

There are of course difficulties here, and the second of the abovementioned questions raises the most serious of them. One might object that talk of the body's "endeavor to be affected in ways which increase its power of self-preservation" is not just innocent functionalism, but is outright teleology and purposiveness introduced at the ostensibly purely physical level. At the very least this kind of talk suggests a suspicious degree of intentionality at the level of physical description. Spinoza's failure to deal squarely with the former issue is rather surprising, especially in light of the clarity with which he attacks teleological thinking in other areas.[17] In his defense it might be noted again that on his view a body's endeavor to persevere in its being was but an extension of the Cartesian principle of inertia. He may have thought (rightly) that physical explanations in terms of such "immanent teleology" were less objectionable than explanations in terms of God's purposes.

The issue remains troublesome, though, for regardless of the teleology question, talk of "endeavor to do that which will preserve my being" is clearly intentional talk, and the status of such intentionality is a problem. I argued above that Spinoza embraces what we would call a token-token identification of functionally described states and fine-structure physiologically described states. What of the intentional elements of his functional descriptions? Does he think that ultimately a story can be told at the micro-level which explains the behaviors of the body without attribution of intentionality at that level? Spinoza is not clear on this point, in part because the concept of intentionality was not clearly formulated in the seventeenth century. Led on by what seemed to him a smooth transition from inertia to a body's "endeavor to persevere in its present state" to a body's "endeavor to do that which is conducive to its perseverance in its present state," he never clearly dealt with the discontinuity marked by the introduction of the intentional element. Had he confronted this issue, I think, he would have held that the intentional element is dispensable in the final micro-level account of the body and its actions and passions. The kind of story which he envisions at that level (as presented early in Part 2 of the *Ethics*) is a mechanical story strongly influenced by Cartesian geometrically-based physics. It is a story of "simple bodies" distinguished from one another by their "motion and rest," characterized by a tendency to maintain their characteristic motion unless and until affected by other simple bodies.[18] This kind of story does not involve intentionality, and if the story of the human body can ultimately be told in these terms, then the intentional element in his functional account of bodily states will turn out to be dispensable.[19]

4. The Inadequacy of the Mind's Knowledge of Itself

I have belabored Spinoza's various accounts of the processes and states of the body in order to be able to deal more expeditiously with the issue of the mind's self-knowledge. A summary review is in order here: Our ideas of perceptual objects are in fact ideas of various affections of our bodies (i.e., they are modes of substance which, when understood through the attribute of extension, are ways in which our bodies are affected by other things, and which, when understood through the attribute of thought, are ideas of those ways in which the body is affected). The passions, mentally understood, are ideas of certain functionally identified bodily states and processes in which the body endeavors to preserve itself or in which the body is strengthened or weakened in its power for such self-preservation. These functionally identified bodily states play a role in the explanation of bodily behavior. They are type-type identical with the passions as identified in folk psychology and are token-token identical with physical states identifiable in terms of micro-structure. If, as Spinoza explicitly holds, the "order and connection of modes" is the same under the two attributes, then (odd as it sounds) our ideas of functional states must be token-token identical with ideas of micro-structure states of the body.

This should be sufficient to indicate the extent to which Spinoza holds that our normal introspection fails to reveal the truth about our own minds. Crick's claim that "we are deceived at every level by our introspection" comes to mind.[20] I do not normally think of my ideas as identical with bodily states, nor do I think of my perceptions or my passions as ideas of affections of my body. But according to Spinoza that is what my ideas are, and my mind consists solely of such ideas. In order to understand how the mind works, we must understand how the body works—and the more fully and accurately we understand the body's workings, the more fully and accurately we will understand the mind's workings. A genuinely adequate understanding of the body and mind will indicate that most of what we have thought to be the case about the latter was terribly confused.

5. Overcoming the Confusion

The account of the passions as ideas of functionally described bodily states yields an explanation of the confusion which attends such ideas and provides the beginning of a remedy for such confusion. Love, hatred, anger, jealousy are all ideas of changes in the body's power to preserve itself accompanied by the idea of an external cause of such changes. But the causal story involved in such ideas is highly unreliable. In an almost Freudian manner, Spinoza recounts processes whereby we can come to be affectively disposed toward things in inappropriate ways. Erroneous beliefs about causal connections can arise as a result of non-rational associations of ideas based on superficial resemblances, accidental spatial or temporal coincidence of things, or even bodily pathology. Understanding of the mind's susceptibility to such irrational affects requires an understanding of the etiology of the passions in terms of the body. More specifically, perceptual knowledge systematically misrepresents certain general features of the world. Our eccentric perspective on things (given that we know them only by their fortuitous encounters with our bodies) gives rise to the misguided idea that there is contingency in nature and "free will" in persons. These erroneous ideas potentiate destructive passions, and they can be overcome only by understanding their source and by focusing upon the law-governed character of nature when known scientifically.

All of these therapeutic consequences follow from an understanding of the passions as confused perceptual ideas of functionally described bodily states. Most of Spinoza's discussion in Parts 3 and 4 of the Ethics proceeds at this level. Understanding of this sort relieves us of the illusion that our ideas, passions, and other psychological states constitute an endogenous and autonomous realm to which we have direct and accurate cognitive access and over which we have direct voluntary control. Such understanding is an important step toward what Spinoza calls "liberation from the passions," but it is not the last word, for such an understanding of the mind and body does not represent the final truth about these—it does not provide knowledge of them "as they are in God." As I interpret the crucial propositions of Part 5 of the *Ethics*, ultimate liberation requires knowledge of the mind in terms of the most basic natural processes; i.e., knowledge of the mind as ideas of the *non*-intentional, microstructural features of the body.

Spinoza's account of the method by which such knowledge can be achieved is one of the least plausible aspects of his entire system. He adhered to a kind of Cartesian geometrical physics, and since he took geometry to be an *a priori* science, he was confident that the covering laws for a genuine science of the body could be had *a priori*.[21] This odd view (by contemporary standards) of how we come to know the basic physical laws should not distract us from the more interesting conceptual issues involved in the claim that coming to know the workings of the body constitutes coming to have knowledge of the mind.

As seen above, the closest Spinoza gets to a fine-structure account of the physical counterparts of ideas is with terms such as 'soft parts' and 'fluid parts'. These terms, it seems, can be cashed out in terms of the degree of relative motion among the "simplest parts" which make up organs, tissues, etc. Spinoza's ultimate story will be in terms of myriad tiny particles moving about, bouncing off one another, communicating their motions one to another, etc., all in accordance with basic laws of geometrical physics. The mind, we are to believe, is identical with these physical processes (in the way in which a mode understood through two different attributes is nonetheless a single mode).

This claim has all the initial implausibility of recent claims to the effect that my pains *are* C-fiber firings and that my belief that Pegasus has wings *is* a process or state in the brain. Pains and specific beliefs, after all, seem to have intrinsic and immediately knowable properties which brain processes do not have. This objection, thought by some to be damning to any identity theory of a reductionist or eliminativist kind, articulates one of the main *attractions* of the identity theory from Spinoza's perspective. Awareness of the mind as ideas of certain physiologically described brain events would indeed lack a number of qualitative features of our present everyday consciousness as we know it.

But as we have seen, Spinoza holds that our everyday consciousness is confused and misleading with regard to the true nature of the mind, and if some of that confusion should be dispelled by gaining a better understanding, so much the better.

For example, our everyday consciousness suggests to us that we are engaged in an effort to maintain ourselves in an environment fraught with threats to our survival. We experience ourselves as subject to conflict, sometimes within ourselves and more often between ourselves and other things and people. All of the passions discussed above arise as a consequence of our perception of our relative success or failure in that effort to maintain ourselves. However, an important characteristic of descriptions and explanations of things at the level of the most basic physical laws is that the notion of "conflict" plays no role, and (assuming that teleology is expunged at that level) there is no mention of "effort." There is no effort involved in neurophysiological processes (so described), and chemicals (or particles) acting in accordance with the basic laws of chemistry or physics do not conflict with one another. Thus, for a mind which knows itself in physiological terms, there is no conceptual foothold for generating the debilitating passions which concern Spinoza as philosophical therapist.

Thus, while Spinoza's attraction to such a view is understandable, it is not clear that the position is consistent with the rest of his program. Commentators have often noted that there seems to be a profound tension between the first level program for overcoming the passions as outlined in Parts 3 and 4 of the *Ethics* and the final version of liberation as described in Part 5. It seems as if the entire conceptuality which underlay the earlier program—a conceptuality involving effort, endeavor, time, passivity, etc.—is undermined in the final view. If the person who truly knows her own mind "as it is in God" does not know herself in terms of these concepts, then what was that first-level program for overcoming the passions all about? The answer which Spinoza must give, I think, is that it was a program based on ideas which, from the perspective of the final truth, are revealed to be confused ideas.

It is not clear that this response makes sense, however, for from the perspective of the final truth, the conflict and passivity which generated the confusion are revealed not really to have been conflict and passivity at all. It is not open to Spinoza to appeal to conflict-induced confusion to *explain* the confusion which gave rise to ideas of conflict and passivity. This difficulty in Spinoza's system is remarkably similar to the difficulty which besets the materialist in the attempt to explain away the mind. As Richard Rorty pointed out back in 1965, "that great dumping ground of unwanted entities, the mind," is not available to the materialist as a way of explaining what the putative "ideas" were which are being claimed not to exist as such. Confronted by the dualist who asks, "What was it, then, the sharpness of which I experienced when I experienced a sharp pain?", the materialist must reply that there was nothing sharp and no sharpness. One might say that sharp pains are confused ideas of C-fiber firings, but "confused ideas" sounds irremediably mentalistic. Shall the eliminativist say that sharp pains are confused ideas and then say that there are not really any ideas at all but only complex brain processes? That, it seems to me, is the sort of thing that he inevitably does say, and it is much like what Spinoza must say.

Spinoza's view that the mind just is ideas of the body is helpful here, for as one comes to think differently about the body, one's mind is becoming different ideas. These newly acquired ideas (ideas of physical processes, physically described) may not have the properties or qualities which require the invocation of notions of confusion to explain. Thus Spinoza could say that the need for an explanation of various properties of putatively confused ideas would not arise for the individual who has come to know him or herself in terms of nature's most basic ways of acting (a "disappearance strategy").[22] A similar strategy is available to the eliminative materialist if he can maintain that due to the theory-ladenness of experience, one's experience would change radically as a consequence of accepting a non-folk-psychological theory of oneself. Perhaps such an experience would not manifest, say, the sorts of phenomenal qualities which seem to some to defy the materialist's account.[23]

My purpose is not to resolve these issues here, but to point out remarkable similarities between the tensions which arise in Spinoza's final position and the difficulties encountered by recent eliminative materialist positions. The problem of intentionality is a final issue which deserves mention. Spinoza defined the mind as the "idea of the body," and as we saw he held that that idea was both identical with and intentionally related to the body whose idea it is. At the level of final understanding, there is a difficulty involved in affirming both the identity and the intentionality, for the physical states of which the mind is the idea are not intentional states. How could the mind have an intentional structure if it is identical with physical states which are not intentionally structured? This issue of isomorphism (between intentionalist and physicalist, micro-structure descriptions) finds no resolution in Spinoza, and recent literature suggests that it may be among the toughest problems for contemporary cognitive scientists. Functionalists can deal with this issue (as could Spinoza, at the level of functional descriptions of the body), but the eliminativist is in the position of needing to make sense of reference and other intentional notions without benefit of the aboutness of mental states. In an interesting historical study of the concept of "*amor dei intellectualis*" (a phrase which Spinoza uses to describe the highest kind of self-knowledge), G. A. Wyneken (1898) describes it as the mind's effort "to purge itself of all traces of subjectivity in order to make of itself, as of all other things, an object." This seems an apt description of Spinoza's therapeutic program and of the tendency (if not the intent) of the Churchlands' program today.

6. Conclusion

In these pages I have been interpreting Spinoza's system in ways which reveal its similarity and relevance to certain issues in contemporary cognitive science and philosophy of mind. Some of the terminology has been anachronistic in application to the thought of a seventeenth century philosopher, though I hope not seriously distortive of Spinoza's intent. One of the most interesting characteristics of Spinoza's philosophy is the fact that it lends itself to plausible interpretation in contemporary terms as well as in terms of Stoicism and Neoplatonism. Some have convicted him of radical incoherence on this account, but while there are certainly problems in his view, such a judgment seems to me unwarranted. He was a philosopher with one foot planted in the Stoic and Neoplatonic traditions and the other striding out into the new "natural philosophy." He shared the ancient view that self-knowledge is the route to salvation, and he shared the modern view that real knowledge consists of an understanding of things in accordance with mathematically (or geometrically) formulated laws of nature.

As I understand Spinoza's views, these two considerations drew him toward an understanding of the mind in terms of basic physical processes. He was a methodological monist (as well as a metaphysical monist) who believed (as quoted above) that there should be "...[one] way of understanding the nature of anything, of whatever kind...viz. through the universal laws and rules of nature" (*Ethics* 3, Preface). Of equal influence in his thinking, however, was the Neoplatonic view according to which apparent "imperfections" in nature must be the result of our nescience or confusion. Insofar as effort and conflict, for example, seemed to him to be imperfections, he was drawn toward an understanding of the human being and the human mind in terms of a theory which has no conceptual place for notions of effort and conflict.

Since he held that the mind *is* ideas, he could hold that the individual who comes to think of himself in terms of the most basic regularities of the activity of nature has a mind which is constituted primarily by the ideas of these most basic "ways in which God acts." Since such ideas make no reference to endeavor or purpose, the extent to which one knows oneself in this way is the extent to which one knows nothing of either. Given Spinoza's view of the timelessness (eternity) of these basic laws, one's mind comes to be constituted in great part by ideas which make no reference to time, and hence one knows oneself in a way which de-emphasizes duration and participates in eternity. Finally, since notions of good and evil are ideas which arise from our regard for the ways in which things affect our efforts to endure and preserve ourselves, such notions are of little significance to someone whose mind consists primarily of ideas which make no reference to effort or to duration.

Coming to know oneself in this way, and hence having a mind consisting primarily of ideas such as these, is Spinoza's version of ultimate liberation from the passions. This kind of "salvation" or "beatitude" has affinities with Stoic "apatheia" as well as with Neoplatonic "self-transcendence" and "union with 'the One'." These days, we have stopped expecting such soteriological results from philosophy—certainly from naturalistic, scientifically oriented philosophy of psychology—although Wittgenstein's similar account of the "happy" man seems not too distant. Spinoza's interesting claim is that these salutary consequences follow directly upon an understanding of one's mind in terms of the most basic physiological processes of the body. Perhaps this may warrant outright dismissal of his views as unrelated to contemporary concerns, despite the conceptual similarities discussed above. On the other hand, one might regard those similarities as grounds for rethinking the possible affective and experiential ramifications of actually adopting an eliminative materialist's perspective on the mind.

Rollins College

NOTES

[1] *Ethics*, 2, P. 29, Cor. In further references to the *Ethics* I will follow the standard practice of listing the part and proposition number of the cited passage. All quotations are from E. M. Curley's translation (1985).

[2] *Ethics*, Part 2, Preface.

[3] This passage lends itself both to a Stoic and a mystical interpretation (as Curley points out in his recent translation).

[4] This suspicion would be reinforced by those influential interpretive approaches to Spinoza's work which view him as a late medieval philosopher or a latter-day Renaissance Neoplatonist. For the former interpretation, see Wolfson's magisterial work (1934). For the latter, see Gebhardt (1921). These interpretations render Spinoza's views irrelevant to concerns of cognitive science only by systematically underemphasizing his commitment to the emerging "natural philosophy" of his day.

[5] *Ethics*, 2, P. 13, Scholium.

[6] See, for example, Stuart Hampshire (1972).

[7] Technically, Spinoza is careful not to allow for the explanation of mental events by means of physical events, for he wants to avoid the embarrassments of Cartesian interactionism (*Ethics*, 2, P. 6). But since ideas are discussed by reference to the physical things of which they are ideas (their intentional objects), adherence to this methodological stricture sometimes requires cumbersome circumlocution.

[8] It is important to emphasize that on Spinoza's view, my idea of Bill when I am looking at him and my idea of Bill when he is across town are *both* "images" of Bill and instances of "imagination."

[9] See *Ethics*, 2, P. 29, Cor.

[10] For impressive evidence of the depth of the Stoic influence upon Spinoza, see Wilhelm Dilthey (1929).

[11] This one-sentence summary hardly does justice to Spinoza's brilliant account of individuation of physical things (especially living things). Those interested in conceptual development in the history of biology should see Hans Jonas (1973).

[12] This principle is Spinoza's extension of the concept of inertia (as formulated by Descartes) to apply to composite bodies in their tendency to maintain their physical integrity over time.

[13] Spinoza actually differentiates between (1) the endeavor when referred to the body alone ("appetite"), (2) the endeavor when referred to the mind alone ("will"), and (3) the endeavor when related to the mind and body together ("desire"). See *Ethics* 3, P. 9, Schol.

[14] I am using (loosely) a term from contemporary cognitive science here. 'Folk psychology' is an appropriate term in that it refers to common accounts of attitudes, beliefs, and behavior—those which employ the notion of "belief that something is the case" and accept uncritically the taxonomy of propositional attitudes, of emotions, etc., which are current in common parlance. In fact, while indeed retaining the intentional idiom (see below), Spinoza's theory of the passions provided something like a reductive account not of emotions as discussed in the streets of Amsterdam but as discussed in traditional learned (Latin) works. He drew heavily on Hobbes and on Descartes's *Passions of the Soul* in formulating his theory.

[15] Spinoza did not, of course, use the term 'functional' or 'functionally defined state', but his accounts of specific affects are clearly and precisely accounts in terms relationally abstracting what we would call "functionally defined states of the body."

[16] Against those who insist upon introducing the mental into the causal story, Spinoza writes:

> They will say, of course, that it cannot happen that the causes of buildings, of paintings, and of things of this kind, which are made only by human skill, should be able to be deduced from the laws of nature alone, insofar as it is considered to be only corporeal; nor would the human body be able to build a temple, if it were not determined and guided by the mind. But I have already shown that they do not know what the body can do, or what can be deduced from the consideration of its nature alone....

> I add here the very structure of the human body which, in the ingenuity of its construction, far surpasses anything made by human skill.... (*Ethics* 3, P. 2, Schol.)

[17] See, for example, his razor-sharp polemic against the attribution of purposes to God in the Appendix to Part I of the *Ethics*.

[18] As I understand it, the discussion of "composite bodies" does not introduce a kind of entity whose account requires irreducible and new principles of explanation. Composite bodies are "constellations" of simple bodies fully governed by the laws of motion of such simple bodies. (The term 'constellation' is from Walther, 1971.) See also my "Self-Knowledge as Self-Preservation" (1986).

[19] This is a controversial interpretation which deserves a much more extended defense than I have given it here. Some of that further defense can be found in my "Self-Knowledge as Self-Preservation" (1986).

[20] Cited in Patricia Churchland (1986).

[21] Actually, the story is much more complicated than this. On Spinoza's view the nomological ways in which God or substance eternally and everywhere acts are "contained in" every extended mode of substance as partially causally constitutive of the mode. (I am taking the "fixed and eternal things" of the *De Intellectus Emendatione* and the "infinite and eternal modes" of the *Ethics* to be the ways in which God eternally and everywhere acts—those "things" which we try to describe when we formulate laws of nature. Curley, 1969, refers to them as "nomological facts.") Since these are contained in every mode of substance, their ideas are contained in, and partially causally constitutive of, every idea of an extended mode. Hence these laws are contained in the human mind *a priori* (in an odd sense, innately) and can be accessed by a process of overcoming the confusion involved in ideas of the imagination. The suggestion that the mind of the person-in-the-street implicitly contains ideas of the fundamental laws of nature sounds wild to our ears. But it is not surprising that Spinoza held this, since he took geometry to be the science of extended nature, and geometry had long been held to be an *a priori* science. For further discussion of this point, see my "Self-Knowledge as Self-Preservation" (1986).

[22] Spinoza does take a position similar to this in his discussion of ideas of good and evil (*Ethics* 4, P. 68). In other passages, however, he says that the "images" of confused cognition would remain, though they would constitute the lesser part of the mind of the individual who truly knows herself and her own mind.

[23] Something like this seems to be Dennett's strategy in "Quining Qualia." The recent exchanges between Paul Churchland and Frank Jackson on the issue of "monochrome Mary" revolve around just this point (Paul Churchland, 1985, 1986; Frank Jackson, 1982, 1986). It seems to me that Churchland ought to bite the bullet and say that Mary, upon con fronting the tomato outside her black and white room, will not experience red qualia, but will experience certain "reflectance efficiencies" in an object or (better) certain neurophysiological goings on in her retina or in her visual cortex. Churchland seems more inclined to pursue a reductionist strategy with regard to qualia (as in his recent *Mind* paper). I have recently argued these issues in "Neuro-Scientific Self-Knowledge and Experiential Change" (read at the Central Division APA, 1987).

REFERENCES

Patricia Churchland (1986), *Neurophilosophy* (Cambridge, MA: M.I.T. Press).

Paul Churchland (1985), "Reduction, Qualia, and the Direct Introspection of Brain States," *The Journal of Philosophy* **LXXXII**, 1, pp. 8-28.

Paul Churchland (1986), "Some Reductive Strategies in Cognitive Neurobiology," *Mind* **95**, pp. 279-309.

J. Thomas Cook (1986), "Self-Knowledge as Self-Preservation" in Grene and Nails (eds.), *Spinoza and the Sciences*, Vol. 91 of the Boston Studies in the Philosophy of Science (Dordrecht: D. Reidel).

E. M. Curley (1969), *Spinoza's Metaphysics: An Essay in Interpretation* (Cambridge, MA: Harvard University Press).

Wilhelm Dilthey (1929), *Weltanschauung und Analyse des Menschen Seit Renaissance und Reformation, Band II of Dilthey's Gesammelte Schriften* (Leipzig and Berlin: Tuebner Verlag), pp. 285-289.

Carl Gebhardt (1921), "Spinoza und der Platonismus" in *Chronicum Spinozanum* (Haag).

Stuart Hampshire (1972), "A Kind of Materialism" in *Freedom of Mind and Other Essays* (Oxford: Oxford University Press).

Frank Jackson (1982), "Epiphenomenal Qualia," *Philosophical Quarterly* **32**, pp. 127-136.

Frank Jackson (1986), "What Mary Didn't Know," *The Journal of Philosophy* **LXXXIII**, pp. 291-295.

Hans Jonas (1973), "Spinoza and the Theory of Organism" in Marjorie Grene (ed.), *Spinoza: A Collection of Critical Essays* (Garden City, NY: Anchor Books).

B. Spinoza (1985), *The Collected Works of Spinoza*, Vol. I, trans. by E. M. Curley (Princeton, NJ: Princeton University Press).

Manfred Walther (1971), *Metaphysik als Anti-Theologie: Die Philosophie Spinozas im Zusammenhang der Religionsphilosophische Problematik* (Hamburg: Felix Meiner Verlag).

H. A. Wolfson (1934), *The Philosophy of Spinoza* (Cambridge, MA: Harvard University Press).

G. A. Wyneken (1898), *Amor Dei Intellectualis: Eine Religionsphilosophische Studie* (Greifswald: Julius Abel Verlag).

FREDERICK KREILING

LEIBNIZIAN RESONANCES

...there would be no more need of dispute between two philosophers than between two accountants. It would suffice for them to take their pencils in their hands, sit down to their slates, and say to each other...: "Let us calculate."
—Gottfried Wilhelm Leibniz

Wovon man nicht reden kann, darüber muss man schweigen.
—Ludwig Wittgenstein

There is so much of Leibniz that relates to our epistemological situation, and to our current concerns as well, that a few brief examples might be helpful at the outset to display the scope of his thought. In physics, Albert Einstein wrote:

This is the gist of the principle of equivalence: In order to account for the equality of inert and gravitational mass within the theory, it is necessary to admit non-linear transformations of the four coordinates. That is, the group of Lorentz transformations and hence the set of the "permissible" coordinate systems has to be extended. What group of coordinate transformations can then be substituted for the group of Lorentz transformations? Mathematics suggests an answer which is based on the fundamental investigations of Gauss and Riemann: namely, that the appropriate substitute is the group of all continuous (analytical) transformations of the coordinates. Under these transformations the only thing that remains invariant is the fact that neighboring points have nearly the same coordinates; the coordinate system expresses only the topological order of the points in space including its four-dimensional character. The equations expressing the laws of nature must be covariant with respect to all continuous transformations of the coordinates. This is the principle of general relativity.

The procedure just described overcomes a deficiency in the foundations of mechanics which had already been noticed by Newton and was criticized by Leibniz and, two centuries later, by Mach: Inertia resists acceleration, but acceleration relative to what? Within the frame of classical mechanics the only answer is: Inertia resists acceleration relative to space. This is a physical property of space—space acts on objects, but objects do not act on space. Such is probably the deeper meaning of Newton's assertion *spatium est absolutum* (space is absolute). But the idea disturbed some, in particular Leibniz, who did not ascribe an independent existence to space but considered it merely a property of "things" (contiguity of physical objects). Had his justified doubts won out at that time, it hardly would have been a boon to physics, for the empirical and theoretical foundations necessary to follow up his idea were not available in the 17th century.[1]

In mathematics, Benoit Mandelbrot, the discoverer of fractal geometry, also attests to the Leibnizian heritage:

J-C. Smith (ed.), Historical Foundations of Cognitive Science, 99–118.

Several basic ideas of fractals might be viewed as mathematical and scientific implementations of loose but potent notions that date back to Aristotle and Leibniz, permeate our culture and affect even those who think they are not subject to philosophical influences.[2]

In the development of probability theory, Ian Hacking notes that Leibniz,

> ...was the first philosopher of probability. He was the first to insist that probability theory can serve in a branch of logic comparable to the theory of deduction. Before coming to Paris, and before mastering the work of "Pascal, Huygens and others" he had tried to develop an arithmetic of probability that was not based on games of chance, and hence was potentially more general in application. He also wrote the first monograph on the theory of combinations, and saw its relation to probability theory. He was the first to try to axiomatize probability as a pure inferential science. He saw how a generalized theory of games should be the foundation for making any quantitative decision in situations where one must act on inconclusive evidence. If we abstract him from his environment, he often seems a twentieth century figure.[3]

In the theory of perception, Leibniz directly noted the crux of an argument that persists in modern disputes:

> Moreover, it must be confessed that perception and that which depends upon it are inexplicable on mechanical grounds, that is to say, by means of figures and motions. And supposing there were a machine so constructed as to think, feel, and have perception, it might be conceived as increased in size, while keeping the same proportions, so that one might go into it as into a mill. That being so, we should, on examining its interior, find only parts which work one upon another, and never anything by which to explain a perception. Thus it is in a simple substance, and not in a compound or in a machine, that perception must be sought for. Further, nothing but this (namely, perceptions and their changes) can be found in a simple substance. It is also in this alone that all the internal activities of simple substances can consist.[4]

Further examples of Leibniz's contributions and provocative insights from calculus to mining, from history to biology, could easily be added. The breadth and intensity of the historical role of the "last man who knew everything" has gained visibility with the rise of the history of science.

1. Leibniz, the Practical Philosopher

That the father of the German idealistic tradition was also an active politician, engineer, and inventor should help dispel notions about the exclusivity of its intellectual or technical or socioeconomic roots. Leibniz knew, after all, that civilization has been an ecological process arising from the interplay of widely diverse human contributions and social institutions. Indeed, the very meanings of the words 'art', 'science', and 'technology' have varied greatly throughout history. Many ambiguities and overlappings occur. A positivist might object to this on the grounds that only science has predictive power, but in our age of probability even that distinction has become somewhat blurred. Moreover, we have forms of science with well-founded paradigms, such as evolutionary biology, in which by and large only retrodiction is possible.

The work of farmers and mechanics, carpenters and fishermen, as well as of a host of others on which civilized life depended and still depends, evoke a tradition that stretches back to antiquity and beyond. Much remains to be studied. For example, we still do not know the origins of the carpenter's toolbox nor, for that matter, the farrier's art of gelding, so important for early agriculture. These basic human skills, essential for the material and even the social foundations of culture, long remained as empirical strongholds, in the sense that they were precise perceptual arts or 'ways of the hand' practiced by men and women who had painfully mastered their craft through an arduous apprenticeship: small wonder that they remained firmly committed to exquisitely traditional practices. During the sixteenth and seventeenth centuries, craftsmen contributed to the scientific revolution in many important ways. They provided the articulation and demonstration of interesting problems which challenged rational and systematic inquiry, such as those related to bridge building, power transmission from both wind and water, lathes, spinning machines, and most provocative of all, the design and manufacture of clocks. Their work led to collections of technical information which might be handled scientifically and to the fascination with use and manufacture of instruments of all kinds.

A typical seventeenth century view is that of Robert Hooke, a man versed in both instruments and theory:

> So many are the links upon which the true Philosophy depends, of which, if any one be loose, or weak, the whole chain is in danger of being dissolv'd; it is to begin with the Hands and Eyes, and to proceed on through the Memory, to be continued by the Reason; nor is it to stop there, but to come about to the Hands and Eyes again, and so, by a continual passage round from one Faculty to another, it is to be maintained in life and strength.[5]

Hooke here perceives the scholar/craftsman relation in terms of a famous type of faculty psychology first proposed by Sir Francis Bacon and widely influential thereafter.

Leibniz was fascinated by the skills of craftsmen, and although he had serious disputes with them from time to time, he showed great respect for them. During the 1680's he writes:

> Concerning unwritten knowledge scattered among men of different callings, I am convinced that it surpasses in quantity and importance anything we find in books, and that the greater part of our wealth is not yet recorded. There are even some sorts of information privately

possessed by certain persons and lost with them. There is no mechanical art so slight and scorned which may not furnish some important observations and considerations; and all callings have certain ingenious skills which it is not easy to be informed about and which may nevertheless lead to consequences of the highest importance. We may add that the important matters of manufacturing and commerce can only be regulated by an exact description of what belongs to all sorts of arts, and that the questions of militia, finance and navigation depend a great deal on Mathematics and Physics particularly. And the chief defect in many scholars is that they occupy themselves only with vague and well worn arguments while there is such a fine field for exercising their minds in solid and real objects to the advantage of the public. Hunters, fishermen, merchants, sea voyagers, and even games of skill as well as of chance, furnish material with which to augment useful sciences considerably. Even in the games of children there are things to interest the greatest Mathematician.[6]

Of course, he well understood the other side of the coin, too, that the craftsman,

...will understand perhaps the common rules about moving forces, the lever, the wedge, the jackscrew, but he will not understand that part of mechanics which I call the science of resistance or rigidity which has not yet been put sufficiently into rules, and he will not understand that the rigid pieces which are to support the moving parts should have a great deal of resistance, else what will yield is these parts rather than the big load supposed to be moved, and the nearer these rigid parts are to the final action, the more they need rigidity. When we understand these things, we are not exposed so vulnerably, and we do not need any longer the enormous masses which workmen of little ability use to assure themselves of rigidity.[6]

Leibniz's years in Paris were most significant for him. When he arrived at a mere 26 years of age, he had barely begun to study mathematics. After Christian Huygens's splendid tutelage (1672-1676), Leibniz devised a form of the calculus, thus paralleling Newton's great feat. Of course, the calculus is of inestimable significance for the history of science and technology. But Leibniz also retained his usual wide span of interests, extending to the work of artisans and industrial production. At one point, for example, he was much taken with the manufacture of mirrors (natural perhaps for a man whose "monads" are held to reflect the universe in varying degrees; I return to this below).

Thus, Leibniz's background was typical of the new bureaucrat of the seventeenth century: a lawyer by training, the new professional who had begun to supplant the cleric in the civil service, but a lawyer with a repertory of technical skills and one who could give advice on everything from mining to theology. Indeed, some of Leibniz's views are characteristic of this 'new man', generally statist in politics and mercantilist in economics. Yet the many duties brought on by his civil service positions have often been deplored as a waste of his talents, and it is highly probable that a man of Leibniz's genius could have produced considerably more science than he did, had he been able to devote full time to it.

However, in view of his actual situation and the outreach of his intellectual interests, it is hard to see what else Leibniz would have done in his youth but accept service in the bureaucracy. Without a post he would have been tied to what he felt was a rather dull life at a university, in fact, a way of life he refused to follow when it was finally offered. In the service of the state he was able to visit the European capitals and come into contact with many of the most brilliant people of his day. Contact with others was very meaningful for him. A good deal of his thought was developed in discussion with others, either in conversation, through his voluminous correspondence, or in journal articles. Perhaps he even foresaw the day when scientific studies would be written by teams, for he continually stressed the value of cooperation for science. But most important of all was Leibniz's intense social conscience, his deep desire to help the Empire and Europe as a whole overcome the disasters that had befallen it during the previous century. The Thirty Years War, one of the most destructive in human history, had ground to a halt in 1648, only two years after his birth, and its devastating effects were still to be seen and felt in many parts of Europe.

Against this backdrop, Leibniz was almost always engaged in several diverse projects at the same time. While involved in the complicated and ticklish correspondence on his plan to unify the Christian churches, for example, he also sought to increase the yield of the Harz silver mines for Duke John Frederick of Hanover. Droughts had prevented the pumps from continuous operation, a great loss to the principality and its typically mercantilist economic policy aimed at self-sufficiency. Suggestions for the use of improved wind power came from the director of the mines, but Leibniz persuaded the Duke that he could produce a better design. In 1679 a contract was signed, according to which Leibniz was to set up and operate one mine and maintain it for a year with a windmill operated pump. If it succeeded, he was to derive an income of 1200 thalers a year. The mining officials protested, however, asserting that Leibniz was a dangerous man to have dealings with.

As it turned out, great difficulties were encountered in getting the windmill to work. When the wind was strong, parts broke down; when the wind was light, the power was weak. Leibniz continued to work on it, though, and produced the windmill design given in Figure 1. Rather than a vertical one, it is, significantly, a horizontal mill, whose function is to capture the force of variable winds. He continued to make new designs, as well. By 1683 he had spent 2270 thalers, while the original contract had called for an outlay of 300. The resentment of the mining officials increased: they protested to the court that Leibniz knew nothing of mining and was deluding himself with the belief that all his

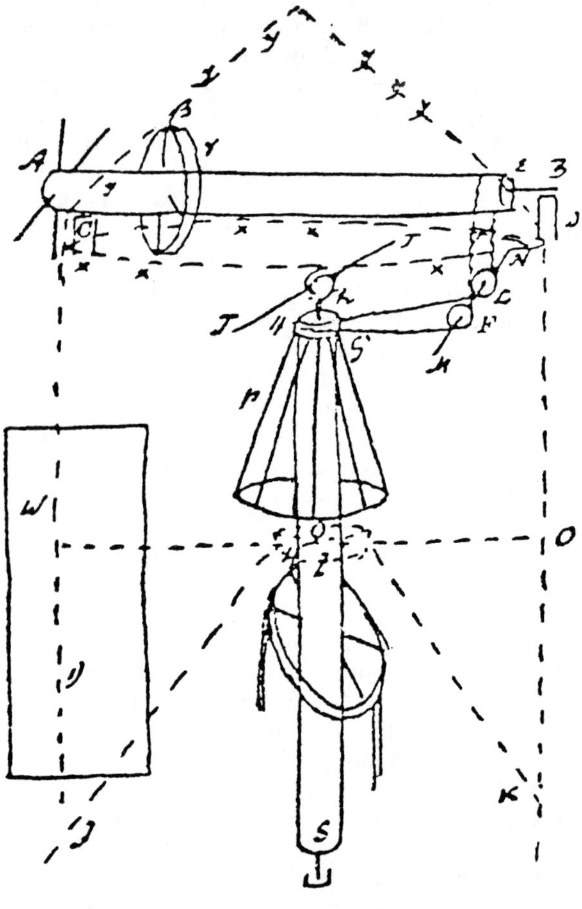

Figure 1. Windmill design by G. W. Leibniz.

mathematics would work in practice. Surreptitiously, they had assigned Leibniz the most difficult mine to work with. In turn, he protested to the Duke, charging the officials with inefficiency and nepotism. Finally, a delegation from the court was sent to look over the mine in 1683, but when they arrived the windmills were not working, because there was no wind, whereupon Leibniz was forbidden to draw any more money for the project. Still, he tenaciously hung on until the autumn of 1685, when he gave up in despair, fearing that his reputation would be ruined.

Later engineering opinion has in fact judged the plan sound. It involved a combination of wind and water power systems that were to compensate for one another through what we would call a pump-storage system. That is, when the wind was at a sufficiently high level, water would be pumped into a reservoir at a higher altitude than the water wheels. Then if the wind died down, the water could be released down a sluiceway to turn the water wheels which would activate the pumps used to remove water from the mine shafts, the aim of the whole process. This is more significant than it may appear at first sight, because it involves the conscious application of the principle of feedback. Although long implicit in nature and craft, this concept would become crucial for many other applications, including computer design in the present century. The failure of Leibniz's work resulted not from his plans, then, but from lack of knowledge of the strength, variability, direction, and frequency of the winds in the Harz Mountains. Deficient meteorology defeated mechanical skill, as Leibniz did not have the broad coastal plains of the Netherlands before him.[7] Still, we see here a revealing episode in the life of this godfather of the Enlightenment, at once engaged in mathematics, engineering, and politics, a potent historical mixture.

2. Leibniz, the Computer Scientist

Leibniz's interest in technology extended to a wide range of topics: barometers, compasses, hearing aids, firearms, telescopes, pocket watches, ship hulls, wagons—at one point he designed his own coach down to the last bolt. He had an incredible sense of detail. Of course he sought general principles, as well. Knowing a great deal about how machines work, and rejecting the possibility of perpetual motion, Leibniz conceived of the law of the conservation of energy, as Weyl notes,

> ...on the basis of the axiom *causa aequat effectum*, by transforming every mechanical change of state, the cause, into a 'standard effect' with but one degree of freedom, namely, the lifting of a given weight. The lifting height is then taken as measure of the energy. Leibniz's idea really strikes at the root of the energy principle. And it will carry over to

all natural phenomena as soon as the lifted weight is replaced, say, by a water calorimeter. (This generalization, however, was not conceived before the middle of the nineteenth century.) Indeed, a non-mechanical change of state may not always be transformable into a mechanical effect, such as the lifting of a weight, but it is always transformable into the heating or cooling of a given standard body.[8]

"*Calculemus*," Leibniz said. This was truly an uplifting and optimistic vision. Leibniz was among the first intellectuals in the Western tradition to perceive the great advantages inherent in mechanical calculation. What better way to resolve our differences, a road to the resolution of bitter disputes, at last? One recurrent strand in his thinking from the *De Arte Combinatoria* (1666) to the *Monadology* (1714) is his quest for a universal language and a reasoning calculus. Over time he created two fundamental concepts: the *characteristica universalis* and the *calculus ratiocinator*. The simplest definition of the first is as a pool of information, or a data bank. The items in it are so coded or labeled that we can know a number of things about them, but not all. We have symbols to designate those properties or processes that we do know. In other words, we have a fact net. For Leibniz the fact net is never complete: only God has a complete fact net, because He is not bound by space-time or otherwise limited. All other points of view are limited in one way or another. The *calculus ratiocinator* Leibniz conceived as a set of operational rules or algorithms that enable us to select out of the pool of information the properties or processes we wish to know something about and to combine them in the hope of seeing new relationships and thus gaining knowledge.

For example, if we take a set U as a universal set or a universe of discourse, subsets can be designated by elective operators x, y,..., so that x(U) is the subset of U, the elements of which have the property definitive of the operator x. If U is the set of fish in a pond, we can choose those that are female by an elective operator x and denote them with 'x(U)'. In the same way all mature fish may be denoted by 'y(U)', red-eyed fish by 'z(U)', and so forth. We may apply the elective operators successively, as well. First we may select all females and from these all mature ones by the symbolism 'y(xU)'. If we first select all mature fish and from these all the females, we have the symbolism 'x(yU)'. In each case the final set is the same. Therefore we can write 'y(xU) = x(yU)', or with the universe of discourse taken for granted, simply 'yx = xy'. This is clearly analogous to the commutative algebraic product. The associative law for products 'x(yz) = (xy)z' is immediately verifiable, because each side denotes the set of those fish that are female, mature, and red-eyed. It is easy to see why Norbert Wiener, a major founder of computer science in this century, considered Leibniz its patron saint. As Wiener wrote,

> ...the step from a system of deduction to a deductive machine is short. The '*calculus ratiocinator*' of Leibniz merely needs to have an engine to become a '*machina ratiocinatrix*'. The first step in this direction is to proceed from the calculus to a system of ideal reasoning machines, and this was taken several years ago by A. M. Turing.[9]

This connection to Turing machines shows, of course, the relevance of Leibniz for current work in artificial intelligence and cognitive science.

Until the Industrial Revolution of the nineteenth century, the clock was the single most complex instrument in general use. With Huygens's invention of the pendulum clock in 1657, it reached a high level of refinement as well as greatly expanded utility. Historically considered, the clock had long been a metaphor for the cosmos, one in use from Nicole Oresme in the fourteenth century through Leibniz's debate with the Newtonians in the seventeenth, to Voltaire and the Enlightenment and still further. Clockwork provided the essential mechanical model for the type of calculator that Leibniz constructed. The functioning of interlocking gears and the transmission of power through rods and pulleys were widely explored. Although the various forms of the gear had been known since antiquity, their use and the means of making them were not often pursued until the rise of Western European technology.

Leibniz had several predecessors, e.g., Samuel Morland, W. Schickard, and the great French savant, Blaise Pascal. They had constructed what are essentially counting machines with interlocking cogwheels to indicate units, tens, hundreds, and so on. The heart of the idea was an arrangement whereby when any one gear made a complete revolution it turned its left hand neighbor one tenth of a revolution. None of these machines ever worked very well for long. There were several reasons for this which were to plague Leibniz's efforts, as well. It was, for example, extremely difficult to cut the necessary gears with sufficient precision out of materials hard enough to be durable. Furthermore, since the machines had to be quite small, the largest of them being no more than 30 cm. in length, the difficulties for the craftsmen in building them were formidable.

Leibniz's machine was unique in that it was designed to perform all four arithmetic operations, while the others could only add, and sometimes subtract. At first, he hoped it would also be able to extract square roots, but this proved too difficult. Although it performed the other functions, it was nonetheless cursed by the construction problems inherent in the earlier models. Still, Leibniz's machine became the forerunner of all later mechanical calculators. The first really effective model, the Arithmometer, made by C. X. Thomas in 1820 in Alsace, made use of Leibniz's design components. Leibniz's calculator has been reconstructed several times in recent years following his instructions to the letter, and it has been made to work. One such model was recently on view at the Hayden Planetarium in New York City.

Leibniz had high hopes for his calculator. It was one of the pieces of work that he submitted to the Academy of Science in France with the unfulfilled hope of gaining a stipend like that of his friend and teacher, Christian Huygens. In his conversations with Peter the Great on forming scientific institutions in Russia, he proposed improving relations with China and suggested that his calculator would make a suitable diplomatic present for the Chinese emperor. He had long had a deep interest in the language and culture of the Celestial Kingdom. Indeed, when he devised a system of binary arithmetic, he was pleased to hear from R. P. Bouvet, a French Jesuit missionary in Peking, that the binary numerals are analogous to the linear signs of trigrams in the *I Ching*, the ancient Chinese *Book of Changes*, and long held to be symbols of human ingenuity.

In the longer run, through the work of people like the logician George Boole in the nineteenth century, the binary system aided in the actual design of the electrical circuits used in contemporary computers. A switch, for example, must be either "on" or "off," and the easiest way to represent this in a string of symbols is by 0 and 1. Furthermore, the transmission of bits of information is often performed by symbolic logic, making use of binary steps indicating "true" or "false." Certainly Leibniz knew nothing of electrical circuitry, but just as surely he recognized the utility and simplicity of the system later employed.

The mechanical problems encountered by the early designers were largely solved through improvements in machine tooling made during the nineteenth century. There were still many obstacles, however. Poor Charles Babbage spent half a lifetime with his "difference engine" and never got it to work! The use of electricity marked a great step forward in the design of calculators, leading ultimately to computers. It facilitated the development of logic machines and ultimately enabled wider use of a tradition in logic and mathematics that stretched back over time. Leibniz, of course, was a prime contributor to this tradition.

Leibniz saw the whole world as a *conspectus* or a collection of beings that he called monads. Their activity consists in perceiving one another on the basis of a pre-established harmony laid down by God. It is clear that he thought of this interaction mainly in terms of optics. The monads are conceived of as having no "windows," so that mechanical interaction becomes at bottom a subtle form of optical interaction. Both of his most original ideas, the *characteristica universalis*, the fact net or data bank, as well as the *calculus ratiocinator*, the direct ancestor of modern mathematical logic, depend upon the transmission of messages, a process central to all contemporary information-processing models, computer programming, robotics, and so on. At one point he even defines causality in terms of perception: "Thus action is to be attributed to a monad insofar as it has distinct perceptions, and passivity insofar as its perceptions are confused."[10]

Nevertheless, Leibniz also clearly speaks the language of physical reality. Indeed, he insists again and again that all natural processes are complexes of mechanical ones and thus can be explained mechanically and mathematically (human perception excepted). Of course, he holds that the principles to be used in these explanations are derived from logic and metaphysics; i.e., that they are essentially ideas. How then does he reconcile his logic and metaphysics with his sense of physical reality?

The answer is that he conceives of monadic aggregates of two basic types. One is composed of inert objects. These are termed ordinary or "mere" aggregates. Living organic entities, and preeminently human beings, are of the other type in being unified aggregates, or monadic hierarchies, which are fully individuated. In man the dominant monad is a spirit capable of reason, reflecting not only the other monads in the universe but also God. Only human beings are capable of self-consciousness (apperception), and human life in its ethical and cultural aspects is separate from the physical realm. However, Leibniz sees parallels. The concepts of plenitude, continuity, and linear gradation appear in his attempts to reconcile what was usually called the "order of grace" with the "order of nature." Indeed, they appear in practically all aspects of his thought. In trying to drain Cartesian dualism of what seemed to him to be its psychological absurdities, for example, he proposes a resolution of the mind/body problem in the form of a parallelism that one might call *psychoneural isomorphism*, the hypothesis that for every dimension of consciousness or mind, there must be a correlated dimension in the nervous system or body. This provides a paradigm case for semantic interpretation of physical processes, a crucial aspect of later computational theory.

Leibniz was deeply taken with symbols, geometric figures, diagrams, and in fact all means of notation useful in mathematical thinking. He called them the "thread of Ariadne" which, like the mythical heroine's thread, could lead one out of the labyrinth of the continuum. He made an extended study of notation, in the course of which he corresponded with many of the leading mathematicians he knew: the Bernoullis in Switzerland, John Wallis in England, and Christian Huygens in Paris. Ultimately, he was responsible for introducing more signs and symbols than any other mathematician with the exception of Leonhard Euler. He recommended the use of parentheses to separate portions of algebraic expressions instead of the vinculum, a line above the terms which had long been in use. He proposed the dot to indicate multiplication (because he felt the St. Andrew's cross was too easily confused with the letter 'x'), the decimal point, the equal sign, and the colon for division and ratio.

The notation of the calculus as we know it today is in large part due to Leibniz. For the process of integration which had been developing slowly ever since the time of Archimedes, he proposed the familiar symbol '$\int$', an elongated 's' signifying "summation." Perhaps the best known of his innovations is the '*d*' for the differential. Isaac Newton, who had invented the calculus independently, used dots and dashes above the letters to indicate what he called "fluxions" and "fluents," and they were difficult to read and to print. It is generally agreed that the development of the calculus in England was hindered until well into the 19th century because English mathematicians remained loyal to Newton's notation, while their Continental colleagues moved ahead into new areas with Leibniz's more expressive system.

Leibniz always sought to make the form of a symbol express its content, for he thought there was a great advantage in signs that express the exact nature of a thing briefly and thereby reduce the labor of thought. Of course, he saw more than technical aids in mathematical symbols. He perceived the link between mathematics and logic. Indeed, he is often held to have founded modern symbolic logic. The *characteristica universalis* and the *calculus ratiocinator* mentioned above connect in an art of combination by which we may discover new knowledge. His metaphysics coincides with his logic, for he asserts that our combinations will always be partial, as we can only imitate imperfectly the divine combination which expresses all that is possible. By laying the groundwork for a universal logical language whose syntax would ensure correctness, he hoped to enhance scientific communication, good ethics, and cultural universalism.

3. Leibniz, the Cognitive Scientist

Cognitive science is a house of many mansions in which adepts of varied persuasions reside. Psychology, philosophy, artificial intelligence, linguistics, neurology, and anthropology can all be found. Naturally, interdisciplinary studies are encouraged. Cognitive theories generally focus on the description of mental structures and processes as distinct from biological or historical factors. Most share faith in the computer as both a tool of scientific investigation and a metaphor for the human mind. Historically, what is known as cognitive science emerges as an effort to answer classical epistemological questions central to the Western tradition since the pre-Socratics, Plato and Aristotle, and the Scientific Revolution of the seventeenth century. The rise of indeterminism, relativity, and quanta in our troubled century has not been without influence either. The distinction between rationalism and empiricism, the debates between continental rationalists and British empiricists, between Locke and Leibniz, lie close to the heart of the matter. Does the mind have innate powers of reasoning that it imposes on the flow of sensory experience, or are mental structures and processes wholly dependent on external *sensa*? Are our actions conditioned reflexes or does a mental core shape our experience?

Generally, behaviorists tend to embrace empiricism, while cognitivists usually hold to some form of rationalism or a blend of both traditions. The term 'cognitive science' first came into use the decade after the Second World War to designate modes of psychological investigation that challenged the then dominant school of behaviorism with an explicit interest in consciousness and in mental structures and processes. Of course, behaviorism has hardly perished. Indeed, its leading exponent, B. F. Skinner, is still probably the most famous American psychologist, and his many disciples continue to practice their craft. Still, it is clear that two new conditions favored cognitive science: the advent of the computer as both a tool for research and a model for the human mind, and the rise of the new Chomskian linguistics. In both, Leibnizian resonances were distinctly audible.

a. The most striking instance of the influence of Leibniz on computer modeling and its epistemology that I am aware of is found in the work of C. West Churchman, a noted theorist in the field. He offers a convincing demonstration that Leibniz's system is a prototype for a particular mode of inquiring system which is in fact the general mode of systems theory assumed to hold in natural science. Unlike the method prescribed by Descartes, the Leibnizian inquirer does not begin with "clear and distinct ideas." Instead it has the capacity of identifying sentences and the ability to apply logic. It can determine which sentences are tautologies, which are self-contradictory, and which are neither. These last Leibniz familiarly calls "contingent." It does this by means of the *characteristica universalis* mentioned earlier, a universal language with a structure precise enough so that the processor, the *calculus ratiocinator*, will not be confused. The store of definitions provides the basis for determining whether a given sentence follows logically from the definitions. If it does, it is a tautology; if not, it is self-contradictory.

Leibniz recognized that the processor need only analyze the form of the sentences proposed to it and that consequently the questions must be posed in the algebra of logic. The processor is thus designed to handle a stream of symbols. Leibniz calls this ability *perception*. Each monad, which is a Leibnizian processor, has this ability to some degree. Perception originates from within the monad, generates the stream of symbols, and identifies its elements. There is a graded hierarchy of monads, each different from every other in its capacity to reflect all other monads and the universe as a whole in the so-called *pre-established harmony*, more felicitously expressed as "the hypothesis of the concomitance or of the accord of substances between themselves."[11] In the higher forms of the monad, and the human mind in particular, the system can operate in different ways on the perception stream. Leibniz calls this capacity *apperception* or consciousness. "Specifically, the system can store segments of the stream in memory, with an appropriate tag to indicate when the segment occurred. Furthermore, the system can retrieve segments from memory and combine them in various ways (by means of imagination)."[12]

The contingent truths belong to what we have called "fact nets." They imply other contingent truths or are implied by them. Any sentence under examination becomes a contingent truth if it can be linked to another prior contingent truth. (Interestingly enough, we do this all the time.) As the fact net grows, certain sentences become crucial, for if they are false, a part of the net, or even all of it, becomes false.

Leibniz's inquirer is essentially a model builder. Because of the infinite combinations that exist, it requires an executive with complete control who must search the stream of perceptions for just those sentences that shape a coherent truth. But what guides the model builder? Leibniz's own answer consisted in finding one model that must dominate every other, because its coherence implies its truth: God, that being who possesses all properties to the highest degree, insofar as these properties can coexist; i.e., the unique guarantor of the optimal system of reality. As Leibniz frequently asserted in his version of the ontological argument, "God exists if He is possible." Since God's goodness is manifest in nature, man can learn the principles of ethics and the guide to human welfare by the scientific study of nature.

Briefly, Churchman's statement of the Leibnizian basis for the design of a system is that it:

1. depends upon innate ideas to control the perception stream;
2. can produce strings of symbols that can be grouped into recognizable units;

3. can classify units as tautologies or non-tautologies;
4. can form nets of units by means of relations or operators;
5. can rank nets by some criterion;
6. has a method of processing symbols and forming the nets such that the system arrives at, or converges to, an optimal net, and *knows it*.[13]

Churchman identifies three types of widely used machines as essentially Leibnizian in design, all of which satisfy most of the conditions listed above: *algorithm machines*, those that automatically derive solutions to particular classes of problems such as finding the maxima of a function or solving certain types of differential equations; *heuristic search machines*, those using approximation or trial and error methods to find the maximum of a function; and *theorem-proving* and *problem-solving machines*, those that can check their data to determine whether the theorem was proved and that have an 'executive' who can discard one approach for another if the first one is not working well.[14]

Two other ideas from Leibniz prove significant in this context. First is an idea that he shares with other rationalists, that no optimal design of a component of a system is possible without a prior concept of the system as a whole. This, of course, poses difficult questions for the modern systems designer. Just think of the complexity of modern machines, systems within massive systems! At some point we know that objects reduce to molecules, atoms to sub-nuclear particles, and so on. Indeed, the very chair I am sitting on at one level of description becomes viewed only as a mass of whirling electrons. But how does that mass which is this chair fit into the increasingly larger systems that make up the world? This, simply put, is the kind of question that frequently occurs when planners attempt to alter one part of a complex system toward some specific end.

Secondly, for Leibniz in particular, there is the idea of a fundamental similarity of all systems in the design of their components. This implies that much, if not all, of the complexity of nature is reflected in all its monadic units. "To translate this into modern terminology, the idea is that in the design and evaluation of any functioning system the same net of considerations are always involved; to ignore any consideration is to design an incompletely functioning entity."[15]

This is certainly one of the most striking examples of the influence of Leibniz on modern thought. It is truly remarkable that his rationalistic philosophy can be readily translated into modern systems analysis and computer design. To be sure, Leibniz defined his monads as 'entelechies', not machines, but they share enough characteristics with machines to make this translation possible. His mathematics gives us the essential clue. Plenitude, continuity, and linear gradation are the prime characteristics in his conception of the universe and the totality of monads, with hierarchical sequence from the lowest form of life to God and with no two alike but each differing from those above and below in the scale by the least possible difference. (Interestingly, in this context he contributed to the discovery of what is known in physics as the principle of least action.)

b. A Leibnizian note is also clearly discernible in linguistics and its contributions to cognitive science. In the 1950's considerable debate was centered on the ideas of Noam Chomsky. They ran directly counter to those of the leading light in the field at that time, the behavioral linguist, Leonard Bloomfield. He had done a great deal of work on methods of notation for exotic languages and stimulated many students in the exploration of phonology. He did, however, consider language a purely physical phenomenon and its use a passage of physiological events, thus avoiding the crucial question of meaning. Of course, there were countervailing voices such as those of Roman Jakobson, Edward Sapir, and Benjamin Lee Whorf. While using analytical techniques, they were also nurtured by the rich philological tradition going back to the eighteenth and early nineteenth centuries, to Jacob Grimm, Franz Bopp (who knew forty languages), Alexander von Humboldt, and ultimately the Port-Royal grammarians and Descartes and Leibniz in the seventeenth century. Chomsky begins with this tradition, as reflected by his early work on Cartesian linguistics,[16] and as a historian I do find it refreshing that Chomsky awakens a sense of history, often so dormant in American intellectual life. Nonetheless, Chomsky is also deeply informed by the work of contemporaries such as Turing, Wiener, Shannon, and von Neumann in his mastery of symbolic logic and mathematics.

Like his earlier predecessors, Chomsky sees language in its creative and logical potential as the essential dividing line between humans and animals. Hardly anyone with any historical background would disagree with this in general, and certainly not Leibniz. As Chomsky (1966) writes:

> ...it seems that after a long interruption, linguistics and cognitive psychology are now turning their attention to approaches to the study of language structure and mental processes which in part originated and in part were revitalized in the 'century of genius' and were fruitfully developed until well into the nineteenth century. The creative aspect of language use is once again a central concern of linguistics, and the theories of universal grammar that were outlined in the seventeenth and eighteenth centuries have been revived and elaborated in the theory of transformational generative grammar.[17]

According to Chomsky, the key to the inviolable rules of syntax is found within the mind in an autonomous innate structure, basically independent but interactive in dimensions of language such as semantics and phonology. He suggests that "the individual is born with a strong penchant to learn language and that the *possible forms* of the language which one can learn are sharply limited by one's species membership with its peculiar genetic inheritance."[18] These forms are known by the direct intuition of the native speaker as part of ideal *competence*, the formal structure of the language, and as opposed to *performance*, the actual process that shapes speech.

In historical perspective, Chomsky's abstract notion of syntactic competence is clearly akin to the notion of proof in mathematics; i.e., an explanation of how combinations of symbols come to establish a proposition, given certain axioms and rules of inference. The search for coherence and formal consistency which characterizes Chomsky's work has been an abiding theme in the history of mathematics. Proofs are always purely formal, as Leibniz recognized.[19] We have no absolute assurance that our formulations will reflect the phenomena under investigation, but their formal consistency encourages our belief that there is a good chance that they do.

Leibniz also offers strong support for Chomsky's innateness principle and is usually cited in conjunction with others from the seventeenth century to the early romantic period. While not members of a single school, they share a constellation of ideas and interests, including a common background in the rationalist philosophy of mind. Chomsky reports that his own concepts of surface structure and deep structure originate in the Port-Royal *Grammaire gènèrale et raisonèe* of Cartesian inspiration, published in 1660. While Leibniz was not in touch with the Port-Royal circles at that time, since he was only fourteen years old, he later conducted a lengthy correspondence with Antoine Arnauld, Jansenist opponent of the Jesuits and a mainstay of the group, in the course of which he sharpened and refined his concept of the monad. Furthermore,, he was intrigued by Pascal's work in probability theory, as well as his calculator. Not only was Leibniz in touch with the earlier writers cited by Chomsky (with whom was Leibniz not in touch?), but like Locke, he started his intellectual life as a Cartesian. Although Leibniz later broke with Descartes on many issues, he never rejected Descartes's views on language. He also exerted a profound influence on W. von Humboldt, A. W. Schlegel, and J. G. Herder, early romantic scholars, all of whom supported the innateness principle.

The main source of Leibniz's ideas on language is that vast compendium of erudition, that register of European thought from antiquity to the seventeenth century, the *New Essays Concerning Human Understanding.* It was written as a critique of Locke's *Essays Concerning Human*

Understanding, a work of great import for the English-speaking world. Locke never got to read it since he died (1704) before its publication, although he had read some excerpts and rejected them. Leibniz set it aside, not wishing, as he said, to criticize a recently departed author; it did not appear until 1765, half a century after his own death. It is a pity, for Leibniz was exquisitely polite and conciliatory and felt that by and large they were not too far apart. Written in dialogue form, it gives us a sense of what a conversation between these two giants might have been like. A discussion of language fills a large part of the work. There is no aspect of linguistic study that Leibniz ignores: the compilation of dictionaries, the analysis of phonology, the search for historical origins, the Adamite *Ursprache*, and of course, the nature of grammar. For Leibniz, language is the clearest mirror of the human spirit, a virtual identification of linguistic with mental processes: "...an exact analysis of the signification of words would show us better than anything the workings of the understanding."[20]

The essential difference between Leibniz and Locke concerns, of course, the existence of innate ideas. Leibniz holds that mathematics, logic, metaphysics, and ethics are full of necessary truths that are independent of the testimony of the senses and inherent in the mind. The necessary truths are considered to be formal or structural, since individual people and diverse cultures have many modes of expression. Everyone may not always perceive them since nature's hand-writing on the soul is not always legible. Although he rejected the doctrine of reminiscences of past lives, Leibniz acknowledges the root of this concept from Plato to St. Augustine and others to Francisco Suarez, a Spanish Jesuit of his own time whose works he had studied. Of course, for Leibniz it is God, not biology, who has impressed these formal exemplars on the intellect. However, he does not dismiss experience for the formal ideas may require appropriate external stimulation to call them into consciousness.

Leibniz goes much further than Chomsky in finding basic teleological principles in the mind, supporting a formal concept of God very much like that from Plato. Ethical ends mean a great deal to Leibniz for they are distinctly the province of the human spirit. Indeed, he is unusually stern with those who would forget about ends.

> As I do not like to judge people harshly, I do not morally criticize our new philosophers who claim to banish purpose in science, but I nevertheless must assert that the consequences of their position seem to me to be quite dangerous,...to go to the length of discarding final causes entirely as though God proposed no end and no good in his activity.[21]

He was concerned that ethical positions might be overthrown by a gross mechanistic determinism. His critique of Spinoza reflects Leibniz's denial of pure necessitarianism. On the other hand, Descartes's interactionist dualism also aroused Leibniz's strong dissent, as did Newton's occasionalism because it demanded miracles. Clearly, in his science Leibniz accepted the generally mechanistic model of the universe that became dominant in his time, but he also maintained its inadequacy in the human sphere.

Leibniz apparently never worried about the negative social and ethical consequences of science and technology. Indeed, he felt the latter were of consummate importance for the successful maintenance of a just society, as befits a seventeenth-century believer in progress. Of course, he could not have foreseen the earth-shattering potential the human race possesses today. Yet his early understanding of the deepest principles of knowledge, practical as well as theoretical, gives his views an enduring relevance to the disciplines of modern cognitive science.

Polytechnic University of New York

NOTES

[1] From Albert Einstein (1950), "On the Generalized Theory of Gravitation," *Scientific American*, Vol. 182, No. 4 (April), pp. 13-17.

[2] From Benoit B. Mandelbrot (1977), *The Fractal Geometry of Nature* (New York: W. H. Freeman), p. 405; cf. James Gleick, "The Man Who Reshaped Geometry," *New York Times Magazine*, December 8, 1985, p. 124.

[3] From Ian Hacking (1986), *The Emergence of Probability* (New York: Cambridge University Press), pp. 57-58.

[4] From *The Monadology and Other Philosophical Writings*, trans. Robert Latta (London: Oxford University Press, 1898), pp. 227-229.

[5] From the Preface of Robert Hooke's *Micrographia* (London, 1665).

[6] Cf. Philip D. Wiener (ed.), *Leibniz Selections* (New York: Scribners, 1951), pp. 45-57.

[7] Cf. Ulrich Horst and Jurgen Gottschalk (1972), "*Uber die Leibniz' schen Plane zum Einsatz seiner Horizontalswindkunst*," *Studia Leibnitiana Supplementa* (Wiesbaden: Franz Steiner), Vol. 2, pp. 35-59.

[8] From Hermann Weyl, *Philosophy of Mathematics and Natural Science*, revised and augmented English edition based on a translation by Olaf Helmer (New York: Atheneum, 1963), pp. 182-183.

[9] From Norbert Wiener (1954), *Ex-Prodigy* (New York: Doubleday), p. 193.

[10] Cf. *The Monadology*, No. 49.

[11] From G. W. Leibniz, *Basic Writings*, *Correspondence with Arnauld* (La Salle, IL: Open Court, 1962), p. 69.

[12] From C. West Churchman, *The Design of Inquiring Systems: Basic Concepts of Systems and Organization* (New York: Basic Books, 1971), p. 30.

[13] Ibid., p. 34.

[14] Ibid., p. 48.

[15] Ibid., p. 41.

[16] Cf. Noam Chomsky, *Cartesian Linguistics: A Chapter in the History of Rationalist Thought* (New York: Harper & Row, 1966).

[17] Ibid., p. 72.

[18] From Howard Gardner, *The Mind's New Science: A History of the Cognitive Revolution* (New York: Basic Books, 1985), p. 193.

[19] Cf. Ian Hacking, "Proof and Eternal Truths: Descartes and Leibniz" in S. Gaukroger (ed.), *Descartes: Philosophy, Mathematics & Physics* (New Jersey: Barnes and Noble Books, 1980), p. 170.

[20] From G. W. Leibniz, *New Essays Concerning Human Understanding*, trans. A. G. Langley (New York: Macmillan, 1896), p. 368.

[21] From Leibniz, *Basic Writings*, *Discourse on Metaphysics*, p. 33.

JOHN BIRO

HUME AND COGNITIVE SCIENCE

In this paper, I want to explore what Hume's science of man and the new *cognitivism* have in common, and wherein they differ. I think that by taking a closer look at their similarities and differences we may understand both better. Indeed, I shall argue that we may learn some lessons from Hume's old science that the new could well take to heart.

1. Stagesetting

Two relatively recent developments make this perhaps surprising coupling of subjects both reasonable and, as I shall urge, profitable for anyone with an interest in either. One is the repudiation of both dominant interpretations of Hume—the classical one, as a pure skeptic, and the more recent one, as a sort of proto-conceptual-analyst—which is the hallmark of the revolution in Hume studies that began in the 1970's.[1] As all revolutions, this one has had its excesses; a more balanced picture will no doubt emerge, one in which the elements at present receiving less emphasis—such as the indisputably skeptical thrust of Book I of the *Treatise*—will take their proper place.[2] However, that it is now possible to see Hume as doing something relevant to modern science is one of the striking and welcome aspects of the fundamental change our view of him has undergone in a relatively short time.

The second development is of course the emergence of the so-called *cognitive sciences*. It is possible that the most notable intellectual development of our era will prove to have been the rise of this cluster of loosely connected inquiries, usually taken to involve at least some branches of psychology, neurophysiology, theoretical linguistics, artificial intelligence research, and computer studies—as well as a miscellany of studies not clearly the property of any one of these disciplines, but criss-crossing the territory loosely demarcated by them. Increasingly, these are being thought of as an emerging *single* science—often receiving the appropriately singular label 'cognitive science'—with perhaps an at least potentially well-defined domain.

J-C. Smith (ed.), Historical Foundations of Cognitive Science, 119–137.
Reprinted from *History of Philosophy Quarterly* **2**, no. 3 (1985), pp. 257-274, with permission of the publisher.

The general aim of this infant science cannot be summed up better than in the words Hume uses to characterize his own *science of man* in the *Treatise*. Its project, he says, is "examining the Mind...to discover its most secret Springs and Principles" (Txviii). While the truth about these "Springs and Principles" "must lie very deep and abstruse" (ibid.), the philosopher may hope to reach it by "anatomizing human nature in a regular manner."[3] In so doing, he is acting as scientist—"as an Anatomist" —as opposed to both a speculative metaphysician and a dogmatic skeptic.[4]

I shall try to make the claim in my opening paragraph plausible by taking a look at three problems for *cognitivism*—the philosophical framework of cognitive science—on which a Humean perspective may yield some insights. These are the problem of meaning and content, the problem of qualia, and the problem of the "exempt agent." But before doing so, I should like to say a bit more about the new science, so that I can make some general comparisons in terms of both substance and method.

2. Hume and Cognitive Science

The leading idea of cognitive science is that the best way to understand both a particular kind of behavior human beings exhibit and the ability manifested in such behavior is in terms of *internal processes* (and states) that underlie the behavior understood as an exercise of the ability in question. The kind of behavior focused on has the twin characteristics of *intelligence* and *intentionality*, typically involved in perception, action, memory, speech, concept-formation, reasoning, and so on. The empirical study of these psychological states and processes proceeds along normal scientific lines: there is no *special* method available, such as introspection or the discovery of *a priori* truths through the analysis of our concepts involving the mental. The mental is to be studied in the same way as the physical, though in part with somewhat different tools.[5]

Reflecting perhaps the eclectic character of cognitive science, there are a variety of competing philosophical theories about its nature and status. For the purposes of this paper, I shall focus on the so-called *representational theory of the mind* (RTM) advocated by, among others, Jerry Fodor. I do so for two reasons. First, while it has important rivals—such as the (alleged) instrumentalism of Dennett, the eliminative materialism of Patricia Churchland and Paul Churchland, and the "syntactic theory" of Stephen Stich—it is, I think, fair to say that RTM has so far both dominated philosophical discussion and, more importantly, informed actual scientific practice.[6] Second, it is the approach closest to Hume's conception of a science of man and the one which, in spite of some real differences to be noted shortly, he would, I am confident, find more congenial than the others.

This second claim may seem surprising in view of the traditional picture of Hume as a sort of super-empiricist, too tough-minded to admit theoretical constructs in the way RTM freely, almost gleefully, does. Indeed it is true that he says: "The only solid foundation we can give this science...must be laid on experience and observation" (Txx); while "we must endeavor to render all our principles as universal as possible, by tracing up our experiments to the utmost, and explaining all effects from the simplest and fewest causes" (Txxi), "'tis still certain we cannot go beyond experience" (ibid.). And this may well seem to be in conflict with the unabashed willingness of modern science in general, and of writers such as Fodor in particular, to appeal to theoretical entities only tenuously connected with observation. Indeed it may seem to conflict no less with Hume's own interest (illustrated in his characterization of the *Treatise* cited earlier, but repeated many times elsewhere) in "secret springs and principles" (as well as "secret operations"—e.g., T104).

But the conflict is only apparent: Hume's empiricism is a more sophisticated one than is often realized.[7] The principles he, as an *anatomist*, proposes as governing human nature are, though *secret*, not *occult*. They are secret in the sense that they don't lie on the surface of phenomena, even to the extent that mere generalizations do: they are *explanatory* principles that have to be discovered through the right method. This is why "a total alteration," indeed "a revolution," is needed in philosophy before it can yield a science of man. But they are not *occult*, as are those principles appealed to by the metaphysicians of whom Hume is so scornful.

Hume's principles are, more or less, those principles of association which in the mental world play a role analogous to Newton's in the physical, and those native propensities and dispositions of the mind that underlie its conformity to the principles of association and explain its reliance on them. Like Newton's principles, they require philosophy to bring them to light. (Note how for this to make sense, "philosophy" cannot mean something radically different from "science" as the traditional interpretations of Hume assume.)[8]

Should we ask, on the other hand, *why* the principles we have thus discovered are the way they are and not otherwise, we would indeed be stepping outside our purview as natural philosophers—i.e., both as scientists *and* as philosophers. Let me illustrate the point by reference to Hume's complaint in the *Appendix* to the *Treatise* about his own account of personal identity. One interpretation of this notoriously puzzling passage has it that what bothers Hume is that his account "leaves completely unintelligible and mysterious the fact that the 'data' [on which his account is based] are as they are."[9] But in fact Hume accepts that to such "mysteries" there is no philosophical answer. Nor is there an answer to the question why the processors of those data, our mental faculties, are the way they are: "We can give no reason for our most general and most refined principles" (Txxii).

The metaphysician pretends to answer these questions by invoking occult principles that "are not founded on [the] authority of experience" (ibid.). But *that* is not the same thing as the genuine discovery of secret principles by the philosopher. The former has to do with the alleged *sources* of these principles, and for Hume these "are, and must forever remain, shut up from human view;" conjectures and hypotheses that purport to found our descriptive and explanatory principles on qualities of the mind or of the world in some sense even more "original" or "ultimate" "ought at first to be rejected as presumptuous and chimerical" (Txxi). It is hypotheses of this sort that are scorned by both Hume and Newton; but there is no eschewal by either of hypotheses in the sense in which modern science in general, and what Fodor calls "speculative psychology"—"the attempt to say how the mind works insofar as answers to that question emerge from empirical studies of language and cognition"—in particular, employs them, with hidden mechanisms postulated as one's theory requires.[10] The dispositions and the propensities that drive the mind in conformity with Hume's principles of association are in the same boat methodologically and ontologically as are Newtonian forces *and* the mental processes postulated by the modern linguist or psychologist to explain cognitive abilities and behavioral regularities.

There are similarities that it would be interesting to explore, as well as some important differences. Before turning to the three specific topics I indicated I wished to explore, a few of these differences need to be at least mentioned.

Hume thinks of his science as at once broader and narrower than does the cognitivist of his. Broader, in that it includes explanations of ideas which are taken for granted and *used* by modern cognitive science, such as those of external existence and causality. It is not that cognitive science has nothing to say about such ideas and about their role in our mental life; it is, rather, that they are not seen as standing in need of legitimation, as they *are* for Hume (albeit in a special Humean sense, of showing their deviation from properties of the mind). To the extent to which such notions can be used uncritically in scientific theories in other domains, they are so used in cognitive science, whereas for Hume, of course, it is just such explanations of such ideas that are fundamental.

But Hume's old science is also narrower than the new, in that in the end its subject matter is always the whole person or mind. True, in explaining the behavior of that whole, it makes reference to processes and operations at the *sub-personal* level (impressions and ideas "doing" things), where many of the most interesting explanations of cognitive science are also pitched. But while its frequent talk of perceptions *doing* things, such as producing or attracting each other, may suggest something different, in the end what does the work is always some propensity of the *subject* whose perceptions they are, and propensities and dispositions are properties of the mind or person as a whole.[11]

Another important difference between the two sciences has to do with their differing view of the *evidence* relevant to their hypotheses. There is in cognitive science much less emphasis than in Hume on introspection as a method; indeed, it is fair to say there is a widespread distrust of it. For example, there is general agreement that if we want to discover which (if any) of a number of possible formalisms mirrors real psychological processes occurring in a subject, *asking* him is usually a poor way of trying to do so. We get much more reliable information from response times, measures of comprehension, error-patterns, etc. Now Hume is not averse to appealing to evidence from the "outside," as it were: he realizes that observing the patterns of judgments people actually make can often tell us what rules they are using better than they themselves can.[12] But for him the rules used are *in principle* open to introspection, in that they are defined over ideas the subject has and beliefs he holds or entertains consciously. This is in sharp contrast to the rules some cognitivist theories invoke to explain, for example, linguistic behavior, where the rules and the objects they govern may both be inaccessible to the subject.

Instead of pursuing these general similarities and differences further, I now turn to those specific matters on which I have suggested Hume may have something to say of relevance to the new science. I shall discuss three such topics, though none in the detail it deserves.

3. Hume's Theory of Meaning and the Problem of Content

One of the central tasks facing any theory of the mental that assigns causal explanatory roles to propositional attitudes, as does RTM, is to give an account of the *contents* of those attitudes. As Fodor argues convincingly, it is plausible to think that folk psychology is right in making its explanatory generalizations in terms of these contents: for example, it is Jesse James's belief *that banks are where the money is kept* and his desire *to get rich without work* that cause him to hold up banks. Without such content-specifying clauses, our commonsense explanations would not get off the ground.

In saying that a belief is a belief *that p*, we make use of the fact that there is some sentence 'S' that means that p; we use that sentence in the content-clause to say *what* Jesse believes. To say that he believes that banks are where the money is kept is to say that his belief has something to do with what the English sentence 'Banks are where the money is kept' means. It is thus likely that we will need some account of *meaning* in order to be able to give an account of content.

Hume is traditionally interpreted as holding a version of the *picture theory* of meaning, according to which our words receive their meaning from the ideas for which they stand, these ideas being derived from, and exactly representing, items of experience, namely, impressions of sense or reflection. Thus ultimately the meaning of a word just *is* a mental image of some sort, such images being private to the mind whose images they are. As has been argued by a number of writers recently,[13] this interpretation, embodying as it does a phenomenalist reading of Hume, is something of a myth. Hume's theory can be seen as, at least in part, a *conventional-cum-historical* theory of meaning, in which the use in a language a term has and the information it carries in virtue of that use are part and parcel of its meaning. Indeed, these features of the term contribute far more to its meaning than any mental image that a user of a term may contingently associate with it.

This is clearest in the case of what Donald Livingston calls past-entailing terms, though it is true in various and in varying degrees of all language. The meaning of a word such as 'father', for example, or 'priest', involves a whole network of properties and relations—most of which have to do with the past and are thus not (directly) observable and not manifest in any image we may have of an individual to whom we apply the term. Thus a particular private image not only does not exhaust the meaning of the term, but in fact has very little to do with it; indeed, if one misapplies the term, as one might, the image one erroneously associates with it presumably cannot be even part of its meaning. (Let me cite just one instance, but a very striking one, of Hume's recognition of the role a term's use plays in giving it meaning. In speaking of the divisibility of space, he says, "'tis certain we have an idea of extension, for otherwise why do we talk and reason concerning it?" (T32).[14] And this in a context where the very question at issue is the *imaginability* of infinite divisibility!)

Livingston marshals much more evidence in the course of his argument that Hume's theory of meaning, *unlike* a picture theory, recognizes and makes room for this conventional and historical dimension of language.

If this is right, Hume's theory is obviously not vulnerable to a private-language criticism and is generally less absurd than has been supposed. Whether it *is* right depends, of course, on whether this is the best way to interpret the various things Hume says in various places about language. This is not the place to try to decide that question; all I want to claim here is that it is a plausible way to interpret them. And if it is, then it is not plausible to think, as most people, and many cognitivist philosophers especially, do, that Hume is saddled with the picture theory and whatever shortcomings it may have as a theory of meaning.[15]

The immediate interest of all this, however, is that it is precisely this public aspect of language that RTM has the most trouble with. RTM combines two theses, both independently plausible, but difficult, perhaps impossible, to reconcile. The first is that only those things that are in a person's head can figure among the (psychologically relevant) causes of his behavior (and thus in a psychological theory).[16] The second is that (among) the chief causes of a person's behavior are his propositional attitudes. The only way to render these two theses compatible is to show that the contents of a person's propositional attitudes can be specified strictly ("narrowly," in Fodor's term) in terms of what is in his head (and perhaps the causal provenance of the latter). But, it is not easy to see how this can be done if the meanings of the terms we use in the content-specifying clauses aren't exhausted by what's in the subject's head.

RTM's proponents usually opt for the first horn of the dilemma created. As between accepting an in-the-head theory of meaning or an outside-the-head theory of content, they prefer to try to make a go of the former: hence their frequent espousal of a Gricean theory of meaning, with its program of reducing public-language meaning to private speaker meaning and that in turn to intentions.[17]

This is not the place to go into the labyrinthine details of various Gricean reductions; my immediate interest is merely to note that if what I have been saying about Hume's theory of meaning is right, there is an irony in the wide-spread assumption that he is a kind of grandfather of the sort of mentalism and individualism that characterizes the Gricean approach. More importantly, if Hume not only holds something like the theory of meaning that I have attributed to him, but is *right* in doing so, then RTM is wrong to the extent to which it must rely on the (wrong) Gricean picture. At the very least, it does not add to the attraction of RTM that it depends on a theory itself so controversial and problematic.[18]

This is not to say, of course, that Hume himself has a good solution to what is in fact one of the central problems for the science of the mind, that of combining two equally essential yet seemingly incompatible features of the mental states and events adverted to in the explanation of behavior.

On the one hand, it seems that RTM is right to insist, along with common sense, that these states and events can figure in the sorts of generalizations we need in psychology only in virtue of their (intentional) content. On the other hand, it is only in virtue of their *causal powers* that they can explain anything, and it seems that these powers depend only on their formal properties and not on their content. Reconciling these aspects of mental representations has been a preoccupation of RTM all along; insisting, as I have suggested Hume may be seen as doing, on a theory of meaning that takes seriously the social and historical dimensions of language makes the problem even more acute.

There is no point in trying to make the problem easier by adopting a bad theory of meaning. So if the kind of Gricean theory on which RTM relies does turn out to be wrong, and especially if it is so for the sorts of reasons that lead Hume away from it, proponents of RTM must re-think one of the two planks in their program. Something has to give: either the insistence on content-based generalization *or* the formality condition must then go. It is not easy to see how RTM could survive the loss of either.

4. The Problem of Qualia and the Case of the Missing Shade of Blue

The fact that Hume's adherence to the introspective method is not as complete as the traditional interpretations would have it lessens the plausibility of attributing to Hume a picture theory of meaning. It also has other interesting consequences, not least for the vexed debate about cognitivism's ability to give a satisfactory account of the qualitative aspects of experience. It has seemed obvious to many commentators that Hume's "first principle" concerning the origin of ideas in corresponding prior impressions and his frequent appeals to introspection are inseparable. That Hume does not separate them is indisputable. That they could not be separated is not. I shall argue that they can be and that if they are, a Humean position results that is of considerable interest for the debate just mentioned.

It is often said that one of the hallmarks of the mental is *consciousness*. (The other being *intentionality*, the property we were concerned with in relation to content.) For some, such as Thomas Nagel, it is "consciousness that makes the mind-body problem really intractable"—it is the sensuous content of our awareness, our "raw feels," that, even more than intentionality, stand in the way of a wholly physical account of it.[19] In fact Nagel has argued that no *objective* account of any sort could do justice to what is essentially a subjective phenomenon. There is a feel to a particular toothache (*my* toothache) that no description can—could ever—convey. Because of this, he thinks, there cannot be a science of experience. But even for those who do not go that far, the problem of explaining how purely physical entities could have the qualitative properties in question often appears intractable. They are then tempted to conclude that for this reason alone physicalism has to be rejected.

Here, too, Hume's official position seems to support the skeptics: his notorious "first principle," according to which we can have (simple) ideas only of things we have previously experienced (had an impression of), seems to entail that you may have some ideas that I cannot in principle share.[20] This, together with a picture-theory of meaning, would imply that I cannot produce or even understand any description of your experience in objective terms; indeed, there cannot be such a description.

There has been much discussion of this sort of issue in the recent philosophical literature.[21] For example, putative cases of absent *qualia* are sometimes taken to show that it is possible for an entity to satisfy all the functional descriptions a well-supported theory provides for it on the basis of behavioral evidence and yet to lack the qualitative content we intuitively associate with the state specified by those descriptions.[22] That is, it seems as if someone could be in the functional state our best theory says pain *is* and yet not *feel* pain (and, of course, the reverse also seems possible: *qualia* without the relevant functional state). If such possibilities are real, that *would* show that functional definitions of at least *some* mental states are at least incomplete, if not altogether misguided, and therefore so are functional accounts of the mind in general.

What could we say about meaning of a sensation term, assuming that it cannot be given by a functional definition? It seems that to understand the term one is required to have *had* the sensation, in which case, an account of the meanings of at least these terms must be at least partly a phenomenological one. Our earlier arguments against viewing Hume's theory of meaning as a picture theory notwithstanding, the copy principle itself may seem to commit him to such a conclusion. The trouble with the notorious case of the missing shade of blue seems to make things even worse. Hume has to dismiss the apparent possibility that I can generate an idea of a shade of blue I have never seen (if I am shown enough *other* shades) as an inexplicable exception to the copy principle.

But perhaps the trouble is that the copy principle has been wrongly interpreted. Without developing it in detail, I shall sketch an alternative interpretation of the principle, one which predicts as normal precisely what happens in the case of the missing shade of blue and supports just the opposite conclusion about the possibility of objective descriptions of experience.

Livingston, as we have seen, objects to viewing Hume as a picture-theorist. But he seems not to fully appreciate the implications of this for the present issue. He says: "Although Hume does not think that having mental imagery is a sufficient condition for having an idea and thus for understanding a word, he does think it necessary."[23] Now as a general claim this is surely mistaken, since as we saw earlier, one's understanding of a term such as 'father' is quite independent of any image one might have of any object to which one may apply it. This is even more obviously so for terms such as '7' or 'justice', say—what imagery could be suggested as necessary for understanding these? So the general claim does not seem very plausible. Yet there does seem to be something in it in the special case of sensation terms. For here the sensation itself is the image, and it seems more intimately connected with the understanding of the sensation term than is one's image of something that (perhaps) happens to be a father is with one's understanding of the word 'father'.

Yet I think the temptation to identify the meaning with the image, and to make understanding depend on having the image (in this case, the sensation) should be resisted even here. Livingston goes part of the way in the right direction when he notes Hume's distinction between having *an* idea of something and having a *just* idea of the thing (T5). Such a distinction would permit us to say that someone has *an* idea of the taste of a pineapple, for example, even though he has never tasted one, if he were able to make certain discriminations associated with our normal use of the expression 'pineapple taste'. He could rely upon various combinations of familiar tastes, much in the way in which we are all forced to do with novel experiences when there is no settled matrix of quality terms specific to the sort of object in question. Consider the predicament of music critics or wine writers: the proof that it can be done is that it *is* done. As with most things, some people are naturally better at it than others; on the other hand, the role of education and practice should not be overlooked.

Still, it is not hard to imagine even the most sensitive, ingenious and practiced judge of tastes realizing on first biting into a pineapple that he had until then not *really* known its taste. (Though note the slipperiness of our language here: in a sense it is trivially true that he had not known *its* taste (the taste of this particular pineapple, one possibly un-pineapple-like for some idiosyncratic reason), though perhaps he did know, as we might also put the matter, what its taste was (likely to be) *like*. In our unreflective talk about knowledge, acquaintance and description are not, and need not be, as well separated as in philosophical discussion.) In terms of the distinction just introduced, we could think of this newly discovered taste as a *just* idea, and it is tempting to think that it is in such a radically different epistemic category that we must accord it a special status *vis-a-vis* the meaning of the expression 'pineapple taste'. Then the claim that having some mental image (more generally, some qualitative experience) is necessary for the possession of at least one category of ideas would have to be admitted.

One might respond to this objection by pointing out that having an idea of some quality and having a *just* idea of it shade into one another rather, and are not as sharply distinguishable as these imaginary cases tend to suggest. One may start with merely an idea, but making that idea more and more detailed and rich can lead finally to what is, to all intents and purposes, indistinguishable from the elusive "just" idea. Let me make an admittedly speculative suggestion. Could not Hume's example of the missing shade of blue be read as describing just such a case? The idea would be that knowing *enough* about *x*'s relations to *y*, *z*, etc., and knowing enough about *y*, *z*, etc., themselves, may give one complete knowledge about *x*, *including* knowledge of its qualitative features. It can't, of course, give one *acquaintance* with *x*. But the question at issue is precisely whether such acquaintance is needed for knowledge, so to assume that *because* it is absent our knowledge of *x* is necessarily incomplete (since it fails to involve a "just idea") would be to beg the question.

The point can be generalized, relying on an interesting interpretation of the copy principle recently put forward by Kenneth Bower.[24] Bower argues that Hume's first principle can be understood in a way that does not require that there be any qualitative similarity between impression and idea. An idea need not be *like* the impression from which it is derived and which it exactly represents, as long as the former contains all the information the latter does. The relation of exact representation may be understood in information-theoretic terms, requiring only that no information that is present in some "corresponding" (and prior) impression be missing from the idea and that the latter contain no *additional* information.[25] If so, the presence of the relation of exact representation need not depend on *pictorial* (or more generally, qualitative) resemblance. Indeed, not all ideas need be image-like at all to satisfy Hume's principle, even if all impressions are.[26]

Even more interestingly, the argument applies the other way around, too. There seems to be no reason why one could not come up with an idea in pictorial form, even though the information embodied in that idea is not present in the same pictorial form in any particular prior impression. The information for "deriving" the idea of the previously unencountered shade of blue is there in the impressions one has had of the *other* shades one has encountered and in the relations among these, and there is no reason why it could not be re-constituted—usually quite subconsciously, at the sub-personal level where most of our perceptual processing takes place—in the form of a novel image.

Admittedly, this is stretching the text—much of what Hume says in *Treatise* I.i unmistakably suggests that he has in mind the stronger requirement of one-to-one pictorial correspondence. But it is worth noting that Hume's own statement of his first principle, often called "the copy principle," does leave room for the sort of interpretation proposed here. After first speaking of the relation between impressions and ideas in terms of resemblance—"every simple idea has a simple impression, which resembles it" (T3)—he switches, when giving his official statement of the principle, to language that, as we have seen, speaks of simple ideas being *derived from* simple impressions (T4). While this statement still involves the notions of *correspondence* and *exact representation*, the terminology of *resemblance* is conspicuously absent from it.

Further, as Stroud points out,[27] Hume himself has a way of explaining the kind of process that might be involved in the missing-shade-of-blue case. One of the most important principles of the imagination is what might be called the principle of momentum: "The imagination, when set in any train of thinking, is apt to continue, *even when its object fails it*, and like a galley put in motion by the oars, carries on its course without any new impulse" (T198). While Hume does not invoke this principle in connection with our present topic, it is one he has recourse to in a variety of other contexts.[28] Here one might suggest that the missing shade is the object that "fails" the imagination, which nevertheless "spreads itself" (T167) to form an idea (in *this* case, an image) of it given its active perception of the other shades. I am not, of course, suggesting that this account is fully adequate as an explanation of what happens in the missing-shade case. What I *am* claiming is that it is one that is available to Hume and one that does not require the resemblance interpretation of the "copy principle."[29]

One reason for being skeptical about my suggestions here may be the traditional coupling of Hume's first or copy principle and introspectionism. On the account suggested here, unlike the usual one involving resemblance, the information used in the construction of a novel idea need not be available to the subject by way of introspection. This is in line with the general cognitivist view that many interesting mental processes are not introspectable and that the information and the processes involved in the construction of *any* idea may be largely unavailable to consciousness. So

Hume's introspectionism and his dismissal of the case of the missing shade of blue go nicely together.

But if we are embarrassed—as we should be—by that dismissal as an "exception scarce worth our observing" (T6), it may be worth considering this reinterpretation of his first principle as one he might in fact welcome. In any case, even *if* acquaintance *were* required for a *just* idea of a quality, the minimal lesson here is that it does not obviously follow that the *meaning* of quality terms is constituted by the object of such acquaintance, rather than by whatever comprises just *an* idea of it. That acquaintance is required for having a just idea (if it is) does not show that it is required for having just an idea. One may be able to come by the latter by way of understanding objective descriptions, something that does not require acquaintance with any particular instances of which those descriptions are true and, *a fortiori*, does not require the having of any particular experiences.[30]

The upshot of the discussion in the last two sections is that neither Hume's commitment to the copy principle, not anything he says either about the relation between having experiences and having ideas or about that between having ideas and understanding words, forces him into positions incompatible with either common sense or with cognitivism. Indeed, if anything, his ideas, properly understood, may be useful in developing an adequate cognitivist theory of the mind. In the next section I shall discuss another topic on which I claim Hume's position has been misunderstood, not least by cognitivists, and one on which he has similarly insightful views to offer them.

5. Exempt Agents, Undischarged Homunculi and 'Hume's Problem'

The last topic I want to look at, if only very briefly, has to do with what has been explicitly dubbed "Hume's problem" by Dennett.[31] But when we take a careful look at Hume, we find that the problem in question is everyone else's problem, not Hume's, and that he in fact comes closest to having anything like a solution to it. Again, I shall take the most developed version of cognitivism, RTM, as my stalking horse, though my argument applies to cognitivism in general.

The problem—somewhat oversimplified—is this. Assume, as some cognitivist philosophers do, that the right strategy for explaining what makes a piece of intelligent behavior intelligent is to *decompose* it functionally, looking for the part or parts that can be said to play a causal role in its etiology *qua* intelligent behavior. This is the strategy even of some experimental work in cognitive science and, more to the point, of those cognitivist philosophical frameworks that endorse the so-called top-down approach.[32] We begin by trying to isolate some subprocess underlying, or some set of events simpler than, the behavior we are interested in, which can be seen as imparting intelligence to the behavior as a whole. What, now, are we to say about these? Do *they* possess the property of intelligence or not? If they don't, how can they lend intelligence to the whole of which they are a part? If they do, where do *they* get it from? We seem to be launched on a regress, or else we must rest content with the same unexplained property at some level. But then why start—why not regard intelligence as a primitive property of the whole? This would be all right, from the point of view of most psychologists, but for many *philosophers* the whole point of the enterprise is to show how it is possible for a purely physical system to exhibit intelligent behavior.

The problem has different guises, but let me focus on the way it presents itself for RTM in particular. We have just noted that a *homuncular* theory is often thought to have a philosophical point only if the homunculi can ultimately all be, as Dennett puts it, "discharged." But for the representationalist there must, it seems, always be an undischarged, and undischargeable, homunculus, one whose intelligence is not itself explained or explainable. For systems of representations require a user or interpreter "external" to that system—representations represent only *to* or *for* someone. Such a user—Dennett calls it an "exempt agent"—is an undischarged homunculus with a vengeance.

Is this a problem for Hume, as Dennett suggests? His gloss on Hume's position *vis-á-vis* this issue is this: "Hume wisely shunned the notion of an inner self that would intelligently manipulate the ideas and impressions, but this left him with the necessity of getting the ideas to 'think for themselves.' This associationistic coupling of ideas and impressions, his pseudo-chemical bonding of each idea to its predecessor and successor, is a notorious non-solution to the problem."[33]

Dennett's response is to suggest that we do take seriously the idea of ideas understanding themselves. The mistake here—shared by many commentators on Hume's theory of the self—is to think that these alternatives are exhaustive. To do so is to overlook the most important aspect of that theory, whose point is to identify the self with a *bundle* of perceptions (related in certain ways). So understood, *it*—the self as a whole—is still the subject of those predications that apply peculiarly to persons, that is, of the *intentional* ones. Hume's purpose is not to deny that there is a self, nor that it is the thing that thinks, desires, has dispositions and suffers passions; his project is to tell us *what* that self is.

Doing so requires spending a lot of time telling us what it isn't, which can give the impression that the thesis is that it isn't anything. But such an impression is mistaken.

Of course this is all very controversial—and all I can do here is to refer the reader to other places where I have tried to present the evidence for this interpretation of Hume's views on the self.[34] Part of the difficulty is that Hume does make *perceptions*—not bundles of them—the subjects of certain quasi-intentional verbs: perceptions are said to *do* things such as producing and attracting each other. And this has encouraged commentators to attribute to Hume what they call a Newtonian account, one on which the real subjects of the intentional activities mentioned by the theory are these sub-personal entities.[35] But on such an account we have to make sense of the notion of perceptions doing other, and much more puzzling, things as well, such as understanding each other or themselves. As Dennett points out, this parallels exactly the representationalist's problem of making sense of representation without an exempt agent: for whom, or to whom do representations represent; who are their interpreters?

The mistake, in both cases, is failing to distinguish between genuinely intentional verbs and others such as those Hume applies to perceptions. These indeed "attract"—as well as "produce," "destroy," "influence," etc.—each other, but in a sense no more intentional than do Newton's bodies. The subject of Hume's serious intentional predications is always the mind or person as a whole. While there may be a mystery about how a mind constituted of such things as perceptions can do the things it does, that is not the same mystery as how *perceptions* could do those things. There is no mystery about *that* for the simple reason that *they don't*.

That they don't—and couldn't—does not mean that they cannot and need not do a good many other things without which the mind could not do what it does. What the cognitivist needs to do is to keep the personal and sub-personal levels clearly distinct, as does Hume, at least most of the time, so that he can make sure that his attempt to explicate intentional behavior at the former level by adverting to processes at the latter does not abort in what *he* has to regard as vicious circularity or regress. For such an explication to succeed, the sub-personal processes must be ultimately describable in non-intentional terms. As Dennett colorfully puts it: we can pay back the intelligence loans we take out with our personal-level, intentional descriptions of behavior only by showing that the processes that underlie that behavior (and any homunculi we postulate as playing a role in them) are *stupid*. If I am right about what might be thought of as Hume's equivalent of a sub-personal theory—his hypotheses about the "behavior" of perceptions and about their contribution to intentionality at the *personal* level—that theory enables him to meet this requirement. It is a leading assumption of cognitivism, and of much modern cognitive science, that it can be met. Maybe it can't be, and maybe it needn't be. But as long as the cognitivist, who thinks it needs to be, is as careful to avoid "Hume's problem" as I have claimed Hume himself was, at least he will not fall into the trap of paying back his intelligence loans in false coin.

6. Conclusion

I have tried to show through a consideration of Hume's views on matters connected with three of the central issues facing the philosophy of cognitive science that he has much of value to say to that science. Philosophers of cognitive science, while recognizing his relevance, sometimes use Hume's theories as a foil against which to measure their own, supposedly more satisfactory, accounts.[36] I have suggested that there is a certain irony in this, in that Hume's theories, properly interpreted, are far from examples of *failed* cognitivist psychology, as some influential contemporary writers have suggested, but are in some respects anticipations of, and superior to, the current theories such writers themselves favor.[37]

University of Oklahoma

NOTES

[1] The spiritual father of this revolution is Norman Kemp Smith; see "The Naturalism of Hume," I and II, *Mind* **14** (1905), and *The Philosophy of David Hume* (London: Macmillan, 1949). Its more recent examples are too numerous to list: useful overviews are provided by N. Capaldi, J. King, and D. Livingston in "The Hume Literature of the 1970s," *Philosophical Topics* **12** (1981), and by D. E. Yalden-Thomson, "Recent Work on Hume," *American Philosophical Quarterly* **20** (1983). A nearly complete bibliography of the Hume literature from 1925 to 1976 is to be found in R. Hall's *Fifty Years of Hume Scholarship: A Bibliographical Guide* (Edinburgh: Edinburgh University Press, 1978) and sub sequent annual supplements in *Hume Studies*.

[2] D. Hume, *A Treatise of Human Nature*, ed. L. A. Selby-Bigge (Oxford: Oxford University Press, 1958). References to this work will be given in parentheses in the text.

[3] D. Hume, *Abstract of a Treatise of Human Nature*, eds. J. M. Keynes and P. Sraffa (Cambridge: Cambridge University Press, 1978).

[4] D. Hume, Letter to Hutcheson, 17th September 1739.

[5] In fact, the tools used are extremely various, ranging from the computer simulation of mental processes, through the minute observation of the brain-activity putatively associated with them, to the study of responses in psychological experiments (such as the speed and pattern of grammatical or perceptual judgments).

[6] I shall refer to various papers by Fodor and by Dennett in the course of the discussion. Some of these are included in recent collections: J. Fodor, *Representations* (Cambridge: Bradford/MIT Press, 1981) and D. Dennett, *Brainstorms* (Montgomery: Bradford Books, 1978). For the other approaches, see Paul Churchland, *Scientific Realism and the Plasticity of Mind* (Cambridge: University Press, 1979) and "Eliminative Materialism and the Propositional Attitudes," *The Journal of Philosophy* **78** (1981); Patricia Churchland, "A Perspective on Mind-Brain Research," *The Journal of Philosophy* **77** (1980); S. Stich, *From Folk Psychology to Cognitive Science* (Cambridge: Bradford/MIT Press, 1983).

[7] See J. I. Biro, "Description and Explanation in Hume's Science of Man," *Transactions of the Fifth International Congress on the Enlightenment*, Vol. 4 (New York, 1979).

[8] For discussions of the relation between Hume's principles and those of Newton, and of their views on scientific method, see J. Noxon, *Hume's Philosophical Development* (Oxford: Clarendon Press, 1973); N. Capaldi, *David Hume: The Newtonian Philosopher* (Twayne, 1975); J. P. Monteiro, "Hume's Conception of Science," *Journal of the History of Philosophy* **19** (1981), and "Induction and Natural Selection," *McGill Hume Studies*, eds. D. F. Norton, N. Capaldi and W. L. Robison (San Diego: Austin Hill, 1979).

[9] B. Stroud, *Hume* (London: Routledge and Kegan Paul, 1977), p. 138.

[10] J. Fodor, *The Language of Thought* (Cambridge: Harvard University Press, 1975), p. viii.

[11] For further discussion of this admittedly controversial claim, see section V below, as well as Jane L. McIntyre, "Is Hume's Self Consistent?", *McGill Hume Studies* (San Diego: Austin Hill, 1979) and J. I. Biro, "Hume's Difficulties with the Self," *Hume Studies* **5** (1979).

[12] This is particularly evident in his discussion of *general rules* (T141 *et pass.*; esp. T149-50); see also T. K. Hearn, "General Rules in Hume's *Treatise*," *Journal of the History of Philosophy* **8** (1970).

[13] Most notably by Donald W. Livingston. See his "Hume's Historical Theory of Meaning" in D. Livingston and J. King (eds.), *Hume: A Re-Evaluation* (New York: Fordham University Press, 1976) and *Hume's Philosophy of Common Life* (forthcoming), esp. Ch. IV; also P. Jones, "Strains in Hume and Wittgenstein," in *Hume: A Re-Evaluation*, and P. Ardal, "Convention and Value" in G. P. Morice (ed.), *David Hume: Bicentenary Papers* (Austin: University of Texas Press, 1977).

[14] This criterion for the possession of an idea has to be set against what is indisputably Hume's official doctrine on the matter: that where impressions aren't, ideas cannot be. So, "when one is born blind or deaf, not only the impressions are lost, but also their correspondent ideas; so that there never appear in the mind the least traces of either of them" (T5). This may be thought to clinch the matter. However, Hume goes on to say that in such cases "we cannot form to ourselves a *just* idea" (ibid., my emphasis); as I shall argue in section IV below, having *an* idea and having a *just* idea are not necessarily the same thing, and a different necessary condition may be associated with each. Observing a distinction of this sort may help explain why, in the face of his categorical statement at T5, Hume could allow that a blind man could understand color words: "We have in this Town a singular Phaenomenon, one Backlocke, a very elegant Poet, born blind. His Writings are particularly remarkable for the Justness & Propriety of their Imagery; tho' he

owns that he has no idea of Light or Colors" (Letter to the Abbe Le Blanc, October 24th, 1754).

15 For example, Jerry Fodor, "Mental Representation: An Introduction" (unpublished), esp. pp. 12-13.

16 This is, of course, not quite right but will (have to) do for present purposes. It is really more a matter of there being *something* in a person's head that in *some* way represents the putative distal causes *and* can, in virtue of so doing, connect these with their effects, even though *its* causal powers are due to its purely "formal" (i.e., non-content-relative) properties. This last requirement is the so-called "formality condition,"—see J. Fodor, "Methodological Solipsism Considered as a Research Strategy in Cognitive Psychology," *The Behavioral and Brain Sciences* **3** (1980).

17 For details, see H. P. Grice, "Meaning," *Philosophical Review* **66** (1957), "Utterer's Meaning, Sentence-Meaning and Word-Meaning," *Foundations of Language* **4** (1968), and "Utterer's Meaning and Intention," *Philosophical Review* **78** (1969). There is a voluminous literature on Grice's and Gricean theories. For references, as well as for criticisms, see J. I. Biro, "Intentionalism in the Theory of Meaning," *The Monist* **62** (1979).

18 For additional discussion of the shortcomings of Gricean theories of meaning, see J. I. Biro, "Meaning, Translation and Interpretation," *Australasian Journal of Philosophy* **59** (1981), and "Meaning and What Is Said" in O. Neumeier (ed.), *Language, Mind and Society* (Wien: Springler, 1984).

19 T. Nagel, "What is It Like to Be a Bat?", *Philosophical Review* **83** (1974).

20 This principle—it has also been called, somewhat misleadingly, "the copy principle"—is as follows: "That all our simple ideas in their first appearance are deriv'd from simple impressions, which are correspondent to them and which they exactly represent" (T4).

21 In addition to the paper by Nagel cited above, see also his "Subjective and Objective" in *Mortal Questions* (Cambridge: Cambridge University Press, 1979); C. Hill, "Of Bats, Brains, and Minds," *Philosophy and Phenomenological Research* **38** (1977); L. M. Russow, "It's Not Like that to be a Bat," *Behaviorism* **10** (1982); J. I. Biro, "Subjectivity, Objectivity and Points of View," read at the APA Pacific Division Meeting, 1983; and Paul M. Churchland, "Reduction, Qualia, and the Direct Introspection of Brain States," read at the APA Pacific Division Meeting, 1984.

22 By N. Block in "Troubles with Functionalism," in N. Block (ed.), *Readings in the Philosophy of Psychology* (Cambridge: Harvard University Press, 1980), and in "Are Absent Qualia Impossible?", *Philosophical Review* **89** (1980). For contrary views, see S. Shoemaker, "Functionalism and Qualia," *Philosophical Studies* **27** (1975), and "Some Varieties of Functionalism," *Philosophical Topics* **12** (1981); D. Dennett, "Quining Qualia"; and William G. Lycan, "Form, Function and Feel," *The Journal of Philosophy* **78** (1981).

23 Livingston, op. cit., p. 218.

24 K. Bower, "Imagery: From Hume to Cognitive Science," *Canadian Journal of Philosophy* **14** (1984).

25 Actually, even this is far too strong. Hume's principle should not be interpreted in such a way as to commit him to the implausible view that there can be no more information in an idea than there was in its corresponding impression. First, there is, at least in the standard cases, the information carried by an idea *that* there was a prior and corresponding impression. Second, there is usually a great deal of collateral or background information associated with an idea that goes far beyond its impressional content. On the other hand, there is also information-*loss*, or decay, as with memories, which typically lack detail and are rather schematic. (They are less "lively" perhaps partly for such reasons.) But for the

present argument these complications can be ignored, since the question at issue concerns information only of a certain kind and turns on whether or not there can be some thing in the idea that *would* be in its prior and corresponding impression, had there been one.

26 And perhaps not even all impressions are, once we give up introspectionism. Some of the—in Hume's sense, "original"—information from which we subconsciously constitute our consciously entertained ideas may be in forms quite different from images as usually understood. Whether this is compatible with Hume's official view of *impressions* is another question.

27 Stroud, p. 34.

28 See Biro, "Hume on Self-Identity and Memory," *Review of Metaphysics* **30** (1976), esp. pp. 32-33, on Hume's use of the principle in accounting for how "we can extend our identity beyond our memory" (T262).

29 F. Zabeeh also argues that the resemblance interpretation is needlessly strong. See *Hume: Precursor of Modern Empiricism* (The Hague: Nijhoff, 1973), pp. 54-55.

30 Even if *some* imagery were necessary for the acquisition of ideas of qualia, that would show at most that without acquaintance with *any* of a particular kind or range of qualia, no other member of that kind or range is knowable. This is a weaker claim than the original requirement of a one-to-one correspondence (not to mention resemblance) and is too weak to support extreme anti-objectivist positions such as Nagel's. Indeed, ultimately it may just come down to the triviality that if one cannot experience anything of a certain sort, one cannot experience any *particular* thing of that sort. Defective-organ arguments and bat-arguments may amount to no more than the belaboring of this triviality. (See Biro, "Subjectivity, Objectivity and Points of View").

31 D. Dennett, "Artificial Intelligence as Philosophy and as Psychology," in *Brainstorms* (esp. pp. 122ff.).

32 Dennett, loc. cit.

33 Dennett, "A Cure for the Common Code" in *Brainstorms*, pp. 101-02.

34 J. I. Biro, "Hume on Self-Identity and Memory" and "Hume's Difficulties with the Self."

35 See, for example, McIntyre, op. cit.

36 Notwithstanding occasional disclaimers such as Fodor's disarming admission in "Mental Representation" that what he there speaks of as "Hume's Theory" is "probably a libel on Hume."

37 I am grateful to Chris Swoyer for many helpful discussions on the topics of this paper.

JOHN-CHRISTIAN SMITH

REID AND THE CONTEMPORARY VIEW OF CONSCIOUSNESS

In this article I derive a contemporary theory of consciousness from Thomas Reid's philosophical psychology. His work on this subject agrees with much in current cognitive studies, but little effort has been made to interpret it in modern terms.[1] This is attributable in part to Reid's own failure to state his views in a precise manner. As a matter of historical fact, Reid may not have clarified his position enough for us to determine textually that he would agree with the details of some particular modern analysis of consciousness. The *general* account of cognition he provides is quite rigorous, however, and its highly systematic, relational framework lends itself to further specific development.[2] What I will claim here, then, is that his theory remains consistent under the following interpretation and that there is at least some significant textual support for it.[3]

My point of departure will be a comparison of Reid's remarks on consciousness with a contemporary view characterized by Paul Churchland. Association of these two approaches follows from their common appeal to an analogy between consciousness and sensation (or reflection and perception). This shared feature leads to conflict with the "traditional view," which both Churchland and I consider phenomenological, and its claims of directness and infallibility for consciousness are shown to be unwarranted under the proposed analogy. However, there is also a certain paradox for most modern analyses which is not a problem under my proposed interpretation of Reid's theory. In the end, I argue that this 'Neo-Reidian' account is thus the most principled and systematic approach to the subject of consciousness.

1. The Contemporary View

Paul Churchland[4] presents the following characterization of "self-consciousness" as part of what he labels the "contemporary view":

In summary, self-consciousness, on this view, is just a species of perception: *self-perception*. It is not perception of one's foot with one's eyes, for example, but is rather the perception of one's internal states with what we may call (largely in ignorance) one's faculty of

J-C. Smith (ed.), Historical Foundations of Cognitive Science, 139–159.

introspection. Self-consciousness is thus no more (and no less) mysterious than perception generally. It is just directed internally rather than externally.

In a similar vein, Robert Audi[5] recommends the following approach:

If we take a cue from the etymology of 'introspection', which derives from the Latin *introspicere*, meaning 'to look within', we might construe introspection as attending to one's own consciousness and—when one's mind is not blank—thereby achieving a kind of inner seeing....[W]e can explore introspectional counterparts of some theories of perception and sensory experience.

What I wish to do here is consider whether Reid would have approved of this strategy and whether his approach to consciousness is generally compatible with the resultant contemporary view.

Reid (IP, I:1:2) calls *operations* of the mind "every mode of thinking of which we are conscious." Some of the operations, the conceptual and belief-forming processes, are activities of distinct *faculties* which are innate "powers" of the mind and which are the original, fundamental components of the human "constitution." These faculties include perception, language processing, memory, and a few others, but *not* a range of "capacities" and "habits" (such as math ability and poetry reciting). Where consciousness, itself, fits into this collection is not entirely clear. Reid says,

Consciousness is a word used by philosophers to signify that immediate knowledge which we have of our present thoughts and purposes, and, in general, of all the present operations of our minds. Whence we may observe, that consciousness is only of things present. (IP, I:1:7)

Now this certainly sounds as if consciousness were a distinct faculty on Reid's view, since any mental activity constituting "knowledge" surely involves the conceptual and doxastic (belief-forming) operations which are *definitive* of faculties for him. Moreover he directly compares this activity with operations of perception *qua* faculty:

Thus, *perception* is distinguished from *consciousness*....consciousness is only of things in the mind, and not of external things....As that consciousness by which we have a knowledge of the operations of our own minds, is a different power from that by which we perceive external objects.... (IP, I:1:6-7)

From one perspective this distinction is agreeable to the contemporary view, which holds to a similar distinction between "self-consciousness" and perception; i.e., that self-consciousness results from a separate "faculty" of "*self*-perception," a "species" of perception, and *not* from the faculty used for perception of external objects. It seems, then, that both views agree in distinguishing consciousness from perception according to the different subject domains of their operations (internal thoughts vs. external objects).

The last quotation, however, contains a hint of a Reidian approach beyond this minimal claim. Although Reid sometimes refers to consciousness *as* knowledge, more often he speaks of it as that *by which* we have knowledge. This is perhaps a rather subtle shift in phrasing, but it turns out that this different way of addressing consciousness is indicative of a deeper analysis. If, as the contemporary view holds, self-consciousness results from a special form of internal perception, then there should be some activity as part of the internal 'perceptual' process which corresponds to sensation or sensory activity in the external perceptual process. Perception of the external is not itself sensation, so something must have the parallel role of sensation in the analogous "introspection."

Such activity would indeed be that *by which* knowledge of internal thoughts and feelings is gained, just as sensation, on Reid's view, is the activity *by which* knowledge of external objects is gained (while the perceptual conception and belief *are* the knowledge). For Reid, even though sensation is an innate "power" of the mind, it is not an operation of a faculty and alone never *constitutes* perceptual knowledge. Sensation does not consist in the conception or belief of external objects, as those constitutive operations belong exclusively to the perceptual faculty. Thus, if consciousness is, on Reid's account, the similar intermediary *by which* we gain knowledge of the operations of our minds, then consciousness does not consist in conceptual or belief-forming operations and, although an innate "power," does not constitute a faculty. It would be, instead, a "species" of sensation, viz. internal, *self*-sensation.

Let me support this interpretation by addressing a few other remarks of Reid's that are relevant to the issue of consciousness. To begin with, there are passages which might seem to controvert my approach, as when Reid says, "...perception is accompanied with the belief of that which we perceive, and consciousness with the belief of that whereof we are conscious" (IP, III:1). Again, this appears to be a comparison granting consciousness faculty status, rather than taking it on a par with sensation. But in fact talk of such accompaniment is also present with Reid's characterization of sensation, as in his claim that "sensation is attended by that conception and belief of the external object" (IP, III:1). It is indisputable that Reid nevertheless held sensation to be neither conceptual nor doxastic, so such talk must simply be taken as elliptical or misleading.

Considerations of this sort must be placed in the general context of Reid's continual struggle to separate the notions of qualities, sensations, and faculties in his account, a struggle which is not always successful. It is not as if he proceeds throughout with a fixed set of definitions (see fn. 3). In the early parts of the *Inquiry*, for example, he confusingly uses sensation terms (e.g., 'smell') for the qualities of objects themselves (an ambiguity he notes but does not avoid) and even calls the individual sensory modes *faculties*. It is only through a developing reanalysis, finally completed in the *Intellectual Powers*, that he eventually draws a sharp picture of sensation as the *non*-conceptual, *non*-doxastic, and therefore *non*-facultative operation with an intermediate role of causally "suggesting" the actual perception (conception and belief) of the quality or object perceived.

This final relational account of sensation is most clearly compared to the case of consciousness in the following:[6]

> We know our own thoughts, and the operations of our minds, *by a power which we call consciousness*; but this is only giving a name to the part of our frame. It does not explain its fabric, nor how it produces in us an irresistible conviction of its informations. We perceive material objects and their sensible qualities *by our senses*; but how they give us this information, and how they produce our belief in it, we know not.
>
> (IP, III:2; my emphases)

Here subsequent knowledge of thoughts parallels perception of material objects, and consciousness is unmistakably held to be the correspondent of sensation (as the respective intermediaries): we know our thoughts "*by*...consciousness;" we perceive material objects "*by* our senses." Thus, sensation and consciousness are not the conceptual and doxastic operations constituting knowledge in their respective domains, but rather they are the "informations"—the 'raw data'—in relation to which the operations of faculties do constitute such knowledge.

Reid also says, "The operations of our minds are attended with consciousness; and this consciousness is the evidence, the only evidence, which we have or can have of their existence" (IP, I:2). Elsewhere in his account, however, such "evidence" talk is similarly the chosen means for describing the role of sensation in relation to the perceptual faculty (IP, II:20, e.g.), so characterization as evidence clearly does not entail that an operation has faculty status. (Perception is sometimes characterized as "evidence" (cf. IP I:1), but that is in the context of explicit inferential processes, as in reasoning on the basis of a perception, rather than the processes *producing* a perception, for which sensation is the direct evidence.) Indeed, the functionally mediating but non-facultative nature of sensations, as distinct from independent operations of perception, is fundamental to Reid's realistic epistemology.[7]

Now if sensation is this sort of mediating event for perception and if consciousness is its internal correspondent, then consciousness must be the intermediary *for* something, as well. It is the evidence *of* the existence of other operations of our minds, but like sensation, it must be systematically related to a faculty for any concepts and beliefs *about* those operations to be formed. Reid's theory provides for this by incorporating an independent faculty of *reflection*, which can be held to take the underlying activity of consciousness as its causal intermediary and other mental operations as its domain of represented objects. Reid says,

> ...by consciousness we know certainly the existence of our present thoughts and passions....But it is to be observed that we are conscious of many things to which we give little or no *attention*....It is in our power, however, when we come to the years of under standing, to give attention to our own thoughts and passions, and the various operations of our minds. And, when we make these the objects of our attention, either while they are present or when they are recent and fresh in our memory, this act of the mind is called *reflection*. We take it for granted, therefore, that by attentive reflection, a man may have a clear and certain knowledge of the operations of his own mind....This reflection is a kind of intuition, it gives a like conviction with regard to internal objects, or things in the mind, as the faculty of seeing gives with regard to objects of sight. A man must, therefore, be convinced beyond possibility of doubt, of everything with regard to the operations of his own mind, which he clearly and distinctly discerns by attentive reflection. (IP, I:2)

In this it is most clear that personal knowledge (i.e., the conception and "conviction" or belief) of the internal operations of one's own mind consists in the operations of the faculty of reflection, not the activity of consciousness alone. An analogy between the perceptual faculty ("seeing") and this reflective faculty is also quite straightforward. It is reflection, not consciousness, that makes operations of our minds into objects of our attention. Without the conceptual and belief-forming operations of the reflective faculty, there would be no concepts or beliefs about the operations of the mind at all.

In what I take to be his most definitive statement on this issue, Reid maintains the following:

> When the operations of the mind are exerted, we are conscious of them; and it is in our power to *attend* to them, and to *reflect* upon them, until they become familiar objects of thought. This is *the only way* in which we can form just and accurate notions of those operations. (CS, 7:Concl.; my emphases)

Analogous to sensation, which is non-conceptual, in consciousness there is no conception of other operations of the mind; i.e., nothing is "presented" to the mind about them in the conceptual format of cognition. This role is reserved for reflection alone since—and the point cannot be overemphasized—reflection is "*the only way*" in which we can make other operations of the mind the "familiar objects of thought."

Now as Reid notes, all mental operations are conscious operations, so there is always some activity of consciousness (self-sensation) accompanying them. But we may not have conceptions or beliefs about them because the attention of the reflective faculty is not directed toward the consciousness. Analogously, we may have sensations (or at least sensory activity) but no conception or belief of an external object because the attention of the perceptual faculty is not directed toward that sensory activity: "There are many phaenomena of a similar nature, which shew that the mind may not attend to, and thereby, in some sort, not perceive objects that strike the senses" (CS, 6:XIII:5; also IP, I:1:11). Of course, on the level of innate "original operations" of the reflective faculty, we would expect reflective conception and belief to follow upon consciousness as closely and regularly as the conception and belief of hardness follow upon a sensation of touch. This explains why many have confused consciousness and reflection, just as they have confused sensation and perception (a confusion Reid is most intent upon clarifying). For Reid both perception and reflection, as faculties, must be fully independent of their causal intermediaries (sensation and consciousness, respectively).

What this suggests, further, is an interpretation of Reid's theory according to which consciousness consists in some internal causal activity produced by all (first-order; see below) operations of the mind. Just as causal impressions upon the sense organs are the bodily realizations of functionally characterized sensations, some causal activity (*self*-impressions) must be the bodily basis for consciousness. This activity would be what becomes the consciousness of other mental operations functionally organized in relation to the conceptual and doxastic operations of the reflective faculty. Audi (1988, pp. 43-44) makes a similar point in advancing an "adverbial account of introspection"—which fits well with the popular construal of Reid's theory of perception as adverbial[8]—and argues that "even a true belief about one's conscious states or processes would not be introspective without being causally connected with them," without a causal connection "something like a sensory impression." The results of this connection are reflective concepts and beliefs concerning the existence of the mind and of the particular mental operations causally producing the self-impressions that realize consciousness.

For the most part, this theory parallels Churchland's contemporary view quite nicely. The correlation is direct and requires little terminological manipulation. Churchland holds self-consciousness to be the product of an internalized species of perception occurring through the operations of a faculty called ("largely in ignorance") "introspection." The Reidian view holds conceptions and beliefs about one's own mental organization and operations to be the product of an internal faculty of reflection distinct from but analogous to the perceptual faculty (trivially, since all faculties share the same internal structure and mediated relations under Reid's view). All we need do is take Churchland's rather vague use of 'self-consciousness' to refer to conceptions and beliefs about one's own mind and to consider introspection and reflection the same, in order to make the parallel complete. Moreover, the Reidian and contemporary views both hold that a functional activity must mediate the connection between operations of the faculty and the other mental operations that are objects 'introspected' or reflected upon. After considerable ambiguity in his discussions, Reid seems finally to reserve this role for consciousness *simpliciter*, while Churchland (p. 77) refers to some sort of "second-order sensations," pursuing the analogy with perception to the end. I conclude, then, that the two views are highly similar and basically compatible, even if not precisely the same.

2. Functional Mediation and the Traditional View

A cornerstone for Reid's entire philosophical psychology is exactly the insistence on non-conceptual, relationally characterized intermediaries in cognition of any kind. The emphasis here must be on "non-conceptual," since it is also a major part of his program to deny that there is an intermediate object of thought such as the "ideas" or "impressions" of his Rationalist and Empiricist opponents. Their notion of an intervening "representational" object that is "immediately" or "directly" perceived (as opposed to the *in*directly perceived material object) implies that a *conceptual* operation of some sort mediates all cognition. It is not the mediation itself that Reid objects to—he fully recognizes that "impressions" and "sensations" precede perception—but rather the claim that such mediation is other than purely functional and in fact has objective representational properties. He maintains, instead, that the intermediary is just a "subjective experience" and does not have "any necessary connection" to anything else (CS, 5:III). Interpreting Reid's theory in a way that introduces a non-conceptual intermediary of consciousness leading to reflection is therefore fully consistent with this general approach.

There is a deeper reason for holding to this uniformity of functional mediation, however. Reid's fundamental concern is to establish epistemic equality for all the faculties, and treating their operations as uniformly mediated allows him to do so. In this way Reid is able to deny the Idea Theorist (Rationalist or Empiricist) a basis for special epistemic valuation of the operations of one faculty over those of another, as often claimed for consciousness or reason over perception. At the outset of the *Inquiry*, for example, Reid asks in response to Descartes:

> ...who is voucher for consciousness? Can any man prove that his consciousness may not deceive him? No man can; nor can we give a better reason for trusting to it, than that every man, while his mind is sound, is determined, by the constitution of his nature, to give implicit belief to it. (CS, 1:III)

A difference among the evidential connections of the faculties (ambiguously consciousness/reflection here) to the objects of their operations would presumably provide just such a basis for their epistemic differentiation and so must be rejected. Later, Reid responds to Hume by holding:

> Reason, says the sceptic, is the only judge of truth, and you ought to throw off every opinion and every belief that is not grounded on reason. Why, sir, should I believe the faculty of reason more than that of perception?—they came both out of the same shop, and were made by the same artist; and if he puts one piece of false ware into my hands, what should hinder him from putting another? (CS, 6:XX)

This rejection of preferential treatment for any individual faculty makes Reid's theory depend upon a uniformity of faculty structure and relations.

If formation of one's conceptions and beliefs about one's own mental operations were unmediated, then this would upset the epistemological applecart. Reid is not always consistent in maintaining the paradigm, but the covering argument can be made explicit as follows:

1. Distrusting one mental faculty requires trusting another.

2. No grounds can be given for trusting one faculty that cannot be given for trusting any other.

3. Therefore, the mental faculties are equally trustworthy.

4. Unmediated cognition in the case of one faculty would make its operations more trustworthy than mediated cognition in others.

5. Some faculties, such as perception, produce mediated cognition.

6. Therefore, all mental faculties produce mediated cognition.

The first premise here brings out Reid's belief that general skepticism in regard to the issuances of all the faculties is not a sane option. Support for the second comes from his claim that the only possible reason for trust in *any* faculty is that "every man, while his mind is sound, is determined, by the constitution of his nature, to give implicit belief to it" (CS, 1:III). Perhaps the most crucial step is at the fourth premise, which amounts to a Russellian-style preference for putative 'knowledge by acquaintance'. If there were such immediate acquaintance with the objects of thought, as indeed held under the doctrine that ideas are those objects, it would be "the most wonderful discovery that this or any other age hath produced" (CS, 2:VI). But according to Reid, that hypothesis is rather "the parent of those many paradoxes so shocking to common sense"

(CS, 2:III) and cannot establish greater trustworthiness of any particular faculty over any other.

This argument opposes Reid to the "traditional view" of consciousness, as Churchland (p. 75) calls it. Traditionally, knowledge through consciousness of ("first-order") sensations and thoughts has been held to be "immediate," in the sense of *not* being mediated as perception is mediated by sensations (or ideas or images). Simply put, our sensations stand between us and the objects we perceive, but nothing further stands between us and the sensations we are aware of. Churchland argues against this view on the basis of possible fallibility in "judgments" concerning one's own thoughts and sensations. If mistakes of this sort are seen to be possible, then that is taken to be evidence of the mediation of the underlying processes. In the second quote above, Reid can be taken as making a similar point. Yet he actually holds that all of the faculties are indeed trustworthy and was quoted earlier as claiming that reflection produces a "clear and certain knowledge" and convinces "beyond possibility of doubt," so there is likely to be confusion over this issue.

One way of approaching the subject is in terms of what Audi (1988, pp. 45-48) calls the *Strong Doctrine of Privileged Access*, which he attributes to Hume but which is presupposed by every advocate of the Idea Theory. This is the biconditional assertion that incorporates both the "thesis of infallibility," holding that if one has an introspective belief then one is truly in the occurrent mental state introspected, and the "thesis of omniscience," holding that if one is truly in an occurrent mental state then one has the introspective belief of it; i.e., one knows it.[9] The question is then whether Reid's rejection of immediacy in the process of reflection commits him to a denial of one or both of these theses. After restricting consideration to cases of "attentive introspection," Audi offers counterexamples to the proposed infallibility, but none of these turns on the putative mediation of the process. Rather, what seems possible to him is that desires or other beliefs might influence or distort the introspective belief, as when needing to believe one is thinking about a problem rather than daydreaming. Also, it seems possible that the introspective process might be preempted by other processes and produce a mistaken belief which can later be rectified by returning to the original process, as in discovering upon reinspection that what was imaged was actually a hexagon rather than an octagon. That neither case results from inadequate mediation shows that fallibility does not *entail* mediation. But in fact the second is just a further case of inattentive—in this case, incomplete—reflection, while the first is a matter of inferential defeasibility or suppression. So neither case demonstrates failure of the Strong Doctrine.

Indeed, the errors in introspective judgment pointed to generally are of an inferential nature, e.g., the inference from belief about a taste-sensation to belief about the perception of lime sherbet in one's mouth (cf. Churchland, p. 77). But this is not the sort of *non*inferential concept and belief formation Reid intuitively associates with reflection. To focus on the possibility of perceptual inferential error of this sort is in effect to adopt the Idea Theoretical model of inference from representational sensations or impressions to objects in the world. At best it shows that we can make mistakes if moving from one reflective belief (about a sensation) to another (about a perception), while it says nothing about the possibility of error in formation of an *individual* reflective belief. In an example from Peter Smith and O. R. Jones,[10] the tortured prisoner is tricked by ice cubes being placed on his back rather than the hot iron. Yet the perceptual belief that he was burned is not the result of his *reflective* belief that he had a sensation of "searing pain," and a mistaken inference from reflection on the perception to belief about the nature of the sensation surely does not demonstrate fallibility of reflection. No insertion of reflection into the perceptual process can do so, nor will any example based on general erosion of the reflective ability. A similar point holds in response to Daniel Dennett's[11] suggestion that any intuition that we are the "sole and perfect arbiters of what we experience" should be discarded due to the possibility of memory decay prior to speech acts. The question is not whether a person can make a mistake in less than fully attentive and retained reflection nor in inferring an objective experience on the basis of a reflective belief about a mental experience. Rather, the question is whether a mistake can be made in forming the belief as to one's own mental experience alone.

Now Reid can agree that knowledge of one's own mental operations is "immediate knowledge," *of a sort*. For he also holds that—just as perception counts as immediate knowledge even though it is uniformly mediated by sensation—there is immediate knowledge gained by way of consciousness, as the functional intermediary for reflection. The difference for Reid is that such mediation does not suggest an *inference* from an intermediate *representation*. As a result, no errors of inference are possible in simple operations of either reflection or perception. There are no properties of impressions or sensations from which the existence and nature of objects (including other mental operations) *could be* nomologically inferred. Instead, there is only an internal "natural knowledge" preserved in inherited regularities ("first principles") of the faculties' operations in response to subjective experiences (CS, 5:III). Knowledge gained in this way is "immediate," in being conceptually (so inferentially) unmediated, although at the same time corrigible in being causally and functionally mediated. As seen in the argument above, Reid accepts the trustworthiness of all the faculties, including reflection, so he must be willing to embrace something like the Strong Doctrine, which the passages cited suggest. But denying any intermediary is too high a price to pay for 'introspective' infallibility and one that we have found need not be paid, since (i) lack of mediation is no guarantee of infallibility (nor of omniscience) and (ii) noninferential causal mediation is no guarantee of fallibility (nor lack of omniscience). In the end, Reid cannot hold that reflective beliefs are absolutely incorrigible, but in being conjoined with the entire set of faculties through a uniformity of functional mediation, the operations of reflection fail only if the entire set fails (CS 6:XX).

This last point is crucial for contrast with phenomenology, which is broadly identifiable with the traditional view (cf. Churchland, pp. 86-87). There are two responses to Husserlian phenomenology that I think are possible from the 'Neo-Reidian' perspective. One is that the putative phenomenological "introspective retracing" of mental contents leads only to an endless interpretation of the activities (self-impressions) realizing consciousness, by focusing on their particular causal properties. Such causal structure is never itself an object conceptually represented by operations of *reflection*—just as the proximal stimulus underlying sensation is never itself perceived—so the phenomenological therapeutical program is rendered specious. I return to this point below in addressing Lehrer's Paradox.

The other response is that the proposed separation of the output of the perceptual faculty from the rest of cognition, through phenomenological "bracketing" of the "natural standpoint," undermines the basis for attribution of content altogether. As argued above, the faculties come as an epistemically balanced *set*, and any attempt to upset this balance by breaking up the set quickly undermines meaningfulness of talk about what the remainder of the set is up to. From this outlook the "phenomenological epochè" can be viewed as a method of passing from reflection on intrinsic intentional content—as derived from natural relations of the cognitive organization of faculties *qua* inherited set—to the assignment of an artificially unrestricted, arbitrary interpretation of internal processes. Claims to knowledge on the basis of this inferential exclusion of perceptual operations are therefore epistemologically spurious.

There is a sense in which the resultant arbitrary interpretation of causal activities under phenomenology is very like the approach taken toward computer programming. As long as the formal, causal properties of sequences of physical events preserve the semantic relations of the assigned interpretation then the program has been implemented. This parallel has led some philosophers to maintain that Husserl is the major forerunner of modern cognitive science.[12] Of course, indefinitely many interpretations may be preserved by exactly the same causal sequence, so whatever content is assigned is not intrinsic to the process and is in fact eliminable in the explanation of input-output relations. This is the notorious problem of content that has plagued current *functionalist* attempts to provide models of cognitive processes. There is insufficient space here to go into this matter fully. Suffice it to say that, while Reid would agree with the eliminability of content in regard to *sensory* operations, the basic Neo-Reidian response to both the phenomenological and the functionalist approaches is that they ignore the natural constraints on interpretation of *conceptual and doxastic* operations deriving from the modular or componential nature of cognitive organization (i.e., as a fixed set of faculties). Thus, to the extent that the computational explanation of cognition introduces considerations of naturally selected modular structure beyond mere functionalist analyses—as seems to be increasingly necessary[13]—Reid emerges as the more significant forerunner of cognitive science.

To sum up, then, Reid's account of reflection incorporates reference to a mediating event of consciousness (or self-sensation), in contrast to the "traditional view" of self-consciousness as an unmediated process. However, if lack of mediation means the process is noninferential, then Reid agrees with the claim. The deeper reason for holding that reflection is mediated stems from a more general theoretical claim that the cognitive faculties constitute a set of epistemic equals. If one is mediated, as perception clearly is by sensation, then they all must be mediated. Knowledge gained through operations of the different faculties is therefore either equally indirect or equally immediate, according to whether the question is one of causally functional or conceptually inferential mediation. Phenomenology is relevant at this point because, as a version of the traditional view, it holds both that consciousness involves some sort of privileged access to one's own mental contents and that perceptual output can be meaningfully divorced from the inherited set of faculties. From a Neo-Reidian perspective, neither claim is any more successful in explaining the content relations of cognitive organization than the waning functionalist proposals of current cognitive science. The Neo-Reidian theory therefore provides a more principled defense of the contemporary view of consciousness/reflection, while suggesting the best direction for further computational research.

3. A Neo-Reidian Solution to Lehrer's Paradox

There are numerous paradoxes which surround the issue of consciousness, and I will now consider one which is particularly troublesome for a any contemporary view and whose resolution will bring out some of the finer points of the Neo-Reidian theory. This is the problem of voluntariness, and as it was first suggested to me by Keith Lehrer, I will label it "Lehrer's Paradox." It is an apparent conflict arising from any theory of consciousness which can be understood initially as saying that every mental operation is conscious and then as making the following claims:

1. Consciousness is itself an involuntary mental operation.

2. Consciousness has no object distinct from it as a mental act.

Lehrer's Paradox becomes explicit when we further claim:

3. (1) and (2) cannot both be true.

Since Reid's theory holds that we are conscious of every mode of thought and that some mental operations are in fact voluntary, (1) and (2) seem to lead us to the self-contradictory conclusion that some conscious mental acts are voluntary if and only if they are involuntary. That is, if consciousness is not distinct from the voluntary mental operation which is the object one is conscious *of*, then that consciousness seems both voluntary and involuntary. I will argue that this paradox results from an ambiguity appealed to in the second claim and that the Neo-Reidian theory remains consistent once this ambiguity is resolved.

However, I will first note that this paradox is related to a further problem concerning the possibility of an infinite regress of consciousness. This regress is claimed to follow on the grounds that if we are conscious of all mental operations and consciousness is itself a mental operation, then we must be conscious of operations of consciousness *ad infinitum*. In order to escape such a threatened regress, one might view the operation of which one is conscious as necessarily complex; i.e., as incorporating the consciousness in some sort of 'piggyback' fashion. This would then support the second of the problematic claims. But we need not strictly identify consciousness with other mental operations in order to avoid the possibility of the regress. As usually presented, the argument mistakes a relational psychological claim for an empirical claim about natural nomological causal activities. I shall address this mistake immediately after clarifying the confusion leading to Lehrer's Paradox.

This paradox threatens a general dispute over the role of consciousness in relation to faculties and mental operations. Since consciousness seems to add nothing beyond the operation of which we are conscious, if that operation is voluntary then the consciousness must also be voluntary, *contra hypothesi*. In general Reid ignores this issue, but that is good reason to expect a solution to follow from some aspect of his overall system. Indeed, this expectation is raised by our earlier reliance on the analogy between reflection and perception found in the Reidian and contemporary views.

As it turns out, though, the appeal to this parallel is the very source of Lehrer's Paradox. Comparing sensation with other mental operations, Reid says it "may be distinguished from all others by this, that it hath no object distinct from the act itself" (IP, 1:12). On the face of it, this would seem to rule out an analogous characterization of consciousness, but further remarks suggest that the major distinguishing attribute of having no object is shared by both activities.[14] We can simply take the distinction to apply to the *level of analysis* that captures sensation of any sort, including our proposed self-sensation, viz. consciousness. In particular, since consciousness adds nothing to our perceptions beyond the perceptual operations, themselves, it also has the requisite property in having no mental acts or objects distinct from itself. This again supports the second claim of the paradox. Yet Reid also clearly holds the first claim, that consciousness has a property which some mental operations do not have, viz. involuntariness. By Leibniz's Law of the Indiscernibility of Identicals, then, voluntary mental operations cannot be identical with operations of consciousness. In consequence the latter appear both distinct and not distinct from represented mental objects, thereby warranting the third claim. Lehrer's Paradox thus seems unavoidable once the initial parallel with sensation is accepted.

Now a Neo-Reidian solution to this paradox is not an outright denial of the parallel, which would amount to embracing the skeptical claim that (1) and (2) are contradictory. Rather, we should be led to deny (3) by keeping in mind a broader objective of Reid's theory and reanalyzing the basic comparison. Reid has good reason to hold that consciousness is analogous to sensation in the relevant respect, since his faculty psychology is a viable alternative to the Idea Theory only if no single faculty has a privileged status resulting from greater trustworthiness of its operations.[15] I argued in the last section that since sensation is a level of mental activity between perception and the world, if no comparable level intervened in the formation of concepts and beliefs about one's own mental operations, the faculty responsible for this latter activity would be in more direct contact with its objects and presumably more reliable. But this is precisely the Cartesian result that Reid excoriates (CS, 1:III). The epistemic equality of the faculties thus requires that a parallel mediating role for consciousness must be presupposed.

Once we accept this analogy, however, the relation of consciousness to faculties and other mental operations produces a reinterpretation of the second claim of the paradox. Since sensation is not conceptual, it is not a faculty-level operation, and neither is consciousness, perhaps explaining why both are "involuntary." Moreover, the evidence for this claim in the case of sensation is precisely the critical attribute that it "hath no object distinct from the act itself." The object of sensation *would be* what is perceived, but as there is no conceptual aspect to the sensory act or operation, sensation does not conceptually "present" an object of *any* sort to the mind. Specifically, it does not represent the object or quality that is ultimately perceived. That is the role of the conceptual and doxastic operations of the perceptual faculty. As Reid says, "...in this sense, [sensation] has no object; the feeling and the thing felt are one and the same" (IP, I:1:12). It is not as if there were an object, pain, as well as an act of feeling it. Rather, there is just the subjective experience of hurting, which is accompanied by a *perception* representing the disorder.

Transfering this analysis to the case of consciousness, we can then see that the paradox has made illicit use of an ambiguity in interpretation of what the "object" is that purportedly is not *had*. If what is perceived is the object that the sensation in fact does not have, then the mental operation reflected upon must be the object that the consciousness does not have. The source of the paradox is that both the act of consciousness and the further act can be made objects of reflection, though in distinct operations of that faculty. It is possible to reflect on our experiences of consciousness as they occur, just as it is possible to reflect on our sensations. In doing so, we make those mental operations the objects of representational acts of the reflective faculty, but this entails an additional activity of consciousness mediating between those objects and those acts. I briefly discuss the nature of "consciousness of consciousness" below.

Here the point is that as an *object* of reflection, neither consciousness nor sensation is an intermediary of the reflective process whose object it is. They are, of course, intermediaries in *other* (reflective or perceptual) processes that have other objects. The ambiguity appealed to in (2) is between: (i) consciousness having "no object" in the sense that an intermediary has no conceptual content and thus is *non*-representational, and (ii) consciousness *being* an object when considered "as a mental act" itself. While (1) and (2) are supported through reference to the operation of consciousness as an intermediary of the reflective process, (3) results from taking the operation of consciousness to be the object of reflection.

Without this equivocation there is no paradox, since (1) and (2) can both be true if involuntary consciousness is itself a distinct mental act which (a) has no representational relations to any other object and (b) therefore does not have voluntary operations as objects it represents (though it may function to "signify" them arbitrarily in relation to operations of reflection). The voluntary operations are, instead, some of the objects conceptually represented by acts of attentive reflection.

The Neo-Reidian solutions to such paradoxes of consciousness ultimately turn upon an account of the faculty of reflection. Much of this account must be inferred from Reid's general theory and especially from the supporting analogy with faculties of perception and language. Since reflection is a faculty, it has conceptual and other capacities of the sort that analogy specifies. In general, reflection must share in all the basic features which individuate the level of faculties and their operations. It is unique to reflection, however, that it takes the activity of consciousness as a mediating functional system. In doing so, an operation of reflection (not consciousness) produces a representation of either the representational content of another faculty's operation or the functional role of a mediating event, such as sensation or consciousness, in relation to such an operation. We can reflect upon the conception and belief (i.e., the perception) of a certain odor's material presence or upon the sensation of smell leading to that perception. The operations of reflection accomplishing this are either original or acquired, as with any faculty. Among these, the original must be so "clear and distinct" that differentiating them from involuntary acts of consciousness is comparable to differentiating perception of hardness from sensation of hardness. Reid had difficulty doing the latter, at least through initial parts of the *Inquiry*, and this confusion spills over into the problem of distinguishing primary experiences of consciousness from original operations of the faculty of reflection. As he recognized by the time of the *Intellectual Powers*, however, uniformity in being mediated by functional "signs" (not ideas or images) is an indispensable feature of faculty relations. Without this uniformity Reid's theory is inconsistent.

4. Conclusion

I will leave much of this Neo-Reidian account for consideration elsewhere, but in conclusion I wish to make some payments on a few of the promissory notes taken out in the foregoing. First, if consciousness is comparable to sensation, then the activity underlying consciousness has its role only in non-conceptual, merely functional relations to operations of the faculty of reflection.[16] That is, consciousness results from an "artificial" organization, according to "original" principles of the reflective faculty, of a causal activity which bears "no necessary connection" to the objects that conceptual operations of reflection come to represent, viz. the contents or roles of other operations (cf. CS, 5:III). Second, since this underlying causal activity is presumed to be the effect or by-product of the activity of all other operations of the mind, reflective organization of it must determine its role relative to the entire system of faculties. In this case there is no difference between the "operations of the understanding," what Reid (IP, V:3) calls "generalization" and "abstraction," and at least original operations of the faculty of reflection. (A third operation of "combination" would seem more readily attributable to *all* of the "inner" faculties, those taking outputs of other faculties as their inputs.) This means reflection is necessary for the formation of *general* concepts, which are then known only through *reflective understanding.*

A third point is that this purely relational nature of consciousness allows for an infinitude of involuntary "acts" such as being conscious of being conscious, etc. This is the case since an infinitude of relations of activities underlying consciousness to the operations of reflection may obtain simultaneously. Of course, actual operations of reflection and actual causal activities underlying consciousness will be finite, but relations of consciousness which *could* be reflectively determined will always be infinite once *any* are determined. This is a complex claim about the nature of a consciousness C′ of some other consciousness C of a voluntary operation V. The point is not simply that C of V is itself a conscious operation of the mind, since two different types of operations of the mind have been posited: functional operations which come to have artificial "signification" (as in sensation and consciousness *simpliciter*) and conceptual operations which have intrinsic representational (or intentional) properties (as in all faculties). Iteration of the sort at issue must take that distinction into account.

Consciousness of a faculty operation obtains by virtue of relations of (additional, subsequent) causal activities which come to be associated with the faculty of reflection. To be conscious of being so conscious, however, is merely for those same activities to be further organized under association with the faculty of reflection. This is not a response to further causal activity, as the first-order significatory *relations* of consciousness cannot themselves have *causal* by-products. Rather, it is just a more complex organization of the same causal activity in a different relation to reflection. The operation of that faculty is still functionally mediated by the causal by-product of the activity realizing object operation, but the organization imposed under that mediation turns out to be systematically distinct. Since consciousness is not the neurophysical activity but a signification relation defined over it, a consciousness C′ of C of V does not depend upon there being a further causal by-product of (the causal by-product of) the operation's activities. It depends, instead, upon the same causal by-product having systematically different relations to the causal activities of the faculty of reflection.

The framework for such relations is determined by the original principled operations of the reflective faculty which makes all possible relations of consciousness, at whatever depth of iteration, implicit in the mind. Of course only some of this infinity of relations will become explicit in connection with occurrent operations of reflection. C of V exhibits an activity which can be functionally organized in relation to operations of reflection so as to constitute a consciousness C′ of C of V, as well, but this may never actually occur. C of V also exhibits an activity suggesting C″ of C′ of C of V, and so on. In this sense only is there consciousness of every significatory relation, so the threatened regress is not a regress of nomological causal activities but rather an innocuous regress of implicit psychological relations. Every operation of the mind is conscious in this sense, but those we reflect upon are normally quite limited in number and complexity. As suggested earlier, however, the forced phenomenological focus on conscious content can be understood as endlessly wandering through this infinity of abstract relations, in a meaningless enterprise.

The final point is that the operations whose contents or roles are the objects represented by acts of reflection do not in any way constitute that activity which "suggests" those conceptions and beliefs in reflection. The restriction of reflection as a response only to the activity of consciousness is a major factor in the contrast between reflection and other inner faculties. These faculties must be differentiated according to the mediating activities they naturally organize, and reflection differs in that its input domain of consciousness is realized by an independent *causal by-product* of the neurophysical activities of the faculties and sensory systems. Such a view is obviously *Neo*-Reidian, but it makes Reid's broad picture into a perfectly symmetrical, consistent masterpiece. Attempts to construe his views otherwise cannot explain how the faculties are individuated or how Lehrer's Paradox and other problems are avoided without leading to inconsistency.

In such discussions of consciousness, there is a strong tendency to conflate it with other operations of the mind. Smith and Jones (1986, p. 210), for example, describe a driver who "wasn't conscious of anything unusual in the road," as if consciousness were a version of perception. To be sure, they go on to agree with Dennett that there is "unresolved messiness in our ordinary talk about consciousness" and that "what is needed is evidently conceptual reconstruction rather than analysis of the inadequate concepts we already have." But their subsequent reconstruction of a "transitive use" of 'conscious' only tends to confuse consciousness with reflection, parallel to the tendency Reid observes toward conflation of sensation with perception. There is a further tendency to think of each mediation by consciousness as depending upon a unique causal activity or impression whose nomic properties determine what is signified. This is to accept a version of (type-type) Identity Theory for states or acts of consciousness, and although "what it is like to be" a conscious human being is probably closely linked with such first-order causal properties,[17] the role of consciousness in production of behavior is functionalistically independent of its particular causal realization. All of these tendencies are Idea Theoretic and should be resisted for now familiar reasons.

It should also be noted that the concept of consciousness is not always the same as a concept of specific awareness. A *general* conception of "being conscious," as opposed to being asleep or anesthetized, picks out some overall interaction of the faculties, especially that between reason and reflection. This is what Smith and Jones (loc. cit.) refer to with the "intransitive use" of 'conscious'. It is what I call 'having one's faculties about one' or 'being coherent'. The *particular* conception of consciousness of some mental operation, like believing that one is imaging blue waters, is generally used for what is now hopefully understood as an operation of the faculty of reflection alone. As for consciousness *simpliciter*, in the Neo-Reidian sense, ordinary talk of the consciousness of animals probably best suggests it. It is not that non-human animals lack consciousness, but rather that they lack the faculty of reflection for organizing the consciousness they have. Selection of this faculty was perhaps the evolutionary leap most central to making us the intentional beings that we are.

The Neo-Reidian account I have offered proceeds from some rather simple intuitions. If I have conceptions and beliefs about my own mental operations, then something at that time functions as a sign of those operations in relation to the faculty providing those conceptions and beliefs. For something to be a sign of this sort is for it to have the requisite role in relation to that faculty, according to its own principles, and when activity has this role we call it *consciousness*. Thus, Reid's philosophical psychology provides us with a basis for understanding consciousness and reflection as different systematic features of the mind, and that basis leads to a theory which is both commonsensical and consistent with the contemporary views of cognitive science.

University of Arizona

NOTES

[1] Reid's works include *An Inquiry into the Human Mind on the Principles of Common Sense* (CS) in 1764, *Essays on the Intellectual Powers of Man* (IP) in 1785, and *Essays on the Active Powers of the Human Mind* (AP) in 1788, which will hereafter be cited as abbreviated followed by chapter, essay, section, etc. I find, in particular, a great deal of correspondence between Reid's faculty psychology and the modularity thesis advanced by Jerry A. Fodor in *The Modularity of Mind* (Cambridge, MA: MIT Press, 1983), which the following generally assumes to be correct.

[2] For an analysis of Reid's view in terms of modern cognitive science, cf. my "Reid's Functional Explanation of Sensation," *History of Philosophy Quarterly* **3** (1986), No. 2, pp. 175-193.

[3] There is considerable confusion over terminology due to Reid's gradual refinement of usages over the 21 years from CS to IP. This leads to apparent inconsistencies in textual evidence. I will illustrate this problem in regard to his uses of the terms 'consciousness' and 'reflection'. Perhaps the most glaring example is the development of Reid's sense of a "faculty," from at first the vague notion of a single sensory mode (such as smelling) to the well-defined concept of a functional mental subsystem of conceptual and doxastic operations. Recognizing this developmental aspect of Reid's short works seems minimally charitable, though some 'Reidian scholars' seem more interested in the letter than the spirit of his works. The latter are inclined to forget that Reid was, after all, a *popular* ("commonsense") philosopher—the most popular of his times—rather than a polished academic. (This may also account for the paucity and rather low quality of Reidian scholar ship over the last two hundred years, although there is little doubt that academics, from Kant to Moore to Putnam, have drawn upon his work.)

[4] *Matter and Consciousness* (Cambridge, MA: MIT Press, 1984), p. 74. Subsequent references to this work will be by page number only.

[5] Robert Audi, *Belief, Justification, and Knowledge* (Belmont, CA: Wadsworth Publishing Company, 1988), p. 41.

[6] Cf. John-Christian Smith (1986), pp. 188-189.

[7] A general treatment of the connection between Reidian psychology and his epistemology can be found in K. Lehrer and J-C. Smith, "Reid on Testimony and Perception," *Canadian Journal of Philosophy*, Supplementary Vol. 11 (1985), pp. 21-38.

[8] Cf. Roderick Chisholm, *Theory of Knowledge*, 2nd ed. (Englewood Cliffs, NJ: Prentice-Hall, Inc., 1977), fn. 14, pp. 46-47.

[9] Only occurrent mental properties—process properties like active conceiving and thinking or phenomenal properties like sensing and imaging—are appropriate objects of reflection; cf. Audi (1988), p. 41. Reid considers all belief to be an active operation of the mind and seems not to have distinguished dispositional mental states as a separate category.

[10] Peter Smith and O. R. Jones, *The Philosophy of Mind* (New York: Cambridge University Press, 1986), p. 215.

[11] Daniel C. Dennett, *Brainstorms: Philosophical Essays on Mind and Psychology* (Montgomery, VT: Bradford Books, Publishers, 1978), pp. 169-170.

[12] Cf., e.g., Hubert L. Dreyfus (ed. with Harrison Hall), *Husserl, Intentionality, and Cognitive Science* (Cambridge: MIT Press/Bradford Books, 1982).

[13] Cf. Jerry A. Fodor, *The Modularity of Mind: An Essay on Faculty Psychology* (Cambridge: MIT Press/Bradford Books, 1983).

[14] A particularly interesting passage in this regard is the following, from the University of Aberdeen's Birkwood Collection, MS. 2131/6/I/18, dated October, 1748:

> I know nothing that is meant by Consciousness of Present Perceptions but the perceiving that we perceive them. I cannot imagine there is anything more in perceiving that I perceive a Star than in perceiving a Star Simply, otherwise there might be perceptions of perceptions in Infinitum....

All such talk of internal "perceiving" strengthens the analogy between sensation and consciousness that I have argued best explains Reid's view.

[15] Cf. K. Lehrer and J-C. Smith (1985), pp. 24-25.

[16] Although Reid maintains that sensation is involuntary in a general way, he also holds that "in most of the operations of the mind in which judgment or belief is combined with feeling, the feeling is the *consequence* of the judgment, and is regulated by it" (AP, V:7).

[17] Cf. T. Nagel, "What Is It Like To Be A Bat?", *Philosophical Review* **83**, pp. 435-450. This problem of the 'raw feel' of experience, which is closely tied to the problem of qualia, finds similar treatment in Sydney Shoemaker's "Functionalism and Qualia," *Philosophical Studies* **27**, pp. 291-315. Also, cf. Smith and Jones (1986), pp. 216-219.

RALF MEERBOTE

KANT'S FUNCTIONALISM

1. Introduction

Kant's transcendental psychology, often maligned,[1] is a cognitive psychology. More specifically, it is a faculty psychology which speaks of capacities and abilities of various sorts which are needed for empirical cognition. The exercise of such capacities and abilities typically consists in mental actions of several types. An activity-characterization of cognitive mental life is the indispensable core element of transcendental psychology.[2] Kant conceives of cognitive mental actions as goal-oriented and as performed by an agent on the basis of the agent's conception of rules governing the actions in question.[3] Such a conception assigns to cognitive activities a high degree of structure.

The capacities of sentience, imagination and intuition, understanding, subsumptive (determining) judgment, and inventive (reflective) judgment will serve as examples of the capacities of cognitive mental life.[4] Such capacities are concretely at work both in the formation and in the employment of general mathematical-intuitive concepts[5] and of general empirical qualitative concepts. Such concepts, once available, are in turn capacities in their own right, falling under some or all of the just-mentioned higher-order capacities. They are, as Kant points out,[6] rules for the performance of various acts of cognitive synthesis and will themselves be subject to general rules of the higher capacities.

Many capacities are acquired and employed in a manner requiring sensation, and others are possessed and employed in a manner wholly *a priori* or originary.[7] Generically, cognitive synthesis is production, through combination, of cognitive mental content, such content not being given in intuition.[8] In this generic sense, all of cognitive activity is cognitive synthesis. The distinction between what is given in and through intuition and what is cognitively synthesized (or produced through cognitive combination) is co-extensive for Kant with the distinction (in cognitive situations) between what passively enters the mental life of a person and what (by way of mental content) is achieved through mental activity.

Not all mental content is itself cognitive mental content since not all is produced through combination. Some mental content is given in and

J-C. Smith (ed.), Historical Foundations of Cognitive Science, 161–187.

through intuition.[9] But content given in intuition will turn out to be necessary for cognitive mental content since necessary for production through combination. The passivity of at least one capacity is necessary for the cognitive activities of other capacities. I am referring to the capacity of sentience or sensation. In the list of capacities, sentience stands out from the rest in that it is necessary for the exercise of cognitive capacities without itself being such a capacity.

We shall see that this non-activity character of sentience is of fundamental importance for Kant's analysis of empirical cognition and its objects. We shall also see that his characterization of sentience leads to a treatment of the nature of sensation qualia significantly different from that of states and products of mental activity. To anticipate, it will turn out that Kant adopts a functionalist (and teleological) psychological account of the cognitive mental life of persons and a mixed, or dual, psychological approach to those portions of mental life which are taken up by sensation and other mental content not produced through combination. We shall see that according to him we need to give both functionalist non-causal and non-functionalist causal psychological characterizations of qualia and other content not produced through combination, with some aspects of qualia and of that content not functionalistically characterizable and yet necessary for cognitive mental life.

If such a reading of Kantian texts can be defended, Kant's would be an important place in the history of cognitive psychology. It would also turn out that he takes sides in debates going on in the theory of knowledge and in cognitive science during the last third of the twentieth century. Kant's theory would be interesting since it would be an instructive mixture of a causal-realist and a functionalist account of sensation qualia and other content and hence of the complete set of conditions of the formation and employment of empirical cognitive concepts.

In part II of this paper, I shall discuss Kant's conception of sensations and of their difference from, and their role in, perception. This will involve a discussion of the content, and of the conditions of occurrence, of beliefs a person has concerning his or her sensation qualia. It will emerge that important conditions have to be introduced in order for perceptions to be fully objective, and these conditions require a conception of sensation qualia that is significantly different from the conception which needs to be adopted when we characterize qualia in less than fully objective perceptions. An adequate account of the nature of cognitive concepts is required to gain insight into full objectivity, and part III will be devoted to a statement of Kant's views concerning the nature of such concepts. Part IV sketches Kant's functionalist analysis of the tokening of concepts in the mental life of persons and the considerations which lead him to the mixed, or dual, approach to sensations and some other content alluded to above. It will turn out that Kant's views on these matters, considered in their entirety, shed light on his conception of empirical realism and transcendental idealism, as well as on his conception of the cognizability of things as phenomenal and their uncognizability when considered in or by themselves. Part V brings this paper to a conclusion with a brief statement of these ramifications, together with some indications of the relevance of Kant's views to recent discussions.

I conclude this introduction by considering some questions of textual support, questions primarily of interest to Kant scholars who will insist on some preliminary, antecedent justification for ascribing to Kant a use of functionalist descriptions of the content of the cognitive mental life of persons.

My main support for parts II - IV are the *Critique of Pure Reason* and the Jäsche edition of Kant's *Lectures on Logic*, together with some use of his *Anthropology from a Pragmatic Point of View* and of his *Prolegomena*. But functionalism, and its analysis, become explicit topics for Kant only in the *Critique of Judgment*.[10] Hence some basic orientation may be in order.

The first *Critique* speaks of talents, capacities, and (mental) activities,[11] and a significant portion of the Subjective Deduction in that work needs to be understood as a transcendental-psychological account of the conditions of ascription and of exercise of pertinent cognitive capacities (given sentience).[12] Supplementing this, the *Prolegomena* contains an attempt to explain the step from not fully to fully objective perceptions (by means of the distinction—variously drawn—between judgments of perception and those of experience), and the *Lectures on Logic* significantly add to the theory of concepts.

The *Anthropology* raises important questions, and attempts some answers, largely negative, concerning the nature of mental life, but we have to turn to the *Critique of Judgment* for a positive account of the sort of description psychological description must turn out to be in light of Kant's negative characterization given elsewhere.

This negative characterization amounts to the following. Psychology is limited to the status of a body of descriptions falling short of the conditions required for a science most properly so called. In terms of the program for a science outlined and defended in the *Critique of Pure Reason*, psychology falls short because it (a) cannot incorporate any proper mathematization of its domain and (b) cannot incorporate conditions of proper predictability and retrodictability of the items in its domain. In other words (so Kant might say), psychology fails to satisfy the demands of the cognitive employment of the imagination (in its geometrical and arithmetical use) and of the categorially governed understanding.[13]

As a result of these claimed limitations, psychology is a body of non-determining descriptions, with the notion of determination here reserved for descriptions which satisfy the requirements for scientific cognition laid down by the first *Critique*. Hence the account of acts of cognitive synthesis, an account which speaks of persons determining various sorts of states of affairs and changes of states of affairs, is itself an account couched in non-determining descriptions. The positive view of such descriptions is developed in the third *Critique* in Kant's theory of teleological descriptions.

Teleological descriptions are there[14] said to be descriptions both of the purposiveness of actions with a purpose intentionally pursued and, more basically, of the purposiveness of various states of affairs in the absence of a purpose intentionally pursued or in the absence of a pursued purpose altogether. Kant's distinction between two species of purposiveness is a forerunner of recent attempts at distinguishing between teleological and functionalist explanations. I am thinking, in particular, of Robert Cummins's efforts in that direction.[15] In conformity with recent usage such as his, I shall reserve the term 'teleological' for the former type of case and the term 'functionalist' for the more basic or generic type of case in which we speak of functions either in the absence of, or in abstraction from, pursued purposes. According to this usage, a functionalist description need not be teleological, whereas every teleological description will require functionalist ones.

As was already suggested in the course of characterizing transcendental psychology as a faculty psychology, Kant's account of acts of cognitive synthesis is a functionalist (as well as teleological) account. At the same time, concepts, essentially possessed and employed in and for the performance of such acts and themselves rules for acts of cognitive synthesis, can (and in some respects should be) viewed in a manner abstracting from their use tokened in persons (and hence abstracting from psychology). When they are viewed in this manner, concepts themselves, in their interconnections, also turn out to be functionalistically describable. Hence Kant speaks of the purposiveness or systematic nature of concepts, where systems are contrasted with mere aggregates.[16] A scientific theory, quite abstractly regarded, turns out to be a system of concepts and propositions.

Hence functionalist (and, when appropriate, teleological) descriptions turn out to provide us both with a psychological account of acts of cognition and with an account (not itself psychological) of general features possessed by concepts if they are to form systems. Both Kant's transcendental psychology and his theory of cognitive concepts (abstractly conceived) are functionalist accounts.

But all of this talk about acts of cognitive synthesis and about cognitive concepts has left sensation qualia and some other content, and their descriptions, out of account. Kant also speaks of an empirical psychology, and despite vagaries in his use of the term "empirical,"[17] at least the following is true of such a psychology. It addresses itself to causally necessitated sensation and feeling. If it were to turn out that transcendental psychology is psychology of acts of cognitive synthesis and empirical psychology is (at least) psychology of sensation and feeling, such a distinction (imperfectly) mirroring that between cognitive mental activity and mental passivity, then one highly significant aspect of the distinction between two kinds of psychology would be that transcendental psychology is wholly functionalist (and teleological), whereas empirical psychology must exemplify the dual or mixed approach alluded to above. (It would also turn out that the empirical is necessary for transcendental psychology.) But even if the distinction between his two kinds of psychology is not as clear in Kant's hands as it ideally should have been, the basic point is and remains that for him many, but not all, aspects of a person's mental life can be functionalistically accounted for.

2. Sensation and Perception

Affection and sensation are central to Kant's account of the formation and employment of empirical concepts. Some empirical concepts are straightforwardly appearance-property concepts such as the concept of greenness. Others are removed from the sensory periphery and, as we shall see, must be linked to straightforward appearance-property concepts by means of categorial requirements such as nomological connectedness. According to Kant, empirical concepts away from the periphery require schematization, and we shall see that appearance-property concepts and the sensation qualia to which they are intimately linked play an essential role in schematization.

Affection and sensation are central because sentience is. This is a recurrent theme in the *Critique of Pure Reason* and in other works of Kant's. Conditions of sentience and those of the understanding are the two main stems of all empirical knowledge that it is possible for human beings to have.[18]

The centrality of affection and sensation is documented by the fact that in the crucial Transcendental Deduction of categorial rules of the understanding, Kant refers to conditions of sensation in order to distinguish between what is *given* to a mind equipped with sentience and understanding and what is actively synthesized by such a mind.[19] What is given, it owes to sentience, and what is actively synthesized is the work of the understanding. Without such a distinction, the problems of the Transcendental Deduction would not assume the form they do. Without such a distinction, Kant could not assume, as he does in the Deduction, that there are species of mental content received by the mind, thereby providing active synthesis with a needed starting point. The nature of this givenness and starting point thus needs to be examined. At the same time, the Transcendental Deduction is incomplete until it has been demonstrated that categorial requirements are applicable to objects of empirical cognition. That is, an incompleteness would arise because the applicability of such requirements to such objects demands their applicability to what is given through sentience.

The centrality of affection and sensation is also documented by the schematization requirement for empirical concepts, already alluded to. Specifically, this is shown by the requirement that there be a homogeneity between what is intuited and what is thought.[20] This centrality is further documented by the fact that one of the categorial requirements itself is a requirement on sensations and straightforward appearance-properties;[21] by the reference to what is empirically given in Kant's initial statement of transcendental logic;[22] and by the deliberately dramatic opening paragraphs of the Introduction to the second edition of the *Critique of Pure Reason*, at B 1-2. The two opening paragraphs deserve to be quoted in their entirety. In them, Kant states a basic assumption and problem of the first *Critique*, to wit, the difference and indispensable collaboration between the senses and the understanding.

> There is no doubt that all our cognition begins with experience. For how should the faculty of cognition be awakened to [its] exercise, were it not through objects which touch our senses and on the one hand on their own produce representations and on the other hand move our capacity of the understanding to compare these representations, to combine or to separate them, thereby working up the raw stuff of sensory impressions into that cognition of objects which is called Experience? *In the order of time*, therefore, there is no cognition in us prior to experience, and with experience all our cognition begins.
>
> But even though all our cognition begins *with* experience, not all [cognition] arises *out of* experience. For it may well be that even our empirical cognition is a composite of that which we receive through impressions and of that which our own cognitive capacity (merely occasioned by sensory impressions) supplies from itself. Such an addition we do not distinguish from that raw stuff until extended practice has called our attention to it and has made us adept in separating it out.

As Kant says in the first of these two paragraphs, the only manner in which the human understanding can be gotten to act is the following.

Objects touch our senses. As a first result of this, objects on their own bring about some representations.[23] A second result of such touching is the exercise of the (active) understanding, with the consequence that the raw stuff of the first result—sensory impressions and their immediate results, if any—are worked up into empirical cognition of objects. (This work is of course the work of cognitive synthesis.)[24]

The second paragraph is equally important. There we are told that empirical cognition may very well turn out to be itself a composite of the raw stuff of the first result and of what the understanding itself contributes in the stage reached as the second result. Then follows a crucial characterization of at least one condition of the collaboration between the senses and the understanding and hence of the condition under which the second result is reached: sensory impressions *occasion* the workings of the understanding. And finally there is a warning, not to be taken lightly: the contributions of the understanding are not easily separated from the raw stuff. Only extensive practice makes us attend to these contributions and allows us to separate them out. Had he been in a polemical mood at this moment, Kant would have added, with great conviction, that none of his predecessors had engaged in the practice long enough.

It is fair to say that answers to the following questions would constitute a statement of Kant's theory of empirical cognition and of the objects of such cognition. (1) What are the objects like that touch our senses? (2) How do they touch? (3) What sort of representations are brought about in this manner? (4) How are such representations brought about? (5) What does it mean to call sensory impressions (and their immediate results, if any) raw? (6) What does it mean to say that sensory impressions occasion the work (acts of cognitive synthesis) performed by the understanding? (Or, what does it mean to say that sensory impressions are worked up or synthesized into empirical cognition of objects?) (7) What are the objects of empirical cognition like? For example, are they the objects in (1)? (8) What are the contributions of the understanding? (9) What then is the nature of the composite which consists of what is given through sentience and of the contributions of the understanding? Answers to these questions should tell us what Kant's theory of empirical cognition is, both from the point of view of the mental content of the representations which is the content of such cognition and from the point of view of the acts (and other mental events) which produce or are those representations.

I begin with (2). According to Kant sensory impressions are sensations brought about by objects affecting the senses.[25] Affection of the senses is causal and necessitating.[26] By means of affection, the senses are passively put into particular sensation states on the strength of causal conditions which involve properties of the objects in (1).

For Kant, sensations are a species of consciousness. Sensations are mental states.[27] They are not physical impressions on physical sense organs, although no doubt physical impressions are necessary for sensations. Kant conceives of sensations as a species of empirical consciousness and assigns to particular sensations particular qualia which their owner will introspect by means of inner sense.[28] In a manner which is not elaborated upon, sensations are inherently transparent to inner sense. They are best thought of as that species of empirical inner awareness that is immediately brought about by the affecting objects. The inner awareness at stake here is itself neither referential nor propositional.[29] Neither is there anything referential or propositional about sensations in any other regard: they also fail to refer to anything different from themselves. They are "wholly subjective" and by themselves carry no semantic content. This gives us a partial answer to question (5): sensations are raw in that they carry no semantic content. But they are qualitative. A particular sensation state is or has an immediately experienced quale.

Kant is still willing to call sensations representations,[30] despite the fact that they carry no semantic content. This gives us partial (albeit only partial) answers to (3) and (4). Sensations are among the representations brought about by objects on their own, and they are brought about by causal affection. (A more interesting answer will be forthcoming shortly in terms of representations which are not sensations.) The answers given so far assume that affection is temporal.[31] This in turn has consequences for matters raised in (1), which obviously bears on (7). I shall postpone discussion of all of this.

Kant also holds that there is semantically laden empirical self-consciousness (or semantically laden empirical apperception), namely, full-fledged inner perception. For such perception to be different from mere sensation requires that it be representational in a manner not to be met with in sensation taken by itself. This must mean that even in those cases of inner perception nearest to mere sensation,[32] there is in inner sense an occurrence of reference. Such reference obtains on the strength of the presence of an immediately experienced quale. Following R. Howell's[33] model, I interpret such reference as perceptual ostension, in this case, perceptual ostension of some sensation state and of a minimal self. Hence this and further species of empirical self-consciousness differ from mere sensation in that they carry semantic content. In the case of perceptual ostension, the semantic content is of a very restricted sort requiring a non-predicative use of a conception, with the conception itself being of a special sort we shall consider in a moment.

The result just arrived at is an outcome of Kant's general theory of both outer and inner sense, a theory which centrally contains an account of perceptual ostension. In the case at hand, any ostended self is ostended as the quale's owner, temporally co-extensive with the experienced quale in question. All Kant can (and needs to) say about the connection between sensations and ostensive inner self-reference based on sensations is that such self-reference arises *naturally* as a result of a sensation. By "natural" I here mean, negatively, that there is no cognitive synthesis and no inference from sensation to self-reference. And I mean, positively, based on the model of affection itself, that there is a causal connection. Our person in question in a causally mediated manner arrives at the semantically laden state in which he or she self-ostends as just described.

As Howell has pointed out, perceptual ostensions (in the present case inner ostensions) can be what he calls impure (and it would seem that they have to be impure for purposes of cognition). This means that in the second stage of this natural and causal tale our person uses an *acquired* conception of the particular experienced quale, such a use being required for the occurrence of self-ostension. But the conception is a very special and restricted one: it is itself ostensive since of the form {this quale}. Its ostensive and hence *de re* and singular character will be of great importance later on.

In addition to inner perception as just described, there is outer perception. Outer perception is perceptual ostension of perceived referents in space. When impure, such ostension will contain a singular *de re* conception of a quality located in one of a person's perceptual fields. On the strength of such an ostended and ostensively conceived quality,[34] our person ostends a referent as being the referent located in the subregion filled by the quality in question. Outer perception is as natural and as causally induced as is inner perception, and our person will ostensively conceive of the quality in question and will believe of it that it is *that* quality which in the case at hand caused *this* experienced quale.

Both inner and outer perception carry some semantic content and must not be identified with sensation. But sensation in both cases is a causal antecedent of perception: it is causally necessary. Another condition of perception is the set of *a priori* semantic requirements that help make perceptual ostension possible, namely, space and time as forms of empirical intuition. The first two paragraphs of the Introduction, quoted above, are silent about the fact that an apparatus of spatial and temporal ostension is required, but in other places Kant makes up for this omission.[35]

Perceptions (of all kinds) are semantic representations. So, completing the answers to questions (3) and (4) is the claim that singular representations (composed of ostensions including ostensive conceptions of qualia and of qualities in space) are brought about naturally. But the rest of the answer to (4) also shows that we have introduced an important complication (above and beyond the original story about sensations). Without our person's *a priori* employment of the forms of space and time (as forms of empirical intuition), no singular representation of the sort now under discussion could arise. Kant does not hold that such representations arise on the strength of only natural necessitation. Only sensations do. For outer perceptions, this means that by touching, objects naturally bring about outer ostensions (as described) only because the form of space as an *a priori* form of such ostension has been introduced.

The single most important result of the part of the story just completed is that persons naturally and without cognitive synthesis come to have singular representations of the sorts described. (Since in the normal course of events they will accept them, we can also call them singular beliefs, certainly after predication has been introduced.) I submit that the account sketched up to this point is the best account of what Kant in the *Prolegomena* wants to mean by judgments of perception, viz. singular beliefs of the sort described. Sensations are "wholly subjective," while perceptions are not. Only sensations are "wholly subjective," and perceptions are not sensations. But although perceptions involve singular conceptions and differ from sensations in this regard, they nonetheless are not wholly objective. They refer to items, but they represent them only ostensively and only with singularity (incorporating only singular conceptions, at least up to this point). Total objectivity will turn out to require generality, and more.

On the positive side, (a) perceptual ostensions are *de re*,[36] and (b) in them (or with them) singular representations, or singular beliefs with propositionality of a very restricted sort, arise in the mind. Both (a) and (b) are of the greatest importance for the rest of the story. Singular perceptual beliefs (with ostensive conceptions of qualia and of qualities) are still passive mental states, or at least passive with regard to the understanding. That is, they are not results of cognitive mental activity, properly so called. More specifically, they are not results of cognitive synthesis, since for Kant the understanding is invariably discursive and non-ostensive.

Kant's claim that sensation qualia are objects of inner sense (as described) does not commit him to the Cartesian-Lockean view of inference from inner qualia to the outer, whatever this outer might be. Kant rejects this view, or at least he rejects the view properly understood. (We shall see that much depends on what *else*, in addition to what has already been discussed, we are going to say about qualia.) In other words, Kant rejects the view that sensations are ostensively accessible immediate objects of inner perception which then provide some evidentially privileged basis for inference from them to other states of affairs. The reasons he rejects this view are two-fold. First of all, sensations themselves cannot enter into any semantic relation, and secondly, the content of beliefs that are singular in the conceptions they employ cannot enter into any inferential cognitive connection, for reasons which still have to emerge. Hence, at the moment nothing is in place that warrants our speaking of epistemic inference. And this then provides a further answer to question (5): the immediate results of sensation states are also raw in that they are not cognitively laden.

I will for the moment skip question (6) (and will for a while longer ignore (1)). The next step in our story concerns generality of conceptions. This means that I must now turn to a discussion of Kant's theory of active cognitive synthesis. Equipped with an answer to (8), we can then begin to deal with questions (6), (7), and (9). An answer to these questions will fill in the rest of the picture, including the matter raised in (1).

3. General Concepts and Synthesis of the Understanding

The introduction of space and time as *a priori* forms of perceptual ostension made it possible to continue the story from sensation to ostension and to singular beliefs about qualia and qualities. An account more complete than I can provide here would also pay attention to intuitive *a priori* constructions of arithmetical and geometrical quantities (and would continue this discussion by also paying attention to the role of categorial requirements laid down in the Axioms of Intuition and in the Analogies for *a posteriori* determination of quantities.) But disregarding quantities, hence concentrating on qualities and qualia, we have seen that perceptual ostensions and singular perceptual beliefs are not "wholly subjective" and yet not wholly objective, either. In the spirit of the opening paragraphs of the Transcendental Deduction, in the B edition of the *Critique of Pure Reason*, both sensations and singular perceptual beliefs are given to the understanding in that, subject to space and time as *a priori* forms employed by a person in ostension, they are naturally and causally produced in a mind and are not the result of acts of synthesis on the part of the understanding.

The sort of consciousness which has been described up to this point is a wholly, or merely, empirical consciousness (including empirical self-consciousness). Nothing like the sorts of conditions Kant calls transcendental conditions of the understanding, required for transcendental consciousness, have been described so far. Central in the move towards transcendental consciousness is the first step which the active understanding undertakes, namely, the construction of general concepts. This in turn leads to the construction of propositions possessed of generality in their predicative slots. All of this is in accordance with, and from a conception of, categorial requirements necessary for the constructions to be of service in empirical cognition. The result is scientific theories of nature, the objects of which are in essential ways theoretical objects. That is, objects in essential regards just are what constructed theories have them be. (I shall return to this.)

I have argued for this view elsewhere (in the explicit form of a reconstruction of the Transcendental Deduction in the B edition of the first *Critique*[37]). In this section of the present paper I shall expand on my interpretation of Kant's conception of general theoretical properties, or general empirical concepts at work in empirical cognition, a conception which drives the Deduction. It has important consequences determining how we need to characterize sensations to allow for an actively constructing understanding (a) to be occasioned by sensation and (b) to apply itself to what is given to it through sensation. Points (a) and (b) are clearly relevant to most of the questions remaining from our original list. Eventually, even question (1) can be answered once we understand Kant's conception of general theoretical properties *and* the resultant requirements on sensation.

Briefly, the basic story of general-concept and theory construction goes as follows. It has to be possible to form beliefs about uniformities and regularities, hence about identities and differences. Cognition certainly requires such beliefs. But identities and differences require generality, beginning with the simplest case of qualitative sameness and difference. Thus we need qualitative types or kinds (of whatever is in outer or in inner sense and of whatever can be inferentially linked to what is in outer or in inner sense).

Kant holds that qualitative sameness and difference are *not given* in any empirical intuition or finite series of such intuitions. The content of singular beliefs alone invariably underdetermines any desired sameness and difference. As he puts it in his *Lectures on Logic*, the generality of general concepts is a formal property of such concepts and is always made or produced by the understanding, and this turns out to be done "as it were through convention."[38] Kinds are chosen by selection, not in a manner that is defined by, or epistemically justified by, what is given in and through empirical intuition, but in a manner that is *a priori* with respect to what is given. Still, the choice is not wholly *a priori* in that sensation occasions the choice. As Kant has it, for empirical kinds the *matter* of the kind is *a posteriori*. But the fact that generality is produced underscores the invariable activity-character of mental events in that part of mental life which is taken up by the understanding. Cognitive activity is for Kant *a priori* (absolutely or relatively).[39]

At least two aspects need to be questioned and accounted for here. (i) What, in light of what we have just learned, is Kant's overall conception of the nature of general concepts, disregarding for the moment the doctrine of occasioning? That is, what is his conception of general concepts viewed just in terms of the activities of the understanding considered by themselves? (ii) What is the very special connection between the *a posteriori* given matter of an empirical general concept and the concept as chosen or produced in its form? What is Kant's conception of occasioning in his doctrine of general-concept formation?

Now a kind is a similarity relation between instances or tokens. Clearly, when the understanding produces an empirical kind (*modulo* its matter), it does so in order to fix contrariety relations between instances of what are to be different, coordinated kinds. Since the choice of kinds is not determined by what is given in and through empirical intuition, if the given is to exercise any constraint, it must be a constraint which does not determine choices. It would seem that the doctrine of occasioning is just such a doctrine of non-determining constraint being exercised by the given.

Also consider that choices of kinds have consequences for other choices (of other kinds): there are constraints in the relations of a kind to other kinds, in particular to other coordinated kinds together with which the kind in question is chosen. And the only type of relation which seems relevant is a relation of functional interdependence of empirical kinds (and of their instances). An empirical kind is the kind it is (*modulo* its given matter) on the strength of the role the kind plays in systems of concepts (in Kant's sense of system as an organized whole). But since its instances turn out to be instances of singular and ostensively accessible qualities making up the given matter (the qualities being directly ostensive or else indirectly ostensive through schematization), the instances turn out to be simultaneously characterized as instances of (directly or indirectly) ostended qualities and as instances of kinds specified from the point of view of roles played by the kinds. As his discussions in the *First*

Introduction to the Critique of Judgment and in the main body of that work show,[40] Kant holds a view of this sort. His notion of similarity driving concept construction is therefore a functionalist notion, and his theory of concept formation is a functionalist theory.

Since, according to Kant, the understanding makes its choices in behalf of its goal of empirical cognition, and since it proceeds reflexively,[41] he also holds the view that we have a proper teleology at work, with the understanding choosing kinds by its conception of epistemic goals to be achieved thereby. The understanding, in short, deliberately constructs theories the objects of which are the objects of empirical cognition. The story from construction of general kinds to that of theories is the story of successive syntheses performed by a deliberating understanding. The categorial principles of such syntheses are the basic rules of the transcendental logic which govern production and use of kinds. Kinds constructed for epistemic purposes are kinds constructed to satisfy these categorial principles. Kinds are constructed natural kinds, kinds established "as it were through convention."

Natural kinds for Kant are thus nominalistic (*modulo* matter), and the resultant modalities of real possibility and real necessity are *de dicto* modalities; i.e., they are modalities relative to the constructed natural kinds. Functional notions of similarity which furnish us with kinds are also *de dicto* notions, in that they do not contain any *de re* reference to anything given in intuition.[42] The understanding produces and works only with *de dicto* descriptions of its objects. In this respect, recent interpreters of Kant who have emphasized the character of Kantian objects of empirical cognition as immanent or internal to theories have been correct.[43] But their emphasis is one-sided since their interpretation neglects the given matter of concepts. Such bias is due to neglect of perceptual *de re* reference and its conditions, presupposed for the understanding. Internalist interpretations also correctly stress truth-value gaps within any particular theory, since only a subset of kinds is selected at any one moment. But from this the existence of any truth-value gap *per se* (for existents) does not follow, since other theories are always possible (though not always compossible).

For Kant, cognizable essences and possibilities of objects of empirical cognition are systems of categorially constructed kinds. With such kinds required for empirical cognition, Kant's account has the consequence that all empirically cognizable properties of objects are constructed empirical kinds. Property idealism, of the sort described, is therefore necessary for empirical cognition and constitutes an essential part of Kant's transcendental idealism.

A more finely grained step-by-step version of the story I have been telling would begin with construction of appearance-properties closest to the sensory periphery and then proceed to properties farther removed from it (not to mention construction of quantitative concepts). Empirical properties which are not themselves appearance-properties need to be schematized. This crucially involves standing in nomological connection to appearance-properties. Furthermore, for truth of their application, theoretical concepts are required to stand in nomological connection to occurrent sensation (as claimed by the Second Postulate of Empirical Thought, at B 266 and B 273-274). Since the entire discussion of the understanding has proceeded so far *modulo* matter, nomological *de dicto* necessities by themselves do not entail in the actual world the truth of propositions possessed of singular referring expressions (or empirical intuitions, according to Kant). After all, the understanding has yet to be *applied to* (and has to be seen as beginning with) what is given in singular intuition.

We have now given an answer to question (8) on our original list in terms of an account of several species of cognitive synthesis, and we have given partial answers to questions (6) and (7). More complete answers to these and an answer to (9) require a specific reply to (ii) above: What are the details of the contribution of occasioning and of the givenness of matter? This is the issue at hand.

For Kant, general appearance-properties need to be constructed just like any other theoretical property. But then, how can sensation serve in such construction, while it also serves as a criterion for truth required by the Second Postulate? It is clear that sensation qualia can be part of the nomological (hence cognitive inferential) network only if they are of constructed general kinds. To be part of the network is to be part of the theory, so qualia kinds would have to be themselves theoretical. But if so, it may seem impossible that sensations occasion and serve as constraints and checks on the theory. Put differently, the matter and the occasioning which Kant speaks of would turn out to be themselves internal to theory, and surely we should not speak of a composite consisting of sensory impressions and of contributions of the understanding. Passivity would not be required (or alternatively, Kant's distinction between passivity and activity would not be), rendering otiose our account of singular beliefs and of the given in intuition.

On the contrary, it would seem that sensation qualia, characterized ostensively and *de re* and with beliefs about them singular, must serve as occasions and constraints and checks. They should not be part of the nomological network. The resolution of this turns out to be that sensation qualia are characterized in both ways. Kant adopts a mixed, dual, approach to sensations. Sensations and what they naturally give rise to are properly non-theoretical when they are non-functionalistically characterized. The next part contains a discussion of such an approach after a brief return to the understanding's activities of synthesis, a return intended to highlight the asymmetry of Kant's treatment of mental activity and mental passivity.

4. Sensations and Kant's Functionalism for Mental Activity

According to Kant, general content of representations cannot be reduced to sensations. (Neither can perceptual reference and singular qualities be reduced in this manner.) Conversely, it also is false that the qualitativeness of qualia can be functionalistically reduced to representing activities. It turns out that there must be a *de re*, ostensive way of characterizing qualia as a necessary (but not sufficient) condition for the formation and employment of general concepts, just as it has already turned out that *de re* ostension of individual qualia has a role to play in the coming about of singular qualitative beliefs. (The two observations are of course closely connected.)

We have also seen that Kant analyzes the content and acts of the understanding into various components (species of cognitive synthesis). In a componential analysis, capacities and acts are described in terms of subroutines, as in the usual fashion of flow charts. Such subroutines are functionalistically described, as can be seen from the case of Kant's conception of the formation and use of empirical kinds. The functionalist notion of similarity in Kant's conception of such kinds and his account of theories complement and parallel his psychological account of cognitive capacities and acts in terms of subroutines. Among Kant commentators, Patricia Kitcher[44] has specifically utilized this view that Kant's conception of acts of (cognitive) synthesis is a functionalist conception, on the grounds that the order of dependence of such acts on one another is a contentual one.

Kant moves on to more and more complex descriptions encompassing one's reference to cognitive states tokened in oneself, such reference being for Kant necessary (and, it can be argued,[45] also sufficient) for rationality and for freedom of rational action. Hence, Kant's account of active mental life is properly teleological. The rational mind is a meta-mind: Transcendental consciousness is apperceptive, and a rational mind acts from self-conscious descriptions of what is tokened in it by way of reasons-connections and by way of beliefs which are part of such reasons-connections.

Elsewhere, I have also argued[46] that according to Kant, all of this at the same time receives a physical realization in the fashion of an anomalous monism of the active mental life. There will in principle be causal-physical descriptions of all tokens of mental activity. Passivity is left out from all of this, however, for two reasons. First, sensation is for Kant not an activity and cannot receive the same sort of componential analysis which he provides for mental activity. Second, there is need to characterize qualia ostensively and non-functionalistically, and this rules out anomalous monism as an account of the relation of qualia to the physical. But sensation also needs to receive a functionalist account, since there needs to be a functionalist notion of quale similarity. Because of a simultaneous need for a functionalist and for a non-functionalist characterization of qualia, Kant as claimed earlier adopts a dual or mixed psychological account of sentience, in contrast to the wholly functionalist (and teleological) psychological account of mental activity considered by itself.

For qualities in general, we get the result that functionalistically conceived similarity does not exhaust empirical properties, because of the singular content of singular qualitative beliefs (i.e., beliefs in the presence of singular qualities, *de re* accessible on the strength of *de re* accessible qualia). For qualia, what is *de re* ostensively accessible must not be identified with the constructed and functionalistically characterized similarity relation between qualia which gives the quale type that is part of the theoretical network. A functionalist characterization of quale types does not characterize a singular quale *de re* accessible in its particularity. But at the same time, *that* quale is a token of a functionalistically characterized quale type, and the *de re* accessible quale is necessary for the formation and employment of functionalistically characterized quale and quality types in theories.

What needs to be explained is the special connection between singular *de re* accessible qualia and theoretical appearance-properties, a connection *via* quale types (and singular beliefs). Roughly speaking, according to Kant we could not get hold of *qualitativeness* in our construction of empirical general kinds, or alternatively, we could not get hold of an important *difference* in different choices of general appearance-kinds, if we did not have *de re* ostensive access to a singular quale with its own particular non-constructed qualitativeness. This sheds some light on why passive sensation is a necessary factor in concept- and theory-construction and hence why it is one of the two overall sources of control in perception, with the other source provided by the legislating and synthesizing contributions of the understanding. But we can do better than this. For Kant the necessity which attaches to functionalistically characterized general content is only *de dicto* necessity. The *de re* necessity attached to the way *this* sensation feels (or, the qualitative character which is essential to its being the quale that it is, to wit *this* quale) is not entailed by, or nomologically connected to, any theoretical general content with its *de dicto* necessity.

We choose or construct some appearance-property in a manner partly based on singular qualia, and we recognize instances of such properties on the strength of an ostensive awareness of a particular quale. If this is the case, then we do so without benefit of either analytic or nomological connection of the appearance-property to or from that particular quale (or to or from *that* particular singular belief in question). Kant's view is that in both formation and employment of general concepts, the singular quale will serve as an ostended paradigm for what will turn out to be tokens of functionalistically characterized quale types. As a (necessary but not sufficient) paradigm, the ostended quale itself (i.e., in its characteristic nature) will not be part of the functionalist conception of the quale-type, since a functionalist conception cannot incorporate such particular qualitativeness. But in order for the ostended quale to be a paradigm, Kant must hold it to be false that *any arbitrary* ostendable quale (in its characteristic nature) could play the role that *particular choices* of empirical kinds introduce. In other words, although *some* ostendable qualia different in their particular nature from the ostended quale in question can and do play the same role and receive the same functionalist characterization, *not all* possible ostendable qualia could play the role of *that* quale, under the different choices of empirical kinds that are made. Ostendable qualia, in their characteristic natures, can play their roles as paradigms for many quale types but not for all types. Kant must hold that the nature of qualia, *qua* ostensively accessible, introduces some limitations on roles to be played by qualia, roles drawn from a chosen range or repertoire of types.

The reason for constructing quale types and for characterizing them functionalistically is the following. Experienced qualia are to serve as a necessary means for forming and employing empirical kinds, beginning with appearance-properties. Sidney Shoemaker has recently discussed this issue in a highly instructive manner.[47] He has pointed out that, in order for qualia to serve in this fashion, they must be connected with empirical kinds in such a way that identities and differences in sensations occurring in perceptions of tokens of such kinds are suitably correspondent to identities and differences of the empirical kinds. In other words, as we have known all along, there must be a way for sensation qualia to be part of the theoretical network. Hence their kinds must be constructed together with other empirical kinds to form a system of concepts. Hence they need to be given a functionalist *de dicto* characterization. As a result, tokens of functionalistically characterized quale types are cognizable along with tokens of all other kinds constructed for the purposes of cognition: tokens of quale types are objects of a theory. At the same time, such tokens are *de re* ostended qualia and *in their de re accessible given natures* not objects of a theory and not cognizable *in this respect*.

In short, without a functionalist characterization sensations could not help in giving rise to empirical kinds, beginning with appearance-properties. But Kant's occasion-doctrine and his view that mental life is a composite of *given* elements and mental activities then mean that, in addition, *de re* ostended qualia in their characteristic natures, these natures not being part of the cognitive network, serve as sensory exemplars of types of qualia constructed by the understanding. Kant holds that both theoreticity of general kinds (including constructed qualia types) and particularity of qualia and singular qualities, only ostensively accessible, are required in order for there to be empirical cognition that is partially controlled and checked by non-theoretical sensory content. No theory can determine the particular nature of exemplars.

So much, then, by way of discussing occasioning and the nature of the composite of passivity and activity, answering our original questions (6) and (9). By way of more of an answer to question (7), objects of empirical cognition possess ostensively *de re* accessible singular qualities, which are token-token identical to instantiations of constructed empirical kinds. Such tokens, then, make up the matter of otherwise constructed kinds, and objects of empirical cognition are composites, of the sort that question (9) is asking about.

One of the best-focused accounts of these views to be found in the *Critique of Pure Reason* is provided by the doctrine of the schematism, with its discussion of homogeneity.[48] Kant's search for an element in perception homogeneous with general empirical kinds is a search for something given in perception and at the same time falling under a constructed empirical kind. I have offered token-token identity of instances as well as functionalistically conceived similarity, on the basis of given qualitativeness, as an interpretation of such homogeneity.

Kant's view holds that particular sensations and their causes causally fix the reference of singular qualitative belief (given the forms of intuition) and holds them to be necessary (but not sufficient) for discursive and non-ostensive content. Particular sensations, non-discursive singular beliefs, and singular causal connections which are at work in the affection relation are all non-theoretical and hence noncognizable in their particularity. As a result, Kant's position is not the Cartesian one according to which there is cognitive inference from what is *de re* introspected. But at the same time, non-functionalistically characterized qualia are an essential part of mental life. They constitute a qualitative part of mental life which is necessary for mental activity. We have *qualia* playing functional roles, and according to Kant they are not reducible to their functions.

5. Conclusion

Question (1) in our original list remains. What are the objects which through causal affection bring about qualia in a passively receptive part of the mind? We have seen that Kant does not hold the Cartesian-Lockean view that *de re* introspectively available qualia serve as the basis of cognitive inference to other states of affairs. In sharp contrast to such a view, Kant's theory of perception is a direct realism. In outer perception a person in a *de re* fashion ostends a referent located in a subregion of at least one of his or her perceptual spaces and ostends it on the strength of a quality's being located in the subregion in question. The quality in turn is believed to be the cause of the sensation quale which obtains as part of the perceptual ostension. The referent, like the quality and the quale, is singular, as is the belief in the presence of the quality. As explained in the subsequent story concerning generality and categoriality, a categorially governed, general phenomenal-substance concept that is theory-dependent can eventually be introduced. The perceptual ostension is thereby subjected to conditions of determination and objectivity, in changing from a not fully objective (though not "wholly subjective") encounter to a fully objective one while remaining *de re* and singular in its referent throughout (though not singular in the concept employed). It is thus not rendered otiose by the introduction of complete objectivity conditions. The change consists of this: the referent of the ostension has its determinable, cognizable properties constructed for it and assigned to it from the person's conception of a theory and hence is now theory-dependent in those cognizable properties.

Affection instills in a receptive mind mental states which are required to be given to the active understanding. Once a theory is constructed, affection gets an active mind to enter the system of theoretical beliefs, as explained above. The truth of a theoretical belief requires that the referent, characterized directly or indirectly by some pertinent singular belief, be in the extension of the constructed concept occurring in the theoretical belief. Hence both with regard to the overall truth-conditions of objective determinations and with regard to the manner in which a person's beliefs enter the system of theoretical beliefs, conditions characterized non-theoretically are at work and serve as constraints and checks. But now it should be apparent how we should answer question (1). Given the resources internal to perceptual ostension and given the non-theoretical characterization of the affection relations, it follows that the objects in question are being characterized in terms of properties which have not been constructed for them. Such objects, and the qualia they instill through affection, considered by themselves and not as objects of a theory,[49] have uncognizable properties. Hence they are uncognizable with respect to such properties.

Cognition, in contrast, requires construction of what is to be attributed to a referent. But we have seen that the tokens of the sorts of uncognizable properties we have been talking about are numerically identical to tokens of constructed, cognizable kinds. Because of such token-token identity, Kant's view has it both ways. The objects which question (1) is asking about are empirically cognizable after all, since *instantiations* of their pertinent properties are cognizable as tokens of constructed kinds. Since type-type identity obviously fails, it remains true that some properties are uncognizable. But there is no uncognizability on the level of tokens.

This must suffice as an indication of the manner in which Kant adheres to a theory-external, existential, empirical realism, while at the same time propounding his transcendental idealism of general-concept construction. Moreover, persons with their given states, the affection relations which they stand in, and objects with the sorts of properties we have been talking about are incorporated into theory by means of the token-identity just alluded to. Double affection vanishes on the level of tokens. All of this is the case despite the fact that outside theory we have no non-ostensive, discursive way of characterizing objects and their properties. This version of a two-aspect view is Kant's reply to Jacobi's famous complaint.[50] Kant would say that without things considered by themselves (independent of theory), one cannot form and enter the system of empirical cognitions. But it is nevertheless false that with them we cannot maintain that system.

Kant's metaphysics of things which are uncognizable by themselves, including the just mentioned type-difference from, and token-identity with, theoretical objects, is an integral part of his theory of receptivity conjoined with active understanding. His descriptions of various cognitive capacities and syntheses constitute a componential psychological analysis of the understanding and its cognitions in terms of mental routines and subroutines. In this regard, Kant anticipates more recent analyses such as those employing Cummins's analytical strategies.[51] Use of particular cognitive concepts is described as the following of routines and subroutines, all of this against the background of a fairly specific psychological theory which Kant believes to hold of us human beings[52] and which postulates capacities of various sorts.[53] Among these capacities is one which enables its owner to construct general concepts (*modulo* their matter). The notion of similarity which is operative in such constructions is a functionalist notion, and with respect to their generality, concepts stand in functionalistically characterized relations to each other. Cognitive acts and states are also functionalistically described since their dependence is a contentual one.

The fact that Kant is working with a fairly specific (and human-specific) psychological theory opens him to the charge of chauvinism, to use Ned Block's apt label.[54] But if he is committed to the view that active mental life receives a physical realization and that anomalous monism correctly describes the manner in which this realization obtains,[55] then tokens of different physical types could in principle display the same functions; i.e., they could all instantiate the functions which are the mental types in question. This could defuse chauvinism. Kant only intermittently speculates about physical realizations,[56] and I do not know how he would settle this issue.

Kant's views on qualia complicate the situation further. He believes that functionalistic descriptions furnish a complete characterization of the acts and the contents of the understanding, taking the latter in separation from receptivity. Hence he subscribes to functionalism for the understanding considered in isolation. But he also holds that, in addition, non-functionalistically characterized receptivity is required for construction. Both qualitativeness and qualia, when introduced in this fashion, fail to be discursively grasped since they fail to be functionalistically characterized. As far as qualia are concerned, Kant shares Shoemaker's view[57] that individual *de re* ostended qualia are not functionalistically definable but that classes of them are. Applied to qualitativeness in objects, this view leads to Kant's reply to Jacobi.

For the purposes of Kant's cognitive psychology, qualia are required to be of a dual nature. In ascribing this two-fold nature to them, Kant agrees with both Shoemaker and Block.[58] The fact that one of the natures of qualia is not functionalistically characterized means that Kant treats them differently, and separately, from his functionalism of human information-flow processes. But since in their non-functionalistic nature, qualia are necessary for the construction and for (a partial) causal reference-fixing of constructed concepts, this non-functionalistic part of mentality is an essential part of mental life, at least of human mental life.

New chauvinistic possibilities are hereby opened up: it is possible (but not certain) that Kant would wish to restrict mental life required for cognition to the composite of sensation and understanding that he has been describing. At any rate, since qualia in important regards escape functionalism, so do some of the conditions of the formation and use of general concepts.

The indicated difference between qualia and information-flow processes is radical. Qualia, in one of their natures not functionalistically characterized, constitute a species of mental life which *qua mental* stands in causal connections to what is physical.[59] Yet they are not themselves physical. Moreover, their failure to be theoretical is even more serious than that of other mental states and processes.[60] They fail, not only because they (and even their functionalistically characterized types) fail to be fully mathematizable,[61] but also because qualia themselves, in their non-functionalistically characterized nature, cannot be non-ostensively and discursively described. They are a species of qualitative mental life which is given to a receptive mind without type-identity in place, whereas the type-identity of cognitive representations is that of the constructed concepts which are their contents. Kant's empirical psychology of sensations is no more than ostensive. Qualia, an essential part of human mental life, not physical and in one of their natures not functionalistically characterized, are an indispensable part of the total realization of the functions which are the mental activities of the understanding applied to what is given to it. This realization, then, is a psycho-physical realization.[62]

University of Rochester

NOTES

[1] Among recent Anglo-American commentators, P.F. Strawson has labeled Kant's transcendental psychology imaginary and an aberration. See his *The Bounds of Sense* (London: Methuen, 1966), p. 32.

[2] In the *Critique of Pure Reason*, see B1, B74-75, B93, B103, B129-130, B158 (note). The first *Critique* will be cited in terms of the standard pagination of the second edition (and of the first edition in case of divergence). Reference to other works of Kant's will be to the Academy (Ac.) edition.

[3] Hence he speaks of causality through concepts and of made or reflected rules. See, for example, *Critique of Judgment*, Ac. V, 172; *First Introduction*, Ac. XX, 196; *Anthropology*, Ac. VII, 197.

[4] They are not independent of each other. Kant often thinks of subsumptive judgment as the capacity to use or apply the understanding, and inventive judgment will generate concepts for subsumptive judgment's use. Sometimes he also invokes reason for the task of generation.

[5] Productive imagination is central in the generation of mathematical-intuitive concepts. I have explored the nature of such concepts in my "Kant on Intuitivity," *Synthese* **47** (1981), pp. 203-228.

[6] Among other places, at A106 and B180.

[7] See the heading of sec. 16 of the Transcendental Deduction, B131.

[8] For an elaboration of Kant's distinction between what is given in intuition and what is produced through combination, see my "Apperception and Objectivity," delivered at the Spindel Conference on the Deduction in B, Memphis State University, October, 1986, and since then published in *The Southern Journal of Philosophy*, Vol. XXV, Supplement (1987), pp. 115-130.

[9] Where "given" needs to be stressed. The fact that for Kant mathematical concepts are intuitive does not mean that they are *given*.

[10] This is not to say that he had not written on teleology before his third *Critique*. See his "Concerning the Use of Teleological Principles in Philosophy," Ac. VIII, 157-184. In the *Critique of Judgment*, see his discussions of the formal purposiveness of nature, in the Introduction (Ac. V, 179-186, 192-194); of the objective purposiveness of nature (Ac. V, 359-384, 389-395, 397-410); and also the *First Introduction*, Ac. XX, 201-221, 232-251. For parts II - IV of this paper, the most important portions of the *Lectures on Logic* are Ac. IX, 91-99, 131-133, 139-144. In the *Anthropology*, see Ac. VII, 131-134, 140-143, 196-199. (The Preface of the *Metaphysical Foundations of Natural Science*, Ac. IV, 467-479, is also relevant.) In the *Prolegomena*, see in particular Ac. IV, 294-302, 318-326. In the *Critique of Pure Reason*, among the most important sections are those on Transcendental Logic, B74-88, B93-106; and the opening of the Transcendental Deduction, B129-139.

[11] In addition to passages cited in note 2, see B172-173. Also compare the discussion of talent and genius in *Anthropology*, Ac. VII, 220-227.

[12] I say "a significant portion" because, as I see it, the Subjective Deduction is to derive categoriality *from* such transcendental psychological descriptions. See my "Apperception and Objectivity," op. cit.

[13] The Preface of the *Met. Found. Nat. Sci.*, op. cit., is particularly illuminating in this regard. The *Anthropology*'s concern with lack of conditions of permanence in inner sense (Ac. VII, 134) already makes the same point.

[14] In addition to passages in that work cited in note 10, also see Kant's summary account in sec. 10 (Ac. V, 219-220).

[15] See his "Functional Analysis," *Journal of Philosophy*, vol. LXXII (1975), pp. 741-765.

[16] *Lect. on Log.*, Ac. IX, 139. Also see the instructive *Lecture on Philosophical Encyclopedia*, Ac. 29.1, 1; 5.

[17] 'Empirical' signifies for him connection to sensation. B207 will serve as an example. But complications arise due to the fact that sometimes he thinks of the empirical as the *merely* empirical, in the absence of transcendental conditions, whereas at other times he is concerned with what is both connected to sensation and subject to transcendental conditions. Empirical reproductions of mental states through association, at A100, can serve as an example of the former, whereas empirical cognition is an example of the latter.

[18] B1-2, B33, B74-75.

[19] See my "Apperception and Objectivity," op. cit. I say "actively synthesized" to capture Kant's notion at the beginning of the Transcendental Deduction, B129-130, that synthesis is activity (on the part of the understanding). In a more generic sense, syntheses can be passive as well.

[20] Found at B176.

[21] I am referring to the Anticipations of Perception, B207-218.

[22] Found at B103.

[23] Norman Kemp Smith's translation does not make this fact altogether clear. The translation of the paragraphs just quoted is my own. Also notice that experience, at the beginning of the first paragraph, is sensation and towards the end of the paragraph, the empirical cognition that is subject of the second paragraph. I have marked the second occurrence by use of a capital 'E'.

[24] And B129-130 can be seen to restate the basic message of B1.

[25] See, for example, "affection" at B93 and B207. B562 speaks of sensibility affecting choice, including necessitating affection.

[26] See B210 and compare "necessitating affection" at B562.

[27] Found at B207.

[28] Kant's theory of inner sense is a complicated matter. Compare the consciousness of apprehension, in *Anthr.*, Ac. VII, 134 (note).

[29] There is nothing in place in a sensation alone that would allow us to say that they refer to themselves or to anything else. Reference for Kant requires a form, or forms, of intuition, but at B208 he holds that sensation by itself contains no such forms. He always distinguishes between sensation and empirical intuition or perception, including inner intuition.

[30] Found at B376.

[31] Sensations are in time, although they do not by themselves incorporate the form of time.

[32] These are cases in which the inner perceptual reference rests, in addition to its form, *only* on representing the particular sense quale involved. In such cases the referent, including the self that can be represented here, is very thin or minimal. In richer cases the structure of inner reference will be more and more complex, and so will the referent.

[33] See his "Intuition, Synthesis and Individuation in the *Critique of Pure Reason*," *Nous*, vol. VII (1973), pp. 207-231.

[34] The ostension of quality and, by means of it, the ostension of the item in space themselves require the presence of a sensation quale.

[35] For example, at B33-35 and at B74-75.

[36] Compare Howell, op. cit. This holds both for ostensive conceptions of qualia and qualities and for ostensions of items bearing such qualia and qualities.

[37] See "Apperception and Objectivity," op. cit. The construction of qualitative concepts is only with regard to their form, differing thereby from construction of *a priori* mathematical intuitive concepts.

[38] See the passages in the *Lect. on Log.* cited in note 10. The mention of convention occurs in sec. 11 (Ac. IX, 97). What induced Kant to qualify his remark with the tantalizing "as it were"?

[39] Compare B2-3 with its discussion of what is (and what is not) wholly *a priori*. Relative *a priority* is *a priority* of form occasioned by *a posteriority* of the matter that Kant is talking about. Particular laws of nature are relatively *a priori*.

40 See Kant's account of organization and of cognition as a system in the passages cited in note 10. I have joined this with his discussion of singularity and givenness of matter.

41 See his doctrine of transcendental apperception and the passages cited in note 3. I should also point out that Kant often assigns the task of system-building to reason, with reason building systems *for* the understanding. If this view is preferred, everything I say in the present paragraph can be understood in accordance with it.

42 And in this resides a deep difference between construction of qualitative concepts and that of mathematical-intuitive concepts. For the latter, see my "Kant Intuitivity," op. cit.

43 I am thinking primarily of Gordon Brittan, who has been defending such an interpretation in a series of papers beginning at an APA Symposium in Long Beach, March, 1984. Brittan is represented in B. den Ouden (ed.), *New Essays on Kant* (Bern: Peter Lang, 1987).

44 Among several pieces of hers, see "Kant's Real Self" in A. W. Wood (ed.), *Self and Nature in Kant's Philosophy* (Ithaca: Cornell University Press, 1984), pp. 113-147, and "Kant on Self-Identity," *Philosophical Review* **91** (1982), pp. 41-72.

45 I have attempted to do so in "Kant on the Nondeterminate Character of Human Actions" in W. L. Harper and R. Meerbote (eds.), *Kant on Causality, Freedom and Objectivity* (Minneapolis: University of Minnesota Press, 1984), pp. 138-163.

46 Ibid.

47 See his "Functionalism and Qualia," *Philosophical Studies* **27** (1975), pp. 291-315. My present formulation is indebted to him.

48 Beginning at B176. Also see the characterization of intuitions as examples, at AXVIII and elsewhere, and the doctrine of the inexponibility (by the understanding) of aesthetic ideas, in the third *Critique*, Ac. V, 233-235 and 342-344. In these last passages, Kant speaks of aesthetic norm-ideas to which no concept is adequate. In my "Kant on Intuitivity," I utilized this fact in my account of his conception of the intuitive character of mathematical concepts. I would now add that aesthetic norm-ideas can be *de re* perceptual-ostensive exemplars of the sort I have been talking about in the present paper.

49 ...although they are still considered to be spatial and temporal. In "Apperception and Objectivity," op. cit., I have further characterized their nature by saying that in themselves they are not spatial and temporal and meaning by this that they have some unconstructed properties which are not themselves properties of space and time and which do not by themselves entail or require any properties of space and time. As before, such unconstructed properties are uncognizable, and objects having them are uncognizable with respect to them. For things to be regarded in themselves is for them to be regarded in at least one of two ways. On the one hand, we can regard them apart from what is constructed for them for non-ostensive use by the understanding. This is the approach I have taken in the present (and in my Spindel Conference) paper. In this way of regarding them, they can still appear to a percipient, in the B34 sense of appearance. On the other hand, we can also regard them apart from conditions of empirical intuition and hence apart from conditions of schematization. In this second way of regarding them, they are objects wholly of non-ostensive, discursive representations, but construction requiring schematization is now impossible. At B304-311, in the chapter on Phenomena and Noumena, Kant is concerned with objects conceived in the second way. The complete story of things regarded in themselves needs to deal with both approaches. All of this of course reflects Kant's basic belief that objects of empirical cognition are subject to conditions both of intuition and of the understanding.

50 See F. H. Jacobi, *David Hume uber den Glauben, Werke* (Leipzig: G. Fleischer, 1815), vol. II, 304. Jocobi holds of cognitive representations that without any presumed real and

given conditions on sensibility, we cannot enter the system that these representations form, but with this presumption we cannot remain within that system either. Kant would reply that something real is given in sensibility but that it is not given as objectively determined.

51 See his "Functional Analysis," op. cit.

52 He speaks of human understanding at BXXXII, B556, and B812. At B33 he says that at least for us human beings it is the case that affection relations are in place. Sensibility is then defined as the manner in which we are affected.

53 These are mental capacities, but I have also suggested that they receive a physical realization. When the issue of chauvinism gets raised below, it must be kept in mind that we may be insisting on states and processes ascribed to human beings in either a mentalist or a physicalist language.

54 See his "Troubles with Functionalism" reprinted in N. Block (ed.), *Readings in Philosophy of Psychology*, vol. I (Cambridge: Harvard University Press, 1980), pp. 268-305.

55 See note 45.

56 See his introduction to S. T. Soemmering's *On the Organ of the Soul*, as reprinted in G. Hartenstein (ed.), *Immanuel Kants Samtliche Werke*, vol. 6 (Leipzig: L. Voss), pp. 455-461. The speculation concerns water in the brain as physically realizing perceptions.

57 See "Functionalism and Qualia," op. cit.

58 Both in their papers already cited. Block stresses the importance of such an ascription. Similarly, Shoemaker admits that there is a way of regarding qualia which is not subject to his functionalistic interpretation.

59 According to anomalous monism, cognitive processes receive a physical realization in order to stand in causal connections to what is physical. Also recall that affection relations are causal.

60 Cognitive states and processes, when described mentalistically, are described non-determiningly.

61 They are intensively, but not extensively (additively), measurable. See the *Anticipations of Perception*, B207-218. Kant requires additivity for the highest degree of theoreticity, met with among physical properties. Failure of additivity highlights the character of qualia as mental. This failure also means that the token-incorporation of qualia into theory falls short of the highest degree of theoreticity.

62 For invaluable comments, I owe thanks to Barry Brown, Lewis W. Beck, and J-C. Smith.

PATRICIA KITCHER

KANT'S DEDICATED COGNITIVIST SYSTEM

1. Introductory

Kant described his great work not as a critique of "books and systems" but as a critique "of the faculty of reason as a whole, in regard to all knowledge to which it may aspire *independently of all experience*."[1] Reason provides knowledge which is independent of experience, according to Kant, not because we are able to acquire knowledge in the absence of experience, but because the knowledge that we acquire in the face of experience is partially a reflection of that experience and partially a reflection of our ways of acquiring knowledge. In exploring reason itself in terms of its *a priori*, or experientially independent, knowledge, Kant proposed to examine how the workings of our thought processes are reflected in the knowledge claims we make. One philosophical point of the work is to reveal that particular aspects of our common wisdom are going to be invariant across peoples and times. Such invariances cannot be established by *assuming* metaphysical principles about the necessary uniformity or continuity of nature. Kant introduced a new method. He tried to demonstrate philosophically important invariances throughout the body of human knowledge, by arguing that they reflect the ways in which anyone with our basic mental constitution will be capable of thinking.

Given this project, any lasting results Kant obtained should be a significant contribution to our understanding of cognition. But Kant has not been allowed to play the role of a pioneer cognitive scientist, because his foremost English interpreters have denied the philosophical—and implicitly, the psychological—importance of his efforts to fathom our cognitive powers. No one questions that the argumentative strategy I describe above is present in *The Critique of Pure Reason*; the question concerns its value. Sir Peter Strawson, who more than anyone is responsible for the low status of Kant's efforts in "transcendental psychology," argues that this material must be put on one side:

> Whenever [Kant] found limiting or necessary general features of experience, he declared their source to lie in our own cognitive constitution;....Yet there is no doubt that this doctrine is incoherent in itself and masks, rather than explains, the real character of his inquiry....

J-C. Smith (ed.), Historical Foundations of Cognitive Science, 189–209.

The idiom of the work is throughout a psychological idiom. Whatever necessities Kant found in our conception of experience he ascribed to the nature of our faculties....

But [the deduction of the categories] is also an essay in the imaginary subject of transcendental psychology....The theory..., like any essay in transcendental psychology, is exposed to the *ad hominem* objection that we can gain no empirical knowledge of its truth....[2]

Strawson's (implicit) hasty argument in the last citation is that since the invariances revealed by Kant's method are to be independent of experience, they are incapable of empirical proof. As I will show below, however, either Strawson is using 'empirical' in so narrow a sense that his point has no force, or he is mistaken that Kant cannot appeal to experience—specifically to the possibility of experience—to make his case. A further problem with the apsychological approach to reading the *Critique*, however, particularly for the crucial deduction of the categories, is that if we do skip or excise all the discussions of our cognitive powers, then we will have very little left to read.

Strawson was able to recommend a bowdlerized Kant, because at the time he was writing (1966), Kant's psychology of cognitive faculties seemed hopelessly out of touch with scientific work in psychology. It was, however, but one year before Ulrich Neisser's manifesto for cognitive psychology appeared. The dramatic rise in the fortunes of cognitive psychology (and later cognitive science) that followed left the Strawsonian approach to Kant itself seriously out of touch with the needs and assessments of neighboring disciplines. As is often the case, philosophers did not notice the change that had come over their neighbors. (This is not specifically characteristic of philosophers. Many psychologists still do not realize that not all philosophers are logical positivists.) Thus many Kant scholars have continued the practice of systematically avoiding Kant's analyses of why certain cognitive powers are necessary for cognition. Given the acknowledged importance of these discussions in the text, this self-consciously Whig history does not seem a reasonable way to fathom the complexities of Kant's thought. Moreover, by hewing to this approach, Kantians have lost any opportunity to contribute to current debates about the nature and limitations of our cognitive capacities. I will try to show that this is in fact a lost opportunity, by arguing that Kant had sensible, well-reasoned views concerning at least two topics of great recent interest: the nature of empirical concepts and the problem of the holism of belief.

2. Kant's Approach

The central claim of Kant's transcendental psychology is that, *pace* Descartes, reason is not a universal instrument. Descartes had contrasted reason with more limited organs: "For while reason is a universal instrument which can serve for all contingencies, these organs have need of some special adaptation for every particular action."[3] Although Kant makes much of the distinction between perceiving and thinking, I believe that he regarded our cognitive powers as analogous to our senses in being specialized both in their mode of operation and in their range of application. Kant proclaims again and again that some things are beyond our ken, that our reason operates in particular ways, and thus that, in certain circumstances, it will be led into predictable errors (A293/B350 ff.). In contemporary terminology, Kant does not regard our cognitive powers as all-purpose machines, but takes imagination, understanding, and reason to be dedicated processes that operate only in specific ways on specific materials.

Besides the view that cognitive processes are dedicated, the most distinctive feature of Kant's approach is his overriding interest in the possibility of knowledge. He wants to determine what we can know and how we know it, and what we cannot know and why we cannot know it (Bxxii ff.). Kant's approach is unusual in the context of contemporary cognitive science, because the only features of cognitive processes that concern him are those that make some contribution to the epistemic status of a belief or a claim to know. He is totally uninterested in the actual physical embodiments of particular mental processes; he has only a raconteur's interest in the effects of culture or individual histories on mental processing; he specifically denies any interest in how children acquire knowledge; and finally, he is not even concerned to defend assignments of particular cognitive processes to particular faculties and frequently changes his mind about how faculties divide up the cognitive work. Kant's sole objective is to determine what our cognitive powers must be like if they are to be capable of producing knowledge.

While this focus will seem narrow in the context of current interdisciplinary work in cognition, I believe that Kant offers a valuable perspective on the problem. Because he is not distracted by other issues, he is able to offer unusually rich analyses of the "task environment" (in Newell and Simon's phrase[4]) of warranted belief. I will argue below that these analyses provide important insights into some currently perplexing issues in cognition. Before turning to the crucial matter of results, however, I will try to meet several *a priori* objections to Kant's method.

One obvious objection concerns Kant's apparent assumption that our cognitive powers do, in fact, regularly produce justified beliefs. Recent, widely reported results by psychologists such as Amos Tversky and Daniel Kahneman have suggested that, contrary to this optimistic assessment of cognition, the cognitive strategies employed by most people do not provide adequate justifications for the beliefs they adopt.[5] Kant can avoid the charge of eighteenth century optimism because his faculty psychology actually includes two related but importantly different types of investigations:[6]

> I. Arguments showing that we must have a certain type of faculty if knowledge, or a particular kind of knowledge, is to be possible; arguments showing that a particular set of faculties is sufficient for knowledge or for a particular type of knowledge. I will call such arguments *epistemic analyses*.
>
> II. Arguments showing that, given an uncontroversial account of the cognitive tasks we actually perform, it follows that we possess one or several of the types of faculties noted in arguments of type I. Because they are concerned with capacities we actually have, I will call this second kind of study *analyses of empirical capacities*.

Kant does not just assume that we regularly enjoy warranted belief and then inquire how various faculties contribute to this wondrous achievement. Such an assumption would not have been granted at the time he was writing. Kant did not need Tversky et al. to inform him about the foibles of human cognition. He already had the skeptical challenges of Descartes and Hume to contend with. Because of those challenges, Kant had to offer analyses of empirical capacities—capacities whose existence would be granted even by these remarkably skeptical adversaries—to demonstrate that we actually possess the raw materials necessary for the acquisition of knowledge. This was all Kant tried to show: We have the necessary capacities to achieve knowledge. He did not believe that human cognition was particularly reliable in practice.

Philosophers have sometimes dismissed Kant's faculty psychology with the complaint that he argues that knowledge is possible by arguing that we must possess a certain faculty, because knowledge is possible only if we do. This criticism fails to distinguish the analyses of empirical capacities from the epistemic analyses. Kant clarifies the critical role played by the analyses of empirical capacities in a passage in the *Critique of Practical Reason*:

But human insight is at an end as soon as we arrive at fundamental powers or faculties, for their possibility can in no way be understood and should not be just arbitrarily imagined or assumed. Therefore, in the theoretical use of reason only experience could justify their assumption. Such empirical proof...is, however, denied to us [in ethics].[7]

In Kant's own estimation, then, the transcendental psychology of the *Critique of Pure Reason* proceeds from uncontroversial empirical assumptions about the cognitive tasks we perform. This is why he regards himself as answering the skeptics, rather than begging the question against them as the complaint against transcendental psychology suggests.

If this account of transcendental psychology is correct then it answers the recent empirical objection that Kant was overly optimistic about our cognitive endowment, as well as the long-standing philosophical objection that his reasoning was question-begging because circular. It also answers Strawson's methodological objection that there simply cannot be empirical evidence in favor of a claim of transcendental psychology. Empirical evidence enters Kant's arguments in two places. First, it is experience that establishes the presence of a particular cognitive capacity, e.g., that we have beliefs of a certain sort or that we make inferences of a certain sort. Kant then reasons that a particular feature of our beliefs reflects the way we think rather than the way the world is, by arguing that we cannot account for the presence of that feature in cognition as an acquisition from the world through our senses. Such arguments rest on further assumptions about the sorts of features that our senses can pick up from the environment. So, for example, Kant's well-known claim that our perception of three-dimensional space is a reflection of our sensory constitution rests on the assumption that the senses cannot transmit three-dimensional spatial information.[8] Similarly, his famous claim that causal reasoning is a reflection of human thought processes rests on Hume's demonstration that certain features of our causal beliefs cannot be acquired from experience.[9]

Are these assumptions empirical? I think that some are clearly empirical. The assumption about our inability to register spatial information in our senses is an obvious example. It rests on observations about the geometry of vision, and it is clearly refutable by possible future discoveries. What about Hume's 'observation' that we can never observe any necessary connection between a cause and its effect? Despite the somewhat misleading presentation in *A Treatise of Human Nature*, Hume did not discover that he saw no necessary connections in the same way that Dalton discovered that some people did not see colors. Hume's view did not result from a series of experiments, though it is based on general knowledge about the limits of human perception. Could we look backward and forward in time, we might be able to establish that a certain effect invariably followed upon a certain cause. Although Hume's claim does not rest on a well-delimited empirical base, it is still broadly empirical. The same pattern is typical of the assumptions employed in Kant's psychological reasoning. To attribute a feature to our thinking apparatus itself, he must make narrow or broadly empirical assumptions that enable him to rule out sensory experience as the source of the feature.

Even if Kant relies on empirical assumptions, Strawson could still protest that his method of generating results is non-empirical. In briefest form Kant's method standardly includes four steps: (1) decompose a cognitive task into its essential sub-tasks; (2) note that certain aspects of this process cannot be attributed to features gleaned from experience; (3) infer that the presence of those features is a reflection of the structure of any thinking thing; and (4) infer that those features will always be present in thoughts of creatures who think in roughly the ways that we do. Plainly this method involves analyzing and inferring, and not observing and experimenting. However, if that is Strawson's criterion for the non-empirical, then the more theoretical aspects of scientific reasoning will fall under the same axe as Kant's work. Under more reasonable standards, no view should be stigmatized as non-empirical just because it rests on inference and analysis rather than well-delimited observations, as long as it is consonant with relevant general observations and could be dislodged by future discoveries. As I have argued, Kant's studies, or at least some of his studies, are empirical on both counts.

The final *a priori* objection I will consider arises from Kant's preoccupation with the problem of knowledge. Kant's only interest is cognition. However, unlike contemporary cognitivists who make cognition the model for all mental processes, Kant does not see cognition in everything. He does not assume that perceiving, remembering, classifying, or concept learning are processes fundamentally similar to acquiring warranted beliefs. Rather, he is the complete cognitivist in that all he wants to know about perceiving, remembering, and so forth, is what such processes can contribute to the subject's ultimate claims to knowledge. For Kant, these processes are not like cognition. Rather, they are relevant only insofar as they are sub-processes of the grand task of cognition.

3. Kant's Theory of Empirical Concepts

In presenting and partially defending Kant's approach, I have given prominence to his accounts of faculties, since this is the most notorious part of transcendental psychology. He is also well-known as the philosopher who recognized that knowledge must involve both intuitions which are perceptions produced by information being taken in through the senses and concepts which are used to classify that information. In his famous aphorism, "Thoughts without content are empty, intuitions without concepts are blind" (A51/B75). Kant labored both to differentiate concepts from intuitions—their confusion being the common error of his Rationalist and Empiricist predecessors—and to explain exactly how both were necessary for the production of warranted beliefs.

The issue of whether Kant succeeded in demarcating concepts from intuitions is beyond the scope of a single paper. So I will consider just one interesting piece of the overall theory, viz. the account of empirical concepts. What do you have when you have a concept? This question allows two sorts of answers. We could try to determine the abilities that are criterial for determining whether or not you have a concept, such as the ability to classify new instances. Or we could try to determine exactly what it is about you that enables you to have those abilities that are criterial for concept possession. Kant tries to answer the latter question. What is it about you that enables you to apply concepts to things?

For Kant, a concept is a rule that allows you to unite separately given perceptual materials together under one label:

> ...a concept is always, as regards its form, something universal which serves as a rule. The concept of body, for instance, serves as a rule for our knowledge of outer [objects], by unifying the various aspects which are thought in it....Thus the concept of body requires in the perception of something before us, the representation of extension, and with this, of impenetrability, form, and so forth....(A106, amended translation)

That is, a concept regulates the combining of perceptually acquired information into a single representation, or information-bearing state. Despite the somewhat misleading implications of this discussion, Kant realized that he could not provide a general explanation of how we apply concepts by claiming that we apply concept *X*, by first noting that we can apply concepts *Y* and *Z*, for that would involve an infinite regress.

Kant explicitly rejects the image-based theory of his predecessors, on the familiar grounds that a concrete image is not up to the task and that an indefinite image makes no sense. We cannot determine that a new instance is a triangle by comparing our present perception with a stored image of a triangle, because the stored image "would never attain that universality of the concept which renders it valid of all triangles, whether right-angled, obtuse-angled, or acute-angled..." (A141/B180). Instead, Kant reasons that we must use a *schema*, that is, a rule for constructing images that indicates the sequence of combinatory steps that the perceptual system goes through in producing images of perceived instances of the concept (A141/B180; see also A103-04 and A77-79/B103-04). Presumably this sequence will have some steps that include a number of possible options. For example, in the case of an equilateral triangle, the equality of the sides can be presented, while this step will not occur in other cases. Concepts are rules, and the part of a concept that permits perceptual recognition is the schema, or the rule for constructing images of concept instances.

Other than arguing for the superiority of a schema approach over the standard image comparison view, Kant does not provide any compelling arguments about why schemata must be the right way to understand concept application for the case of sensible concepts. The schema hypothesis seems to permit the latitude required by the concepts we actually use, and it does not involve a regress, because it explains how concepts are applied on the basis of perceptual processes, not conceptual information. I would just add that I think the proposal is quite suggestive. Kant's idea is that it is not the perceptual data themselves that are the key to conceptualization, but rather the processes gone through by our perceptual system in dealing with instances of the concept that are directly connected to our conceptualization. This is plausible in itself, because to understand how we conceptualize, it seems important to consider how the data appear to our cognitive systems, rather than what they are like in themselves. Proposals like this have recently been developed and have met with some success. So, for example, the 'motor theory' of phone perception claims that we classify acoustic signals as different phones on the basis of considerations about how we, as speakers, would articulate the sounds. We would classify a range of acoustic signals as *ba*'s, for instance, not on the basis of their intrinsic acoustic similarity, but on the basis of whether we could produce all of the sounds by the same articulatory movements.[10]

The part of Kant's theory I would like seriously to defend concerns the fixed or changing nature of the rules that embody concepts. Whether one thinks of images, feature lists, or schemata as the relevant tools for directing concept application, a question still arises concerning the stability of the rules. Once you have learned the concept cat, for example, do you now possess a method which offers a necessary and sufficient condition for determining whether everything you meet is or is not a cat? Kant's equation of concepts with rules seems to imply that he thinks of concepts as completely codified and so fixed. In fact, his view is just the opposite.

Because of the regress problem, recognition of instances through a concept cannot in all cases be a matter of consciously going through a checklist of features. Still, it would be possible to draw up a list of features that seem to be crucial to different concepts by working backwards from the instances. Kant repeats the point many times that synthesis must precede analysis: we must actually abstract and use a rule for producing images of instances before we can ask ourselves what (perceptual) features are, in fact, combined in various concepts (B130, e.g.). The problem is that this sort of analysis cannot be performed on empirical concepts:

> Since the synthesis of empirical concepts is not arbitrary but empirical and as such can never be complete (for in experience ever new characteristics of the concept can be discovered), empirical concepts cannot be defined.[11]

For Kant, as for his Empiricist predecessors, an empirical concept is one that has been abstracted from experience. We have the concept cat because, having encountered a number of cats, we have abstracted, among other things, a schema for generating perceptual images of cats. The crucial epistemological point about empirical concepts is that there can be no question about their legitimate applicability to the world. Since we abstract similarities in image processing from our dealings with instances, we cannot merely be deluding ourselves into believing that a class of such objects exists (cf. A84/B116). The use of empirical concepts is warranted by the very process that produces them.

From an epistemological point of view, when should the abstracting of similarities stop and the similarities among instances be analyzed and codified in a definition for future use in classifying new instances? Kant denies the possibility of defining empirical concepts, because he recognizes that since we continue to have experiences there can be no good moment to stop. For whenever you stop, you are limiting your evidence, or at least favoring some evidence over other evidence. Suppose you stop and codify your concepts on your 21st birthday. If you strictly adhere to your definitions, you will deny similar cases membership in the extension of the concept. Since those cases are potential evidence about the adequacy of your classifications, you can no longer claim that your concepts are warranted on the basis of your total evidence. Thus your 'empirical' concepts would lose their experiential warrant.

To avoid this problem, you might consider codifying your concepts on your 21st birthday but taking a liberal attitude towards your conceptual system. While you use your definitions to classify new instances, you are prepared to reflect on new cases and to allow considerations about theoretically important similarities to lead you to reject a concept as you have defined it, in favor of some more satisfactory concept. This liberalization will not commit the epistemological error of limiting the evidence. Further, this model may well reflect the practice of scientists in dealing with critical concepts in their disciplines.

If we probe a bit deeper, however, there are still epistemological problems with this liberal definitional model. First, we should notice that it is evidentially biased towards old cases. New cases can get a hearing, but they must prove their importance as data, whereas during the acquisition of the concept, any case was freely admitted to the data base. Favoring old evidence has some epistemic virtues in the context of organized scientific research. If a datum or a theory has been abroad in the scientific community for some time then there is reason to hope that it has withstood attempts to deny it. But this is surely not the case with data which is old only in the lifetime of a particular individual. Yet, in the model this is the type of old data that is given pride of place.

The second epistemological shortcoming of the liberal definitional model is that new data are not admitted automatically, but only on the basis of conscious reflection. Conscious reflection about the adequacy of our concepts is epistemically virtuous, as Kant would be the first to stress. For the perceptual similarities that lead us to group instances together may turn out to be theoretically unimportant. However, the demands put on conscious reflection in this model seem too high. Classification is standardly not conscious but automatic. Given the regress problem, some of it must be automatic. If classification proceeded automatically via necessary and sufficient conditions, then given that our use of concepts is to be well-grounded in the data we encounter, this process would have to be subject to constant correction by conscious reflection. While this is possible, it is hard to believe that we are sufficiently conscientious that our concepts would not gradually slip out of alignment with the data experience has provided, so that our continued use of them would no longer be warranted. If empirical concepts are to be reflective of reality because of the way they are produced, far better to have a process that allows for automatic continuous updating. As Kant realized, we need such a process to explain how we manage to acquire concepts at all, so he assumed that this process continues to mold our concepts to fit the data we continue to experience.

In recent years, there has been considerable debate about how concepts are represented in us and, in particular, about whether they are represented by necessary and sufficient conditions that apply to all instances in the extension of the concept. Edward Smith and Douglas Medin conclude their lengthy review of the literature with the observation that while the balance of the empirical evidence argues against the presence of necessary and sufficient conditions, that evidence does not establish any one view of concepts as correct and rule out all the others.[12] We cannot turn to Kant and expect to discover that the true theory of concepts has been lying, undetected, in our midst for two hundred years (although I believe that his schema hypothesis does offer a plausible and novel hint about how we might recognize instances). What Kant's analysis of the necessary conditions for knowledge shows, I think, is that concept theorists have a choice: if they are going to conceive of concepts (or some subset of concepts) as reflective of reality because of the ways in which they arise, then they do not want to use a necessary and sufficient conditions model. Indeed, if they envision our conceptual repertoire as extremely adaptive to changing environments, then they may not want to appeal to concepts at all, but rather to something like conceptual worms that continually evolve through time. Kant's analysis also has the rather startling implication that concept-learning is something of a fiction. Insofar as this process is standardly conceived of as taking place over a relatively fixed span of time, before which the subject lacked the concept and after which he's got it, then concept-learning does not take place. To meet the requirements for empirically warranted belief, concept-learning must be, in the educational cliche, a 'lifelong experience'.

Kant's analysis would provide some guidance for theorizing about concepts and concept-learning, if contemporary researchers share his views about empirical concepts and warranted belief. I think these assumptions are in fact ubiquitous, if not always explicit. The only reason psychologists worry about concepts is that they need to explain feats of cognition. In addition to acting on the environment, we also represent to ourselves the way the environment is. Since we have an elaborate machinery for representing our surroundings, considerations of natural selection suggest that we often succeed in forming tolerably accurate representations.[13] The world is a varied and surprising object to represent. As David Marr argues in the case of vision, it is hard to explain how any all-purpose visual system could succeed in its task, unless it is able "reliably to derive properties of the world from images of it...."[14] The same holds, *mutatis mutandis*, for cognition. Given the very wide range of situations we need to think about, how can our cognitive system succeed unless it is able to derive the representations or concepts it will use to classify objects together as similar from actual similarities among the objects it encounters?

For these very good reasons, I believe, contemporary theorists are, like Kant, Empiricists about empirical concepts. They assume that such concepts are applicable to reality because of the process by which they are derived from reality. Further, I believe that current researchers share Kant's global assumption that a theory of concepts should be part of a general explanation of successful cognition. Given these implicit starting assumptions, Kant's analysis then suggests that the model of necessary and sufficient conditions may have much less initial plausibility than philosophers and psychologists have traditionally assumed.

4. The Holism of Belief

Kant offers a two-tiered theory of belief that appeals to two separate faculties, 'understanding' and 'reason'. We have already seen parts of the theory. The understanding is the "faculty of concepts" (A68/B93, e.g.). Since concepts are rules that include schemata for recognizing when the perceptual system processes instances, the understanding is the "faculty of rules" (A126) and the faculty that carries out "recognition [of instances] in a concept" (A103). The understanding is also the faculty that unites concepts in a judgment (A68-70/B92-94), as in 'the body is heavy' (B142). One of Kant's most distinctive views, which I will not try to evaluate here, is that the understanding operates with a specific set of rules for uniting concepts in judgments, the 'transcendental concepts' or 'categories'. So, for example, the understanding uses the rule for combining concepts associated with the concept of negation to produce the judgment, 'the soul is not mortal' (A72/B97).

For my purposes, the crucial fact about the conceptualizing and judging activities performed by the understanding is that they are virtually unconscious and under no voluntary control.[15] According to Kant, the combining of different representations (either perceptual information in a concept or concepts in a judgment) is carried out by "the power of imagination, a blind but indispensable function of the [understanding]" (A78/B103).[16] I take the metaphor to imply that the imagination is not under conscious guidance. If it were, we would face the regress problem about concept application and a similar problem in judging how to make judgments (cf. A132-3/B171-2). Still, if diverse perceptual information and conceptual information were not somehow combined or collated, it is hard to see how we could conceptualize or judge. So we must have some power or powers that carry out this requisite unconscious combining.

As the faculty of concepts, the understanding plays a critical role in generating perceptual beliefs, beliefs about what we are seeing, hearing, feeling, tasting, or smelling right now. It is also subject to the law of association: "...representations which have often followed or accompanied one another finally become associated...[so that] even in the absence of the object, one of these representations can...bring about a transition of the mind to the other" (A100). Again, this function of the understanding operates blindly, without conscious guidance. Since it is governed by the law of association, the understanding can supply us with beliefs about objects and properties that co-occur in nature.

Not all of our beliefs arise through perception or association. If I meet someone who sees red lines when looking at a horizontal grid and green lines in a vertical grid, I may infer that he has been involved in too many demonstrations of the McCollough effect. Just about 300 pages into the *Critique of Pure Reason*, Kant turns to the topic of reason. Reason is the faculty with the power of inference. As he notes,

> Since we have constantly to make use of inference, and so end by being completely accustomed to it, we no longer take notice of this distinction...between what is immediately known and what is merely inferred. (A303/B360)

Nevertheless, the distinction is real and important because serious mistakes can arise when we err about which faculty is responsible for a certain aspect of knowledge (A261/B317). The role of reason is "to reduce the varied and manifold knowledge obtained through the understanding to the smallest number of principles...and thereby to achieve the highest possible unity in knowledge" (A305/B361, amended translation). That is, the role of reason is to make our total set of beliefs as systematic as possible (A645/B673).

Kant is offering a critique of reason, so he focuses on the delusions we are led into by our reasoning practices. For example, since we always seek a theory which has the widest possible scope, we unjustly assume that there must be some one principle that actually governs the workings of the whole world (cf. A571/B599 ff.). Despite Kant's negative emphasis, it is obvious that inference, and so reason, must also play a positive and less grandiose role in production of warranted beliefs. It is no good looking at the moon as it rises and judging it to be of a certain size and then looking at it at its zenith and judging it to be of a different size. These two judgments must be compared and contrasted, if the erroneous one is to be rejected. While Kant does not stress these simple contributions of reason to knowledge, he explicitly states, "Knowledge is [essentially] a whole in which representations stand compared and connected" (A97; Kemp Smith's interpolation). Also, he is clear that it is reason that introduces the highest unity or systematization into our knowledge (A299/B355).

Only intuition (perception) and understanding are (original) sources of knowledge (A294/B350). Reason cannot supply any new data, but it still plays a crucial role in determining our beliefs. As Kant describes our reasoning processes, they are conscious and under (some) direct control:

> In every syllogism I first think a *rule* (the major premise) through the *understanding*. Secondly, I *subsume* something known under the condition of the rule by means of *judgment* (the minor premise). Finally, what is thereby known I *determine* through the predicate of the rule, and so *a priori* through reason (the conclusion)....
>
> If, as generally happens, the judgment that forms the conclusion is set as a problem—to see whether it does not follow from judgments already given...—I look in the understanding for the assertion of this conclusion, to discover whether it is not there found to stand under certain conditions according to a universal rule. If I find such a condition, and if the object of the conclusion can be subsumed under the given conclusion, then the conclusion is deduced from the rule, *which is also valid for other objects of knowledge*.
>
> (A304-05/B360-61)

So, to use Kant's favorite, if shopworn, example, if I want to figure out whether Caius is mortal, I consult what I already know, and discover that Caius meets a condition–he is a man–which enables me to find a general principle, 'All men are mortal', which can serve as the major premise of a syllogism that will deliver the conclusion that Caius is mortal. This process is under considerable voluntary control. In the first case, I deliberately arrange given premises so that a conclusion will follow. In the second, I deliberately look for premises that will enable me to deduce the desired conclusion. Although I can set myself the task of looking, I do not have voluntary control over the search itself. So I cannot guarantee success through sheer force of will, even if I do possess the requisite information.

Over thirty years ago, W. V. O. Quine wrote a landmark article, "Two Dogmas of Empiricism,"[17] that presented a doctrine that has since been called the 'holism of belief'. Quine made two closely connected points, which had been anticipated by the early twentieth century French historian of science, Pierre Duhem. One point is that in evaluating the evidence in favor of a belief, any piece of evidence whatsoever is relevant. Every one of your beliefs is (potentially) relevant to the credence placed in any one of them. The second point is that when a new piece of evidence comes in and seems to contradict previously held notions, we need to amend our system of beliefs by invoking considerations of *overall* simplicity, conservatism, fecundity, and the like. Thus fixation of belief is holistic, because any member of the system can matter to any other and because the system is revised by invoking global considerations of least overall change.

Quine's doctrine of the holism of belief has itself become a dogma of contemporary philosophy. It is frequently cited and, to my knowledge, never seriously challenged. The problem with this doctrine, from the point of view of cognitive science, is that it implies that we can never have a science of belief fixation. To develop a rigorous theory of a psychological process, we must first isolate the relevant variables. If

Quine is right about the holism of belief, however, such isolation is not possible. For any of the indefinitely many beliefs people seem to possess could be brought into play in explaining any case of the fixation of belief.[18] One might hope to introduce some order by partitioning the beliefs into subsets and developing some principle by which entire subsets could be handled at once. However, this suggestion fails to appreciate the genuinely radical nature of holism. For holism amounts to the claim that there are no privileged relevance relations among beliefs, and such relations would be required both to carry out an effective partitioning of beliefs and to enable us to dismiss entire subsets as not worthy of further search.

In the final part of *The Modularity of Mind*, Jerry Fodor offers roughly this argument from Quine's premises to the conclusion that a science of central psychological processes—those involved in fixation of belief—is impossible.[19] However, I think that Fodor is probably not alone in suspecting that Quinean holism has grave implications for a science of belief. What I will try to show is that if we appreciate Kant's points about understanding and reason, then it turns out that the holist's model of belief contains two serious confusions.

Kant's first and better known point is that the systematizing principles of reason are not "constitutive" but "regulative" (A644/B672, e.g.). That is, our reasoning is correctly guided by the principle that we should seek a theory with, for example, the greatest overall simplicity. It does not follow that we can ever find such a theory, however. *A fortiori*, it does not follow that we actually go through an indefinitely long chain of reasoning that allows us to decide on the theory that provides the simplest account of *all* our data. The systematizing principles of reason represent epistemic ideals. They are normative principles about the best ways to systematize our beliefs, not empirical accounts of what actually happens when we reason (A642/B670 ff.). This is true for Kant's principles, and it is true for Quine's principles. These principles were discovered by careful reflection on the best ways to systematize our beliefs to guarantee that they are maximally warranted. Neither Kant nor Quine ever tested any subjects to see what processes they go through in modifying their beliefs. But if the principles are merely normative, then they hold no terrors for the development of cognitive science.

Even if we do not always, or indeed ever, reason by bringing all potentially relevant information to bear on the belief to be accepted or rejected, we do sometimes reach conclusions by combining remarkably disparate sources of information. Darwin's reading of Malthus's *Essay on Population* and his observations on the Galapagos is one of many textbook examples of the value of wide cross-fertilization of ideas. Since we do reason and since reasoning sometimes involves the utilization of *prima facie* wildly unconnected pieces of information, there still seems to be a problem for cognitive science. How could these actual, not ideal, pieces of reasoning be explained?

Here, I think Kant's reflections about the two-tiered structure of belief formation reveal a crucial conflation in the holist's case against a science of belief. If Kant is right, then the concepts of the understanding are 'holistic' only in the sense that all of one's dealings with instances will have shaped the current schemata for recognizing instances and the current associations among concepts based on co-occurrence of instances. This is not a worrying holism, however, because there is no reason to believe that the processes on the first tier cannot be explained scientifically. Not having access to individuals' particular perceptual histories, we may not be able to predict their particular associations, but presumably we can determine the laws that would enable us to make such predictions, given the requisite facts. Similarly, there is no reason to believe that we would not be able to formulate principles relating encounters with particular types of instances to specific features in the development of concepts. So, for example, it might turn out that instances that were encountered under pleasant or painful circumstances were particularly influential in determining the content of the subsequent concept, as Locke hypothesized.

It is when we turn to the second tier of belief formation, where inference enters, that we must be guided by the normative principle that anything we know could be relevant to our attempt to determine the degree of credence we should put in a claim. So what do we do? We "look in the understanding" to find other claims that might help us construct a solid chain of reasoning with this claim as its conclusion. Many factors will affect this phase, including past associations, current ideas that are running through our heads, and most obviously, the ways in which we conduct memory searches. I take it that no one believes that these processes cannot be understood scientifically. That is, we may reasonably hope to find general principles governing the structure of memory, or memory systems, as well as general principles about how memory searches occur, the availability of reasonably current material, and so forth. Again, because individual psychology is a historical science, we cannot expect to be able to predict the results of a particular individual's particular search. In this respect cognitive science as applied to individuals is in the same position as geology or evolutionary biology. Again, however, none of this speaks against the possibility of a *bona fide* scientific theory of the processes involved in this phase of reasoning.

If materials are found, then we can actually engage in reasoning about the claim. Kant believed that in doing so we follow good principles of Aristotelian syllogism. However, recent empirical studies have suggested that we use much more dubious principles in revising our previous beliefs on the basis of new evidence.[20] Further, Kant believed that when we reason, we are conscious not only of the claims under consideration but also of the types of inference we employ. Recent studies also suggest that this is not always, and perhaps not even often, true. Nevertheless, there is no *a priori* reason to believe that we cannot determine the rules or heuristics that are used in inference.

If we do not look at reasoning as a totality but break it down into phases, as Kant did, then the claim that there can be no science of belief fixation seems much less plausible. Interestingly, like Quine, Kant did not believe that terms (concepts) could be defined, and he did believe that the fixation of belief should be guided by systematic considerations. However, in Kant's treatment these two views do not place the prospects for cognitive science in danger.[21] What Kant's position shows is that the move from Quinean holism about belief to Fodorian skepticism about a science of belief fixation rests on two confusions. First, and most clearly, this reasoning slides from normative considerations to descriptive claims.[22] Second, it also trades on conflating the roles of understanding and reason in belief formation. If Kant is right, then there is a certain kind of limited holism involved in the processes by which we quickly and unconsciously modify the concepts we use in judgments. By contrast, when we try to apply holistic principles in conscious (or more conscious) reasoning, then while our goal is to be as systematic as possible, our execution falls far short. This process is holistic in intent, but in actuality, it is slow, cumbersome, and as we would put it, highly resource constrained (because we cannot hold too many items in mind at once). As a result it is not very extensive. But while such a process may be limited in its epistemic virtues, it does have the virtue of being *prima facie* explicable. Thus, in Kant's theory both tiers of the formation of belief—and so the entire process—would be amenable to scientific explanation.

Philosophers will rush to object that, on my reading, Kant's account of belief formation avoids the dire implications of Quinean holism only by invoking an illegitimate distinction between perceptual beliefs and inferential beliefs, or between observational beliefs and theoretical beliefs. I will begin my defense of this move with two *ad hominem* arguments. Arguing *ad hominem* is normally quite inappropriate, but in this case it enables me to make the historical situation clear. The doctrine that there is no serious distinction between beliefs based on observation and those based on theory is as much a dogma of contemporary philosophy as the holism of belief. However, there has been one prominent dissenter from this overwhelmingly popular view, viz. Jerry Fodor.[23] Since Fodor lacks this reason for moving from Quinean holism to the impossibility of a cognitive science of belief fixation, I think his inference does rest on the two confusions cited.

The second *ad hominem* is about Kant. To anyone trained in the history of philosophy, objecting that Kant did not sufficiently appreciate the subject's contribution to his own beliefs is like complaining that the Pope is insufficiently Catholic. Kant's Copernican revolution in philosophy consisted in changing the fundamental question of epistemology from 'How can our ideas conform to objects?' to 'What must objects be like if they are to conform to the conditions we require for knowledge of them?' (Bxvi). Kant's awareness that perception is not a matter of simply registering what is before you right now is evident in the views already presented. On Kant's general model, how you describe what you perceive will depend on all your previous encounters with instances, on the associations formed in those encounters, on particular information that you now bring to bear (e.g., that penguins are properly called 'birds', even though they are not much like other birds), and on whatever global considerations have shaped your views. At least originally, the latter sorts of influence will be via conscious reasoning. However, given Kant's views about association, this information could become part of the conceptual rules that are fluently and unconsciously applied to new instances. This would happen if, for instance, when seeing a penguin, you remind yourself that scientists classify them as birds, or if you simply rehearse the penguin-bird connection enough in short-term memory.

Given the historical record, then, the general complaint that Kant does not recognize the influence of past experiences and beliefs on perception cannot be maintained. So the objection must be about the specific ways in which he thinks of this influence. To resist the Kantian way out of the problem of holism I sketch above, the claim must be that Kant's crucial division between the fluent, unconscious processes of the understanding and the explicit, laborious processes of reasoning is somehow incorrect.

Now such an objection would be question-begging unless based on an independent source of evidence. In fact, for the last 25 years philosophers have appealed to the enormously influential work of Thomas Kuhn to make the critical point. In *The Structure of Scientific Revolutions*, Kuhn maintains, "What a man sees depends both upon what he looks at and also upon what his previous visual-conceptual experience has taught him to see."[24] Kuhn means this in the strong sense that a perceptual report will be influenced by high-level theoretical considerations that are unconsciously employed in making the report. That is the assumption that enables him to reach his famous doctrine that the adherents of different high-level theories will not be able to appeal to a common body of observations to mediate their dispute. For example, Aristotelians will look at a swinging stone and see a heavy body moving towards its natural state of rest, whereas Galileans will see a pendulum.[25] If Kuhn is right, then seemingly it is a mistake to divide belief formation into two separate stages, one automatic, unconscious, and involving only a limited holism and the other laborious, conscious, and completely holistic in intent.

So we have the following doctrinal dependencies: the view that a cognitive science of belief is impossible rests on Quinean holism about belief; Quinean holism about belief is widely supported in part because of the fall of the observation-theoretic distinction; one of the major factors in the fall of this distinction was Kuhn's work on scientific revolutions. What, then, were the major sources of Kuhn's reflections? I think that philosophers have not always appreciated that Kuhn rested a significant portion of his argument for the 'paradigm' paradigm on (then) contemporary research that seemed to show that perception is permeated by unconsciously employed high-level beliefs.[26] Kuhn appeals to this work to support his well-known claims that previous conceptions of objectivity in science, including the existence of a neutral data base, stable data, raw data, and brute or immediate experience, cannot be maintained.[27] There is more than a little irony here. Insofar as this chain of dependencies is accurate—and I believe that it does reflect crucial, though not exhaustive, reasons for the support enjoyed by the various doctrines—the philosophical argument against the development of a particular empirical science depends on the largely unconscious acceptance of previous empirical results. A second, related irony is that Kuhn became well-known for the view that what you accept as facts depends on your current theories.

I am in no position to determine whether or not high-level theoretical beliefs unconsciously influence the production of perceptual beliefs. Recent work by David Marr and his colleagues suggests that this is not the case.[28] But my point is only that this is an empirical issue that cannot be decided by philosophical fiat. Hence I am unwilling to give up the Kantian solution to the problem of holism, without more thorough empirical investigation of this issue.

Kant was deeply concerned with the possibility of obtaining knowledge or justified belief. To be truly justified in our beliefs, we must appeal to the broadest possible evidential base and to considerations of overall systematicity. What Kant's analysis of the differing contributions of understanding and reason shows is that we can honor these epistemic requirements and still hope to understand how we actually form beliefs.

For Kant, we have the ability to achieve a systematic understanding of the natural world, only because we are capable of reasoning about the information we have picked up. Kant's admiration for reason is surprisingly balanced, however. To him, the remarkable thing is that we can reason at all, not that we do it particularly well.[29]

University of California, San Diego

NOTES

[1] Norman Kemp Smith (trans.), *Immanuel Kant's Critique of Pure Reason* (New York: St. Martin's Press, 1968), Axii, amended translation. Following standard practice, I will refer to this work in the text and cite page numbers using Kemp Smith's notation of 'A' for paging in the first edition and 'B' for paging in the second edition. I will also indicate in the text if I have amended the translation.

[2] P. F. Strawson, *The Bounds of Sense* (London: Methuen, 1966), pp. 15-16, 19, and 32.

[3] Elizabeth S. Haldane and G. R. T. Ross (trans.), *The Philosophical Works of Descartes*, vol. I (Cambridge: Cambridge University Press, 1972), p. 116.

[4] Allen Newell and Herbert A. Simon, *Human Problem Solving* (Englewood Cliffs, NJ: Prentice-Hall, 1972), p. 55.

[5] For a review of the literature, see Richard Nisbett and Lee Ross, *Human Inference: Strategies and Shortcomings of Social Judgment* (Englewood Cliffs, NJ: Prentice-Hall, 1980).

[6] I first developed this analysis in "Discovering the Forms of Intuition," *Philosophical Review* **86** (April, 1987), pp. 205-248.

[7] Lewis White Beck (trans.), *Critique of Practical Reason* (Indianapolis, IN: Bobbs-Merrill, 1956), p. 48.

[8] See "Discovering the Forms of Intuition," op. cit.

[9] Immanuel Kant, Lewis White Beck (trans.), *Prolegomena to Any Future Metaphysics*, (New York: Bobbs-Merrill, 1950), pp. 24-25.

[10] Sally P. Springer and Georg Deutsch, *Left Brain, Right Brain* (San Francisco: W. H. Freeman, 1981), pp. 201-202.

[11] Immanuel Kant, Robert Hartman and Wolfgang Schwartz (trans.), *Logic* (New York: Bobbs-Merrill, 1974), p. 142.

[12] Edward E. Smith and Douglas L. Medin, *Categories and Concepts* (Cambridge, MA: Harvard University Press, 1981), p. 182.

[13] Given that we seem to have a complex system for representing the environment, it may be reasonable to assume that this structure is reason ably well-adapted to the task. However, as Stephen Jay Gould and R. C. Lewontin have forcefully pointed out, this inference is not always warranted. See their "The Spandrels of San Marco and the Panglossian Paradigm: A Critique of the Adaptationist Programme" in *Proceedings of the Royal Society of London*, B 205 (1979), pp. 581-598.

[14] David Marr, *Vision* (San Francisco: W. H. Freeman, 1982), p. 23; see also pp. 32-36 and 340.

[15] Kant sometimes suggests that we are aware of the combining activities of the understanding. See A104, A108, B134. The odd point about this claim is that Kant usually makes it and then immediately retracts it. I believe that it reflects the views of a contemporary psychologist, J. N. Tetens. In any case, the important point is that even if we do watch this process, we do not direct it.

[16] In the text, Kant describes the imagination as a power of the soul. However, I learned from Manley Thompson that in Kant's own copy of the *Critique*, he crossed out 'soul' and substituted 'understanding'.

[17] Reprinted in *From a Logical Point of View* (New York: Harper Torchbooks, 1961), pp. 20-46.

[18] The question of the number of beliefs we actually possess—as opposed to those we would assent to upon hearing—is itself quite vexed. For a discussion, see Daniel C. Dennett, "Brain Writing and Mind Reading" in his *Brainstorms* (Montgomery, VT: Bradford Books, 1978), pp. 45-46.

[19] Cambridge, MA: MIT Press/Bradford Books, 1983.

[20] See Nisbett and Ross, op. cit.

[21] In fact Kant had his own reasons for believing that a science of psychology was impossible. First, Kant limited the domain of psychology to what we would call psychophysics. He then argued that psychology could never be a science, because a science must be quantitative but the qualities we encounter in sensation cannot be quantified. See Immanuel Kant, James Ellington (trans.), *Metaphysical Foundations of Natural Science* (New York: Bobbs-Merrill, 1970), p. 8. While Kant rejected a science of psychology qua a science of psychophysics, he spent a great deal of time trying to work out the necessary contributions of various faculties to thinking; i.e., doing what we would call theoretical cognitive psychology. I argue for this view of Kant's attitudes toward psychology in "Discovering the Forms of Intuition," op. cit.

[22] Although I think it is an interesting question, I am not going to try to determine here whether this slide is in Quine or in his interpreters.

[23] Jerry A. Fodor, "Objectivity Reconsidered," *Philosophy of Science* **51**, 1, pp. 23-43.

[24] Thomas S. Kuhn, *The Structure of Scientific Revolutions*, 2nd ed. (Chicago: The University of Chicago Press, 1970), p. 113.

[25] Ibid., p. 119.

[26] Ibid., pp. 62-64, 85, 112-135.

[27] Ibid., pp. 121, 125.

[28] See my "Marr's Computational Theory of Vision," *Philosophy of Science*, March, 1988.

[29] I am grateful to Philip Kitcher for helpful suggestions about an earlier draft.

RONALD MCINTYRE

HUSSERL AND THE REPRESENTATIONAL THEORY OF MIND

Husserl has finally begun to be recognized as the precursor of current interest in intentionality—the first to have a general theory of the role of mental representations in the philosophy of language and mind. As the first thinker to put directedness of mental representations at the center of his philosophy, he is also beginning to emerge as the father of current research in cognitive psychology and artificial intelligence.

—Hubert L. Dreyfus

So writes Dreyfus in his introduction to *Husserl, Intentionality, and Cognitive Science*.[1] These provocative comments launch a most interesting discussion of Husserl's relationship to important recent work in philosophy of mind, especially that of Fodor and Searle. If Dreyfus is right, Husserl himself is the author of a proto-Fodorian theory of mental representations, and the tasks he conceived for transcendental phenomenology anticipate modern-day research projects in artificial intelligence and cognitive science. But Dreyfus is a critic of such efforts: indeed, he believes that Heidegger's reasons for rejecting the very possibility of transcendental phenomenology are basically right. Thus, his ultimate goal in comparing Husserl with "modern mentalists" such as Fodor is to show that both can be tarred with the same brush.

In this paper I shall be reexamining these comparisons from a standpoint that is more sympathetic toward Husserl and that attempts to be more neutral toward contemporary "representational" theories of mind. I have discussed Searle's views in relation to Husserl's elsewhere,[2] and so my focus here will be on Fodor and Husserl. As a contributor to the Dreyfus anthology and an advocate of the general line of Husserl interpretation represented in it, I am interested in dissociating that interpretation from Dreyfus's strong *computationalist* reading of Husserl.[3] However, I agree with Dreyfus that there are some remarkable points of agreement between Husserl and contemporary representationalists; my strategy will be first to push these as far as I plausibly can (or perhaps even a bit further in some instances) and only then to draw out the points of disagreement. By doing so, I hope not only to sharpen these points of agreement and disagreement but also to show where and how Husserl's views on meaning and intentionality would suggest modifications in the representational approach to an understanding of mind.

J-C. Smith (ed.), Historical Foundations of Cognitive Science, 211–233.
Reprinted from *Topoi* **5** (1986), pp. 101-113, with permission of Kluwer Academic Publishers.

Although I shall disagree with Dreyfus's characterization of Husserl as an advocate of a formalist or computationalist version of cognitivism, then, I am also concerned to show that Husserl and contemporary cognitivists share much common ground. In particular, I shall argue, Fodor and Husserl share a methodological principle that marks them both as opponents of "naturalistic" psychology, and Fodor seeks an understanding of the nature of mind that shares some of the goals of Husserl's "transcendental" phenomenology. Furthermore (on the interpretation I favor, at any rate), Husserl's noematic *Sinne* can be seen—up to a point—as a version of what Fodor calls "mental representations," having both formal (or "syntactic") and representational (or "semantic") properties and so forming a kind of "language of thought." Nonetheless, I shall argue, Husserl differs in important ways from Fodor and other contemporary representationalists on each of these points. These differences culminate in an importantly different conception of the intentional, or representational, character of mind and the role of meaning in our mental life.

1. Methodological Solipsism and Phenomenological Epoché

In his much discussed article Fodor (1980) endorses a thesis Putnam first called "methodological solipsism." As described by Putnam, methodological solipsism is "the assumption that no psychological state, properly so-called, presupposes the existence of any individual other than the subject to whom that state is ascribed."[4] Fodor characterizes it, somewhat more broadly, as the "Cartesian" view that "there is an important sense in which how the world is makes no difference to one's mental states."[5] Although many of our mental states are intentional and so stand for or represent things as being external to the mind, these mental states themselves—on this assumption—have a kind of intrinsic character of their own, which is just as it is even if there actually exists no mind-independent world at all. And if that is so, then a theory of mind *per se*—one designed to effect an understanding of this intrinsic character of mental states—ought to be one that even a consistent solipsist could accept. The point is not to affirm solipsism, of course, but only to proceed as though it were true, so that the resulting account of mind will presuppose nothing about the natural (especially causal) relations between the mind and its actual environment, or anything else about the "natural" setting in which minds are embedded.

Now, Husserl's methodology, which he calls "phenomenological reduction," takes its departure from this very thesis about the independence of mind from "natural" reality. "...*No real being*," he says, "*is essential for the being of consciousness itself*."[6] Hence, Husserl's version of methodological solipsism: "...Let us imagine...the whole of nature, physical nature above all, 'annihilated'....My consciousness, however much its constituent experiences would be changed, would remain an absolute stream of experience with its own essence."[7] Indeed, Husserl thinks, we each have a kind of first-person knowledge of the intrinsic features essential to mind (or "consciousness," as he prefers to say) that is independent of the truth or falsity of our beliefs about the world of nature. And so he thinks that a properly philosophical (or phenomenological) account of mind should be consistent with what he calls "epoché" or "bracketing"—i.e., that it should appeal only to the internal features of mind that we know after an epoché, a suspension, of all our beliefs about extra-mental reality.

The fact that methodological solipsism, or epoché, is so controversial, with decriers ranging all the way from Heidegger to Wittgenstein to Skinner to Putnam to Dreyfus, makes its endorsement by both Fodor and Husserl a significant point of agreement. However, it should be noted that even its contemporary proponents disagree about just where it leads, and Fodor and Husserl endorse it for rather different reasons.

Fodor is interested in "mental causation," the causal role that mental states play in behavior. And as he observes, this role often seems more dependent on how the world is represented to us in our mental states than on how the world actually *is*. For example, Oedipus's desire for Jocasta produced radically different kinds of behavior, first courtship and later self-directed violence. Why? Not because of any significant change in Jocasta—she was his mother all along—but because of a change in how Jocasta was represented to Oedipus in his mental states. This reason for endorsing methodological solipsism is further reinforced by Fodor's commitment to a computational account of mental processes and mental causation. "Computations" are operations on formal, or syntactic, elements *internal* to a system, and so these operations and the behaviors they produce are independent of any relationship those elements bear to the rest of the world. Accordingly, he notes, computationalism has no chance of being a true theory of mind unless the assumption embodied in methodological solipsism is true.

Husserl emphasizes two different considerations. First, the representational character, or intentionality, of mental states itself displays a certain independence from the reality of what is represented. Thus, a mental state may represent or be "directed toward" an object or state of affairs that does not actually exist at all; and, where what is represented *does* exist, the properties it is represented as having need not coincide with those it actually has. There is a crucial difference between Husserl and the computationalists here, we shall see. Husserl's other main consideration is epistemological: what we know about the representation of reality in our mental states is epistemologically prior to what we know about the nature of reality itself, since we have no access to reality except via our mental representations of it. Thus, Husserl thinks, a philosophical understanding of the foundations of our beliefs about natural reality must ultimately derive from a study of mental representation, and so that study itself cannot, on pain of circularity, be made dependent on the truth of those beliefs. This view leads Husserl to a more radical version of methodological solipsism than that described by contemporary representationalists and results in his "transcendental" version of phenomenology.

2. Functionalism, Computationalism, and Transcendental Phenomenology

I intend to pass over some large differences between Fodor and Husserl, but I do not mean to suggest that those differences are trivial. For example, Husserl believes that epoché, the suspension of our naturalistic beliefs, can almost immediately deliver up the data for proper philosophizing if it is properly carried through and followed by a special kind of introspection, or "reflection," on the contents of one's own consciousness. He takes this phenomenological reflection to be indubitably reliable, and the pronouncements issuing from it are not supposed to be mere speculative or inductive generalizations but necessary or "eidetic" truths about consciousness. Claims such as these mark radical differences between the methods Husserl characterizes as uniquely phenomenological and those employed by contemporary cognitivists.

A second difference is perhaps less radical than it first appears, though. Methodological solipsism, as Putnam described it, assumes the existence of no individual except "the subject" of the mental states in question. But what is this "subject"? Husserl characterizes transcendental phenomenology as the study of "transcendentally purified" or "absolute" experiences of the "transcendental ego," as opposed to the "real" or "empirical" experiences of the "psychological" or "empirical ego." Such distinctions suggest a heavy dose of metaphysics that Fodor and many other contemporary philosophers would be most loathe to swallow. Appearances notwithstanding, however, I want to argue that there is a major point of agreement between Fodor (and contemporary representationalists generally) and Husserl here.

The point of agreement is this: neither Fodor nor Husserl—neither cognitive science nor transcendental phenomenology—claims to offer a *naturalistic* theory about how mental processing actually takes place in human minds or brains. Rather, the goal of each is to find abstract general analyses of what is involved in various kinds of mental activities, analyses that apply with equal validity to any sort of entity capable of that kind of mental activity, no matter what its actual physical make-up and no matter what physical processes actually enable it so to perform.

For Fodor and proponents of artificial intelligence this point should be readily apparent. Their claim is not that human minds or brains are *physically* like inorganic computers or that the processes in which human thought is carried out are physically similar to those involved in computer processing; rather, they claim that the same "play-by-play accounts" (as Cummins calls them[8]) are descriptive of both certain mental capacities of humans and certain information processing capacities of computers. Research in artificial intelligence is concerned with finding these "play-by-play accounts," articulated in flow charts or computer programs, and it deals with these abstract objects rather than with the specific physical make-up of the hardware that may "instantiate" them. As such, these research efforts exploit the ontological neutrality characteristic of functionalist theories of mind. What is essential to mentality, functionalism says, is not the kind of substance that is capable of having mental states, but certain sorts of logical or structural (standard functionalism says causal) relationships of mental states to others and to sensory "inputs" and behavioral "outputs." As Fodor says:

> Functionalism, which seeks to provide a philosophical account of this level of abstraction, recognizes the possibility that systems as diverse as human beings, calculating machines, and disembodied spirits could all have mental states. In the functionalist view the psychology of a system depends not on the stuff it is made of (living cells, metal, or spiritual energy) but on how the stuff is put together.[9]

For this reason, functionalists are widely given credit for having made a major advance over both behaviorism and physicalism as well as dualism. But Husserl explicitly articulates such an "ontologically neutral" approach to the understanding of mind that predates functionalism by a half-century. He sees that a consistent anti-naturalism in fact requires it, for naturalism includes not only beliefs about individuals "other than the subject" but also beliefs about the subject herself, insofar as subjects are psycho-physical natural organisms in causal contact with other things and occupying the very same world of nature as they. He accordingly urges that the method of epoché, if rigorously applied, must yield an account of mind that is independent of the truth or falsity of *all* our naturalistic beliefs, including these beliefs about the actual psychological or physical nature of human subjects themselves. Thus, with the phenomenologist's imagined "annihilation" of nature, Husserl says in a passage I earlier quoted elliptically, "there would be no more animate organisms and therefore no more human beings. I as a human being would be no more....But my consciousness...would remain an absolute stream of experience with its own essence."[10] Husserl's phenomenological descriptions of this remaining "consciousness" and its "absolute" experiences are therefore not intended as naturalistic accounts of the "empirical ego," the ego as naturally embodied in us human beings, or of its experiences as "real" psychological or physical processes. Rather, they are intended as distinctively philosophical accounts of "transcendental" features of mind: transcendental inasmuch as those features constitute mentality itself (at least of the sort we humans have), no matter how they are in fact actually realized in us or in whatever other beings they are. It is the subject of experience so transcendentally described that Husserl calls the "transcendental ego," and its mental states or experiences understood at this level of abstraction constitute what he terms "pure" or "absolute" experience.

Thus, as Smith and I have argued,[11] Husserl's doctrine of the transcendental ego and its pure experiences is primarily a methodological or an epistemological, rather than a metaphysical, doctrine. It is not the view that there is a second ego standing behind and manipulating the activities of the empirical ego; rather, it is the doctrine that there is an *ontologically neutral* level of description of the ego and its activities that is *methodologically* independent of any natural description of what the ego and its experiences are in fact like.[12] Like the functionalists and the computationalists, then, Husserl seeks abstract accounts that would capture what is common to various mental capacities, no matter how different in their natural make-up the entities having these capacities may be. In a passage written in 1925 (an especially telling passage, because Husserl is here explaining with approval the aims of his *Logical Investigations*, written 25 years earlier), he explicitly says just this:

> ...Whenever something like numbers, mathematical manifolds, propositions, theories, etc....come...to be objects of consciousness in subjective experiences, the requisite experiences must have their essentially necessary, everywhere identical structure. In other words whether we take us men as thinking subjects, or whether we imagine angels or devils or gods, etc., any sort of beings that count, compute, do mathematics—the counting, mathematising, internal doing...is, if the logical-mathematical is to result from it, in a priori necessity everywhere essentially the same....A realm of unconditionally necessary and universal truths [describes] the...psychic life of any subject at all insofar as it is to be thought, purely ideally, as a subject that knows in itself the mathematical.
>
> The same holds [not only for mathematics but] for all investigations of psychic correlations relating to objects of every region and category....Precisely thereby a novel idea of psychology is presented....Instead of the fact of human subjects of this earth and world, this psychology deals...with ideal essences of any mathematising and, more generally, of any knowing subjectivity at all.[13]

Terminology and the concern with unconditional necessity aside, one can see that Husserl's emphasis here is not on ontological embodiment but on an "everywhere identical structure" that he takes to be exemplified in experiences of the same kind. And it is just this emphasis that I claim he shares with the functionalists.

Of course, Husserl cannot himself *be* a functionalist of the standard "causal-role" sort; i.e., he cannot explicate mental states in terms of their *causal* relations to one another and to the world, for causality (in any naturalistic sense) is "bracketed" by phenomenological epoché. But the computationalist version of functionalism also abstracts away from causal relations among mental states, turning instead to certain inferential relations among mental representations as a way of accounting for these causal relationships. However, we should not be too quick to assume that Husserl must, therefore, be a computationalist: that depends on whether these "everywhere identical structures" are to be articulated in computational terms. And on that issue, Husserl's remarks just a few pages later ought at least to give us pause:

> Since we have all formed the concept of *a priori* science in mathematics,...we tend understandably to regard any *a priori* science at all as something like a mathematics; a priori psychology, therefore, as a mathematics of the mind. But here we must be on our guard....By no means does this type pertain to every kind of *a priori*.
>
> The psychic province...is of a completely different essential type.... By no means is the entire science of the type of a mathematics.[14]

3. Mental Representations and Noematic Meanings

So far I have argued that Husserl and Fodor are in basic agreement on two key points: that mental states have an intrinsic character of their own that can be explicated without reference to extra-mental things, and that what is essentially mental in this intrinsic character is properly explicated at an ontologically neutral level of abstraction. Fodor advocates computationalism as a theory compatible with these two claims, and he characterizes it as but one special version of a more general theory that he calls the "Representational Theory of Mind." By the "Representational Theory of Mind" (abbreviated 'RTM'), Fodor means a theory that attempts to explain the important features of mind by appeal to a system of internal "mental representations." Whether Husserl is sympathetic to computationalism or not, he shares a great deal with Fodor if he, too, is an advocate of RTM.

According to RTM (see Fodor, 1980), each mental state is essentially a relation to a mental representation, which (purportedly) stands for or represents some, usually extra-mental, thing or state-of-affairs. Representational relations between the mind and the extra-mental world are therefore "mediated" relations: each is a composition of the relation between the mental state and its associated mental representation and the relation (if any) between that mental representation and an appropriate extra-mental item. But RTM also holds that there are relations among mental representations themselves. Mental processes, naturalistically speaking, are causal relations among mental states; however, according to RTM, these causal relations are mirrored in the relations that obtain among the mental representations corresponding to these causally related mental states. Thus, at the "transcendental" level of abstraction, mental processes can be explicated in terms of the relations that obtain among their associated mental representations. (And if these relations are computational, they can be captured in appropriately devised computer programs; hence, artificial intelligence.) For adherents of RTM, methodological solipsism is an invitation to ignore mind-to-world relations and to focus instead on this system of mental representations and the relations among them.

Now, there is certainly at least a structural similarity between this description of RTM and Husserl's approach to the intentionality of mind. According to Husserl, each mental state is essentially a relation to an entity he calls a "noema," one component of which—called the "noematic *Sinn*—(purportedly) stands for or represents a thing or state of affairs, usually something extra-mental.[15] Intentionality, or representation, is again a "mediated" affair: a mental state represents an object only "via" its noema. (N.B.: This is not to say that noemata are the immediate *objects* toward which mental states are directed. Husserl insists that the represented or "intended" objects of mental states are ordinary sorts of entities; the noema is introduced to explain how mental states come to represent these ordinary things.) Husserl also holds that mental processes can be explicated as relations among noemata themselves: indeed, that explication is precisely the task of transcendental phenomenology. Phenomenological epoché, for Husserl, is then an invitation to ignore the *de facto* relations of mind to the world and to focus instead on these noemata and the relations among them.

In fact, Husserl's views can be pushed even closer to Fodor's than this. Fodor characterizes the system of mental representations for an individual person as a "language of thought," and—with some important differences—this is also an apt description of Husserl's conception of noemata. Fodor believes that mental representations have both "syntactic" and "semantic" properties, in the same sense that the elements of a natural language do. Sentences, for example, differ from one another in "shape" as the individual words they comprise are different and/or differently arranged: thus,

'Marvin is melancholy' is syntactically different from 'Marvin is happy'. Similarly, Fodor holds, the belief that Marvin is melancholy and the belief that Marvin is happy are relations to mental representations that differ in syntactic structure. But the expressions in a natural language also have "semantic" properties, paradigmatically meaning, reference, and truth-value. Fodor conceives mental representations as having these same kinds of properties. Thus, to believe that Marvin is melancholy is to be related to a mental representation that "stands for" Marvin, that "represents" him "as" melancholy, and that is "true" or "veridical" if and only if Marvin is melancholy. The relations among mental representations that explicate mental processes, Fodor therefore holds, are the same sort of syntactic and semantic relations that obtain between sentences. Finally, Fodor believes, the syntactic and semantic properties of natural languages are inherited from their more fundamental counterparts in systems of mental representations.[16] Natural language is in this sense the "expression of thought": the translation of the medium of thought, mental representations, into a public medium of linguistic communication.

Fodor's mental representations, then, are mental symbols, complex sentence-like combinations of simpler word-like elements, having meaning and truth-value (and presumably tokened in the brain, in a way that some as yet untold naturalistic story will eventually explain). Now, Husserl's noematic *Sinne* are not mental *symbols* in this sense. That is, they are not word-like or sentence-like entities that *have* meanings; rather, noematic *Sinne are* meanings (hence, "*Sinne*"). But despite this important point, to which I shall return, noematic *Sinne* are like Fodorian mental representations in several significant respects. These similarities derive from the fact that noemata, too, are conceived in analogy with language.

Just as speech, for example, consists in temporal sequences of meaningful sounds, so thinking (or any mental process), on Husserl's view, consists in temporal sequences of meaningful mental states or events. Indeed, Husserl thinks, the meanings we express in speech or writing are essentially the same entities—noematic *Sinne*—that make meaningful mental episodes possible: the purpose of language is to express what is "in our minds," so that others may represent to themselves the same object we have in mind and in the same way; and in order for that to take place, he thinks, the meanings we express must be the very same noematic meanings that determine the representational character of these "thoughts."[17] Hence, while Husserl also holds that "language is the expression of thought," his version of this thesis differs from Fodor's. On Fodor's version, one might say, we think in mental "words" that get *translated* into a public language when we speak or write, while on Husserl's version we think in "meanings" that get *expressed* in a public language.

According to Husserl, then, the meanings of expressions in a natural language are derivative from their more fundamental counterparts in systems of noemata. Given this view of the relation between noemata and linguistic meanings, it is not surprising that Husserl thinks of noemata as having syntactic and semantic properties. Frege, for example, considers *linguistic* meanings to be syntactically structured abstract entities: just as a sentence consists of syntactically distinct parts put together in syntactically permissible ways, he thinks, so the propositions expressed by that sentence consists of correspondingly distinct meanings put together in correspondingly permissible ways. Husserl similarly thinks of noematic Sinne as structured abstract entities, mirroring the syntactic structures of the linguistic expressions that would express them. To think "this is white," for example, is to be related to a noematic *Sinn* structured into two distinct meaning-components: an "X"-component (as he calls it) expressed by 'this', which represents the object being thought about, and a "predicate-sense" expressed by 'is white', which represents the property predicated of the object as represented.[18] The semantic, especially referential, properties of noematic *Sinne* are similarly reminiscent of Frege: like Frege, Husserl holds that *meaning determines reference*. Thus, he thinks, the representational or intentional properties of a mental state are determined by its noematic Sinn. I will be discussing this point in the next section.

There are good grounds, then, for construing Husserl's noematic *Sinne* as a version of what Fodor calls "mental representations" and so construing Husserl as an early advocate of the Representational Theory of Mind. Noematic *Sinne* constitute for Husserl a "medium" in which mental processes take place; this medium is syntactically and semantically characterizable and so fundamentally language-like; mental states represent extra-mental things by virtue of how these noematic *Sinne* relate to the extra-mental world; and mental processes can be understood, at an ontologically neutral level of abstraction, in terms of relations among these noematic *Sinne* independent of the actual relations that obtain between mental states and the extra-mental world. Mental representations, at least *prima facie*, play these same roles for Fodor.

On the other hand, Husserl's view of noemata as *meanings* rather than as symbols *having* meaning may be enough to show that they cannot be characterized as mental representations in any legitimate Fodorian sense. If so, Husserl is not an advocate of RTM at all. However, I think such an easy dismissal would only hide deeper differences between Husserl and Fodor. After all, it seems trivially easy to modify Husserl's theory into a genuine version of RTM: simply postulate a system of truly

Fodorian mental representations and let noematic *Sinne* be the meanings of these mental *representations* rather than of mental *states* themselves. (The result would conform to what Bach characterizes as "conceptual" rather than exclusively "formal" methodological solipsism and would be a version of what Stich calls "strong RTM.")[19] What I hope to show is that the resulting version of RTM would still—so long as the meanings of mental representations are noematic *Sinne* as Husserl conceived them—be radically different from contemporary, especially computationalist, versions. Part of the reason why this is so comes out in an argument Husserl himself gives against the Fodorian-like view that symbols or "signs" mediate the relations between mental states and their objects. A symbol functions as a symbol, Husserl notes, only by virtue of being itself the object of a mental state, in which it is apprehended (interpreted, represented) *as representing* something other than itself. Thus, that apprehension would have to be via a second symbol that represents the first, and so on *ad infinitum*. The "sign-theory," Husserl says, fails to explain mental representation, and for the very same reason that the traditional "image-theory" of ideas cannot.[20] The comparison is interesting because Fodor also, for a different reason, rejects the "image-theory" in favor of a computationalist version of the "sign-theory." What saves this latter theory from Husserl's objection is that computationalism is precisely designed to show how mental symbols can do their work without functioning *as symbols*, i.e., independently of their semantic or representational properties. Husserl's rejection of mental symbols in favor of noematic Sinne is based on the very opposite view: the view that the *meanings* of these symbols, not just the symbols themselves or their "syntactic" features, would have to do the work of explaining mental representation. What is seriously at issue between Husserl and computationalists is the notion of meaning itself and its role in mental representation.

4. Meaning, Intentionality, and Mental Representation

The Representational Theory of Mind, as Fodor characterizes it, is but a framework—albeit a rather specific and controversial one—for discussing some of the traditional problems about mind. In this section I want to discuss the problem of mental representation, or intentionality, itself within this framework. Dreyfus assumes that the contemporary notion of mental representation is just an updated version of the Husserlian notion of intentionality, but we shall see that this identification is by no means self-evident: the problem of intentionality as Husserl conceived it and the problem of mental representation currently so-called seem to be radically different problems.

For this discussion, let us assume that noematic *Sinne* can, despite the qualifications we have already noted, be characterized as mental representations and that, for both Fodor and Husserl, the problem of mental representation is a matter of the "semantics" of mental representations. The problem is then subject to various possible solutions: there are numerous approaches to the semantics of *linguistic* representation and the number of approaches to the semantics of *mental* representation is surely no smaller.

In Fodor (1980), Fodor proposes an account of mental representation modeled on the so-called "causal theory" of linguistic reference. According to that theory, the fundamental relation between language and the world is causal: for each (actually referring) name, there are complicated causal chains connecting its various occasions of use to some unique item in the world, that item being thereby the "referent" of the name; other forms of reference (e.g., the reference of definite descriptions) are derivative from such causal forms. Similarly, Fodor sees the representational or "referential" properties of mental representations as *causal* relations: "what makes my thought about Robin Roberts [for example] a thought *about Robin Roberts*," he says, "is some causal connection between the two of us."[21] That is, a mental state is "about" Robin Roberts just in case it is related to a mental representation that itself stands in an appropriate causal relation to Robin Roberts himself; this causal relation is the "semantic" relation—the relation of representation or intentionality—that relates the mental representation to that which it represents. Fodor gives few details about how this theory is supposed to work, and (his more esoteric reasons aside) it is easy to see why. Since world-to-mind causality cannot be explicated independently of how *the world is*, the account itself is incompatible with a thorough-going endorsement of methodological solipsism.

One would suppose this result to show either that methodological solipsism cannot provide an adequate theory of mind (since it cannot account for intentionality) or that the causal account of intentionality is incorrect (since it is incompatible with methodological solipsism), but Fodor draws a different conclusion: because mental representation, so understood, falls outside the realm of what can be explicated by the methodological solipsist, mental representation itself is not a strictly "mental" feature of mental states. If Fodor is right, mental *representation* is no proper concern of the *Representational* Theory of Mind! This rather odd result for Fodor the representationalist and methodological solipsist pays off for Fodor the computationalist, though. Computing machines can make no use of the representational properties of the symbols they employ, but—on this view—they are not thereby deficient in anything essentially "mental."

This result, if unmodified, contrasts sharply with Husserl's views on intentionality. The methodological solipsist, like the practitioner of phenomenological "epoché," "brackets"—makes no use of—anything extra-mental: the world of nature, our minds conceived as natural entities, and the causal relations between them are "bracketed" by this methodology. For Fodor this means that mental representation itself, being a causal relation, is included in these "bracketed" items. But Husserl, throughout his entire career, consistently maintained that intentionality is the *primary* feature his methodology of epochè is designed to explicate. To take but one, quite pointed, example:

> If I perceive a house,...a relationship of consciousness is contained in the perceptual experience itself, and indeed a relation to the house perceived in it itself....Of course there can be no talk of external-internal psychophysical causality if the house is a mere hallucination. But it is clear that the momentary experiencing is in itself not only a subjective experiencing but precisely a perceiving *of* this house. Therefore, descriptively, the object-relation belongs to the experiencing, whether the object actually exists or not.[22]

For Husserl, then, the intentionality of a mental state is a feature inherent in the mental state itself, independent of its *de facto* (especially its causal) relationships to extra-mental things or states-of-affairs. Let us be careful to note, however, that it is the "object-relation," and not the *object*, that "belongs to the experiencing."

In fact, though, Fodor's and Husserl's positions are not quite contradictory, due to an ambiguity in the notion of mental representation, or intentionality, itself. Smith has employed a useful distinction between the intentional, or representational, *relation* achieved in a mental state and the intentional, or representational, *character* of the mental state itself.[23] Consider the following case. I peer under my bed, spot a coiled rope, scream "Snake!", and flee the room. Just what did I see? What was my visual representation a representation "of"? In one sense, of course, what I "saw" was a coiled rope, and in that sense my experience was "of" or "about" the rope: the rope was what my visual experience was actually related to, via mental representation. In Smith's terminology, then, my experience was *representationally related* to a rope. And Fodor's causal story captures at least part of what is involved in this relation: the rope was related to my mental representation by virtue of being the distal stimulus that gave rise to it. In another sense, though, I "saw" a snake: I am generally disposed to fear snakes, not ropes, and I feared this object only because I took it to be a snake. Phenomenologically speaking, my visual experience had the intentional, or representational, *character* of being "of" or "about" a snake. And for that sense of representation

Husserl's solipsistic theory sounds right: my mental state had that representational character even though what it was actually related to was a rope, and it could have had that same character even if there had been no appropriate distal stimulus at all. Intentional, or representational, *relations,* then, concern the way mental states and mental representations actually "hook up" with the world; and of course Fodor is right in thinking that those relations (whether simply causal or not) are not independent of how the world is. But if Husserl is right, mental states and mental representations themselves have an intrinsic representational *character*, which makes them *as though* actually related to extra-mental things whether they are so or not.

The problem of intentionality for Husserl, then, is not to explain how mental states actually relate to the world but to explain how they have the phenomenological or "internal" *character* of relating to anything at all. Husserl "solves" this problem by appealing to a "semantics of reference" quite different from the causal account. A mental state is intentional in character by virtue of its relation to a noematic *Sinn*. How so? Because noematic *Sinne* are *meanings*, and, Husserl apparently thinks, it is simply an intrinsic and irreducible (though not completely unanalyzable) property of meanings to represent. Husserl in fact holds a strong version of the familiar Fregean thesis that meaning determines reference. Speaking of linguistic meaning (which he calls '*Bedeutung*'), he says:

> Reference to the object is constituted in the meaning [*Bedeutung*]. To use an expression meaningfully [*mit Sinn*], and to refer expressively to an object (to form a presentation of an object), are thus one and the same. It makes no difference whether the object exists or is fictitious or even impossible....[24]

Thus, even in the absence of any actual referent, Husserl apparently thinks, the meaning of an expression not only makes it meaningful but gives it a referential character as well; and he takes just the same view of noematic *Sinne* and intentionality: "*The phenomenological problem of the relation of consciousness to an objectivity* has above all its noematic side. The noema itself has an objective relation, specifically through its particular '*Sinn*'."[25] In the final analysis, then, Husserl says of noematic *Sinne* essentially what Searle says of mental states themselves: they have "intrinsic," as opposed to "derived," intentionality (but, n.b., Searle's own account of intentionality explicitly rejects the invocation of meanings, especially as abstract entities). In Searle's view, mental states have "conditions of satisfaction" and so are intentional, whether any states of affairs actually "satisfy" them or not, simply because that is a fundamental property of the kind of entities that mental states are.[26] Meanings or noematic *Sinne*, similarly, are conceived by Husserl as intentional, not because of any relations they bear to anything *else* (e.g., not because they are "interpreted" by someone or caused in some particular way) but simply because they are a sort of entity whose very nature is to *be* representational. On this view, the noematic *Sinn* itself will not, of course, be the sole determinant of which object a mental state is actually related to (causally or otherwise), but its intentional character will determine which object it must be related to in order to be "satisfied."

If mental states really do have such an internal or phenomenological intentional character, then modern mentalists cannot simply give the problem of intentionality over to the extra-mental "natural" sciences. But it is also difficult to see how *this* problem of intentionality could be solved using only the functionalist or computationalist resources to which contemporary representationalists usually restrict themselves. Indeed, the problem here is not unlike the widely recognized one of accounting for phenomenal qualities, such as pain, in functional terms. It seems obvious to many that such phenomenal qualities are *primitive* features of mental states and so cannot be reduced to causal roles, computations, or to anything else. And, although Husserl's view is apparently much less obvious to most, he believes essentially the same is true of intentional character.

What I should *like* to do now is contrast Husserl's view of intrinsic intentional character with actual representationalist, especially computationalist, accounts of representational character. Unfortunately, however, contemporary representationalists seem not to consider the problem of intentional character, at least not in any direct way. Indeed, I suspect that computationalists are more wont to deny the existence of intentional character than they explicitly admit. For one thing, since it is very counter-intuitive to suppose that machine states or the symbols in computer programs have intrinsic intentional character, a deep commitment to the computer model of mind would surely tempt one to deny that mental states or mental representations have it either. (Thus the Churchlands argue, though not on behalf of computationalism, that "our own mental states are just as innocent of 'intrinsic intentionality' as are the states of any machine simulation."[27] But few computationalists have been so candid.) In the second place, most contemporary representationalists have been deeply impressed by Putnam's famous "Twin-Earth" arguments.[28] Those arguments purport to refute the Fregean thesis that meaning determines reference and to show, more generally and contra Husserl, that nothing intrinsic to mental states can suffice to determine which object a mental state represents. Accepting that conclusion tempts one to decide that intentionality or mental representation is *entirely* an "external" matter and that the problem of "internal" representational or intentional character, which Husserl's appeal to noematic *Sinne* is supposed to solve, has simply disappeared. Clearly Putnam's arguments, and others like them, raise important issues, but let me merely suggest that these issues have not been conclusively resolved to Husserl's detriment. To cite but three examples, Bach, Searle, and Smith have offered independent accounts of how "indexical" mental contents can determine (or in Bach's case, partially determine) the object of mental states in Putnam-like cases.[29] Furthermore, however it is to be explained, there *is* a "mental side" to intentionality that is as much a "phenomenological fact" of our mental life as are consciousness and self-awareness; good philosophy demands that there be limits on the degree to which theory can do violence to these facts. In the next section, accordingly, I want to consider whether a computationalist or formalist theory of mind might yet be rendered compatible with intentionality as Husserl conceives it.

5. Was Husserl a Formalist?

Husserl never underestimated the richness and complexity of our mental life; hence he characterized transcendental phenomenology, his attempt to explicate this mental life, as "an infinite task." But Husserl also never wavered from his conviction that this richness and complexity is, at bottom, "rationally" understandable. Indeed, he thought, the very concept of consciousness as intentional, meaningful experience requires the imposition of some sort of rationale on what would otherwise be but an inchoate welter of meaningless sensations. It is the "noema" of a mental state or experience that places it within the context of such a rationale, by relating it, in rule-governed ways, to what Husserl calls a "horizon" of past experiences and future possible experiences of the same object or state-of-affairs.

For example (considerably simplified), suppose I see a particular object as a tree. The noematic *Sinn* of this experience then includes the predicate-sense "tree," and it is by virtue of this sense that I perceive the object as a tree rather than something else. But, Husserl holds, this sense does not do its work of characterizing or prescribing the object in isolation from the rest of my mental repertoire. I believe that trees come in different varieties, that trees are physical objects and so are three-dimensional, and so on. Within the context of such beliefs, the sense "tree" foretokens or "predelineates" a range of further possible experiences in which the object before me would be characterized in further possible ways: as an oak or an elm, for example; as black or brown on the side now hidden from me; and so on. In this way, Husserl says, the *Sinn* relates the present experience and its object to an indeterminate or non-specific, and open-ended, horizon of possible experiences. But despite the indeterminacy of this horizon, Husserl believes, it has a rational, coherent structure: the *Sinn* of the present experience, in conjunction with the *Sinne* of relevant background beliefs, limit in rule-governed ways the kinds of further experiences that can belong to it. To understand *this* experience and its intentionality is ultimately to understand how its *Sinn* is related to the *Sinne* of these background beliefs and to the *Sinne* of its horizon of possible further experiences.[30]

To explicate an experience noematically or phenomenologically is, then, to uncover these relations among noematic *Sinne* and the rules that describe them, and so to unfold its inherent rationale. If Husserl's belief—that for every conceivable human experience there is such an internal rationale, independent of that experience's actual relations to the natural world–is a "cognitivist" belief, then without doubt Husserl *is* a cognitivist.

But Dreyfus is not content to characterize Husserl as a cognitivist in the general sense I have just agreed to; he thinks Husserl was a *formalist*, and so at least an incipient *computationalist*:

> Whether in fact Husserl held what Fodor calls the computational theory of mind—i.e., whether according to Husserl...the predicate-senses [in the noematic *Sinn*] do their job of representing objects...and of unifying diverse experiences...strictly on the basis of their shapes (i.e., as a syntactic system independent of any interpretation)—cannot be so easily determined. There is, however, considerable evidence...that Husserl thought of the noemata as complex *formal* structures [and] there is no evidence which suggests that he ever thought of the rules he was concerned with as *semantic*.[31]

Dreyfus raises two issues here: do noematic *Sinne* represent objects and unify experiences strictly "as a syntactic system," and are the rules that describe mental states or mental processes purely non-"semantic"? Since I think, contra Dreyfus, that the first issue is "easily determined," let me turn to it first.

The advocate of computationalism who does not deny such notions as meaning and intentional character is free to try to explain them in terms of something more congenial to the formalist program. What Fodor calls "functional-role semantics,"[32] for example, attempts to explain at least some of the "semantic" properties of mental states in terms of their causal relations to other mental states (and to causal inputs and outputs). Although a critic of this effort, Fodor suggests that computationalism could make use of it by recognizing an isomorphism between the *causal* network of mental *states* and an appropriate network of purely *formal* or syntactic relations among mental *representations*. Any "functionally" explicable semantic properties of mental states or mental representations would then be, if not reducible to, at least replaceable by formal relations among mental representations. Accordingly, if intentional character were such a semantic property of mental representations, it too would be effectively explained in strictly syntactic formalist terms: to understand the intentional character of a mental representation (or a noematic *Sinn*, if this were Husserl's view) would just be to understand its formal or syntactic relations to other mental representations (or noematic *Sinne*). Dreyfus seems to think it is at least debatable that Husserl held some view like this.

I have already agreed with Dreyfus that noematic *Sinne* have syntactic properties and so stand in certain formal relations to one another. And, as we just saw, *Sinne* "do their job of representing objects" only within the context of a network of other *Sinne*, the *Sinne* of mental states comprised by the horizon of the given experience. But that does not mean that their representational properties are *reducible* to the formal relations among the *Sinne* in this network. For one thing, Husserl always characterizes this network in terms of "semantic" relations among *Sinne*, i.e., in terms of their intentional character: he even *defines* the horizon as consisting of experiences *directed toward the same object*.[33] More importantly, since Husserl holds that intentional character is determined by meaning, this reductionist view of intentional character requires a most peculiar account of meaning. It combines the Husserl-Frege thesis that meaning determines reference or intentional character with the radically anti-Fregean view that meaning reduces to syntax. So far as I can tell, not even contemporary representationalists hold this mixture of views.[34] And I simply do not know of *any* passages in Husserl's writings that suggest he ever thought that meaning is in any way reducible to syntax. Not only are there powerful systematic considerations to the contrary; we have already seen what are only just a few of the many passages that argue for a quite different, Frege-like theory of meaning.[35] Furthermore, Husserl himself sometimes explicitly addresses the question of whether meaning and intentional character can be reduced to relations among merely formal elements or meaningless "contents," always arguing that they *cannot*. For example, he rejects the "sensationalist" view of consciousness, the view that consciousness consists of nothing but sensations and complex relations among them. And on what grounds? On the grounds that sensations are "meaningless [*sinnlos*] in themselves" and so "could give forth no 'meaning' [*Sinn'*], however they might be aggregated."[36] It is hard to believe that Husserl could offer *this* argument against sensationalism while *also* believing that the meaningfulness, and hence the intentional character, of noematic *Sinne* could be reduced to formal relations among them based solely on their shapes. And he in fact says, just a few pages later:

> ...Transcendental Phenomenology...must come to consider experiences, not as so much dead matter, as "complexes of content," which merely are but signify nothing, mean nothing, as elements and complex-structures, as classes and subclasses...; [it must] instead master the ***in principle unique set of problems*** that experiences *as intentional* offer, and offer *purely* ***through their eidetic essence as "consciousness-of."***[37]

Now, the fact that meaning and intentional character are not reducible to syntactic relations among formal structure also relates to the second issue Dreyfus raises, the issue of whether the "rules" that concern Husserl are "semantic." Indeed, insofar as "non-semantic" just means "formal," I find it hard to understand why Dreyfus thinks there is "no evidence" here. Husserl always *describes* these rules as rules for relating experiences on the basis of their *intentional character*, not on the basis of the "shapes" or the "formal structure" of their associated noemata. (For example, he says, each category of object "prescribes the rule for the way *an object* subordinate to it is to be brought to full determinacy with respect *to meaning* and *mode of givenness*.") And Dreyfus offers no comment on such remarks of Husserl's as this: "...Transcendental theories of constitution arise that, *as non-formal*, relate to any spatial things whatever...".[39] As a mathematician and logician Husserl was quite familiar with the notion of a "formal" or "syntactic" theory; yet, he held that even the laws of logic apply to the phenomenological description of experience in ways that are not purely "syntactic."[40] And he in fact goes to some lengths to distinguish phenomenology, as an "eidetic" (or *a priori*) discipline, from *formal* eidetic disciplines such as mathematics. Consider:

> Since the mathematical disciplines...represent the idea of a scientific eidetic, it is at first a remote thought that there could be other kinds of eidetic disciplines, non-mathematical, fundamentally different in their whole theoretical type than the familiar ones. Hence,...the attempt, immediately doomed to failure, to establish something like a mathematics of phenomena can mislead [one] into abandoning the very idea of a phenomenology. But that would be utterly wrong.[41]

> We start from the division of essences and essential sciences into material and formal. We can exclude the formal, and therewith the whole aggregate of formal mathematical, disciplines, since phenomenology obviously belongs to the material eidetic sciences.[42]

> Transcendental phenomenology...belongs...to *a fundamental class of eidetic sciences totally different* from that of the mathematical sciences.[43]

Just *why* is Husserl unwilling to consider phenomenology a "formal" science? There are probably many reasons, but let me suggest just one. I have been urging that a system of Husserlian mental representations would be one whose operations are carried out, not on their formal properties alone, but by virtue of their meaning and representational character as well. *Syntax*, on the other hand, is described by Husserl as dealing only with pure "forms" obtained by abstracting away from such meaningful "content."[44] Accordingly, even those kinds of operations that can be formalized and thus *described* syntactically—e.g., logical and mathematical operations—are not *carried out* syntactically in ordinary thought and experience. In a similar vein, Dretske (1985) has recently argued that even mathematical thinking, such as adding numbers, is not the same thing as manipulating formal symbols. The symbols being manipulated by a person who is adding must represent numbers, and so have meaning, *for that person*, Dretske urges, and she must manipulate them as she does at least partly because they mean what they do for her. Even if the same manipulations are performed, but as purely formal operations on symbols that mean nothing to the system performing them, the performance is at best a simulation—not a true instance—of adding. I see Husserl's views on the formality issue as very much like these.

Of course, to deny that human thought and experience are purely formal or computational does not entail opposition to research in artificial intelligence. The position I attribute to Husserl, and it is also mine, does assert that artificial intelligence is "artificial" precisely because it is only formal and so devoid of what is truly "mental." If that is so, then computationalism is false as a theory of *mind*, and so is what Searle denounces as "strong AI"—the view that computers, and humans as well, are minded simply by virtue of their ability to do certain kinds of syntactic manipulations.[45] But that still leaves open the possibility of "weak AI"—artificial intelligence as the project of *simulating* various cognitive mental capacities by constructing formal analogs of them. Husserl himself, in the midst of the discussion from which I have been quoting, seems to leave open this very sort of possibility:

> The pressing question is admittedly not answered thereby whether within the eidetic domain of [phenomenologically] reduced phenomena (either in the whole or in some part of it) there could not also be, *alongside* the descriptive, an idealizing procedure that substitutes pure and strict ideals for intuited data and that would even serve—as a counterpart to *descriptive* phenomenology—as the basic medium for a mathesis of experience.[46]

Dreyfus has shown that at least one major advance in artificial intelligence, Minsky's notion of "frames," turns on ideas that were first developed by Husserl—without the heuristic benefit of the computer.[47] And I suspect there is much more in Husserl's careful descriptions of experience that would help construct his anticipated "counterpart" to a science of the (real) mind. Nonetheless, phenomenology remained for

Husserl a *descriptive* discipline, descriptive of intrinsically *intentional* experiences, as they are experienced.[48]

California State University, Northridge

NOTES

1 Dreyfus (1982), p. 2.
2 See McIntyre (1984).
3 My views, in fact, are in basic agreement with those of Emmett (1983), although she seems to attribute Dreyfus's views to the contributors as well.
4 Putnam (1975), p. 220.
5 Fodor (1980), p. 64.
6 Husserl (1913), section 49, p. 115; my trans.
7 Husserl (1913), section 54, pp. 132-133; my trans.
8 Cummins (1983), p. 34.
9 Fodor (1981), p. 114.
10 Husserl (1913), section 54, pp. 132-133; my trans.
11 See Smith and McIntyre (1982), pp. 96-99.
12 See Husserl (1954), section 72, p. 264.
13 Husserl (1925), section 3(e), pp. 27-28; with trans. changes.
14 Husserl (1925), section 4, pp. 36-37.
15 Cf. Husserl (1913), esp. sections 87-91, 129-133. The interpretation of Husserl's notions of noema and noematic *Sinn* I appeal to here is developed in Smith and McIntyre (1982), esp. Chs. 3 and 4.
16 Fodor (1984), pp. 8-11.
17 See Husserl (1900), I, section 7; Husserl (1913), section 124; Husserl (1929), section 3. For development, and some qualifications, of this view see Smith and McIntyre (1982), pp. 170-187.
18 Cf. Husserl (1913), sections 130-131; Smith and McIntyre (1982), pp. 194-219.
19 See Bach (1982), pp. 123-127; Stich (1983), pp. 127-148.
20 Husserl (1913), section 43, pp. 98-99; section 90, pp. 224-225. Cf. Husserl (1900), I, section 23.
21 Fodor (1980), p. 300.
22 Husserl (1925), section 3(d), pp. 22-23; with trans. changes. Cf. Husserl (1900), V, section 25, p. 603; Husserl (1913), section 36, p. 80; Husserl (1931), section 14, pp. 32-33.
23 Cf. Smith (1984).
24 Husserl (1900), I, section 15, p. 293; with trans. changes.
25 Husserl (1913), section 128, p. 315; my trans.
26 Searle (1984).
27 Churchland and Churchland (1981), p. 140.
28 Putnam (1975), esp. pp. 223-227.
29 See Bach (1982); Searle (1983), Ch. 8; and Smith (1984).
30 See Smith and McIntyre (1982), Ch. 5.

[31] Dreyfus (1982), p. 10.
[32] Fodor (1985), pp. 85-88, 96-99.
[33] See Husserl (1931), sections 19, 20.
[34] McGinn (1982), for example, advocates an explication of one component of meaning in terms of "intra-individual causal" or "cognitive" role, but he denies that reference is determined by *this* component of meaning. However, since by reference McGinn means *de facto* reference *relations*, it is not clear what is supposed to happen to referential or intentional *character* on this theory of meaning.
[35] For example, the whole of Husserl (1900), I.
[36] Husserl (1913), section 86, p. 213; my trans. Cf. Husserl (1900), V, section 14, p. 565.
[37] Husserl (1913), section 86, pp. 214-215; my trans.
[38] Husserl (1913), section 142, p. 350; my trans. and my emphasis.
[39] Husserl (1931), section 21, p. 52; my emphasis.
[40] Cf. Husserl (1913), sections 59, 134.
[41] Husserl (1913), section 71, p. 164; my trans.
[42] Husserl (1913), section 72, p. 165; my trans.
[43] Husserl (1913), section 75, pp. 173-174; my trans.
[44] See Husserl (1913), section 134; cf. Husserl (1929).
[45] Searle (1984), pp. 28-41.
[46] Husserl (1913), section 75, p. 174; my trans.
[47] See Dreyfus (1982), pp. 17-19.
[48] My thanks to David Woodruff Smith and Frank McGuinness for their invaluable help with the issues addressed in this paper.

REFERENCES

Kent Bach (1982), "De Re Belief and Methodological Solipsism," in Andrew Woodfield (ed.), *Thought and Object: Essays on Intentionality* (Oxford: Clarendon Press), pp. 121-151.

P. M. Churchland and P. S. Churchland (1981), "Functionalism, Qualia, and Intentionality," *Philosophical Topics* **12**, pp. 121-145.

Robert C. Cummins (1983), *The Nature of Psychological Explanation* (Cambridge: MIT Press).

Fred Dretske (1985), "Machines and the Mental," *Proceedings and Addresses of the American Philosophical Association* **59**, pp. 23-33.

Hubert L. Dreyfus (ed.) (1982), *Husserl, Intentionality, and Cognitive Science* (Cambridge: MIT Press/Bradford Books).

Kathleen Emmett (1983), "Phenomenology and Cognitive Science: Should You Be Reading Husserl?", *Cognition and Brain Theory* **6**, pp. 509-516.

J. A. Fodor (1980), "Methodological Solipsism Considered as a Research Strategy in Cognitive Psychology," *The Behavioral and Brain Sciences* **3**, pp. 63-73; reprinted in Dreyfus (1982), pp. 277-303.

J. A. Fodor (1981), "The Mind-Body Problem," *Scientific American* **244**, pp. 114-123.

J. A. Fodor (1984), "Mental Representation: An Introduction," photocopy; colloquium presentation at the University of California, Irvine, 1984.

J. A. Fodor (1985), "Fodor's Guide to Mental Representation: The Intelligent Auntie's Vade-Mecum," *Mind* **94**, pp. 76-100.

Edmund Husserl (1900), *Logical Investigations*, trans. by J. N. Findlay, (New York: Humanities Press, 1970). [Original German edition, 1900-1901.]

Edmund Husserl (1913), *Ideen zu einer reinen Phanomenologie und phanomenologischen Philosophie* (Drittes Buch); 1950 edition (*Husserlian V*), ed. Marly Biemal (The Hague: Nijhoff). [Original German edition, 1913. English trans. by F. Kersten (The Hague: Nijhoff, 1982).]

Edmund Husserl (1925), *Phenomenological Psychology*, trans. by John Scanlon (The Hague: Nijhoff, 1962). [Lectures delivered by Husserl in 1925.]

Edmund Husserl (1929), *Formal and Transcendental Logic*, trans. by Dorion Cairns (The Hague: Nijhoff, 1969). [Original German edition, 1929.]

Edmund Husserl (1931), *Cartesian Meditations*, trans. by Dorion Cairns (The Hague: Nijhoff, 1960). [Original edition, in French, 1931.]

Edmund Husserl (1954), *The Crisis of European Sciences and Transcendental Phenomenology*, trans. by David Carr (Evanston, IL: Northwestern University Press, 1970). [Original German edition, 1954, posthumous.]

Colin McGinn (1982), "The Structure of Content," in Andrew Woodfield (ed.), *Thought and Object: Essays on Intentionality* (Oxford: Clarendon Press), pp. 207-258.

Ronald McIntyre (1984), "Searle on Intentionality," *Inquiry* **27**, pp. 468-483.

Hilary Putnam (1975), "The Meaning of Meaning," in his *Philosophical Papers*, Vol. II: *Mind, Language and Reality* (Cambridge: Cambridge University Press), pp. 215-271.

John R. Searle (1983), *Intentionality: An Essay in the Philosophy of Mind* (Cambridge: Cambridge University Press).

John R. Searle (1984), "Intentionality and Its Place in Nature," *Synthese* **61**, pp. 3-16.

David Woodruff Smith (1984), "Content and Context of Perception," *Synthese* **61**, pp. 61-87.

D. W. Smith and R. I. McIntyre (1982), *Husserl and Intentionality: A Study of Mind, Meaning, and Language* (Dordrecht: Reidel).

Stephen Stich (1983), *From Folk Psychology to Cognitive Science: The Case against Belief* (Cambridge: MIT Press/Bradford Books).

ROBERT CUMMINS

THE INTROSPECTIONISM OF TITCHENER

Unquestionably the most significant introspectionist program in the United States was the "structuralism" of E. B. Titchener. Titchener was concerned to establish the claim that the "new psychology" imported from Germany had made psychology a rigorous empirical science. Lacking a nontrivial account of science, Titchener supported his claim by emphasizing the analogies between psychology as he saw it and an established experimental science—viz. physical chemistry. To understand Titchener's vision of psychology, therefore, we do well to examine his model briefly.

The core of physical chemistry in Titchener's time was the periodic table. The periodic table allowed one to explain analytically an enormous number of the chemical properties of compounds. It provided a list of chemical elements—i.e., components whose further analysis is not theoretically significant for the explanation of properties in the intended domain—together with a specification of the chemically important properties of those elements. With these resources, it was possible to derive laws of composition—which compounds are possible—and laws of instantiation for a large number of the empirically established chemical properties of substances. Titchener's idea was to provide for psychology what the periodic table provided for chemistry, thereby making it possible to explain the properties of mental events and processes by analyzing them into psychological "elements"—mental events that could not be further analyzed. Since Titchener's elements are not things or substances but events, his program requires some account of the origin of elements. He needed a general recipe for answering such questions as: Why do we have just these elements present (in consciousness) at this time rather than some others? Two kinds of answer were available: either a current element is the effect of one or more previous elements or it is the effect of extramental stimuli. Titchener allowed both possibilities, but he concentrated mainly on the second, probably because (i) extramental events are more open to experimental manipulation, and (ii) under the influence of empiricist philosophy, Titchener believed that perception is the most significant source of events in the mind.[1]

J-C. Smith (ed.), Historical Foundations of Cognitive Science, 235–242.
Reprinted from Robert Cummins, *The Nature of Psychological Explanation*, pp. 119-127, with permission of The MIT Press.

The object of psychological theory, then, is to explain the origin and properties of the contents of consciousness—e.g., feeling anger, the visual image of a pouncing cat, or the experience of voluntary action. Suppose the feeling of anger has properties P, Q, and R (as revealed by introspection). We are to proceed by analyzing this feeling into its elements—call them x, y, and z. Then, appealing to the properties of these elements and the laws of composition, we endeavor to show why anger must have the properties P, Q, and R—that is, we endeavor to derive an instantiation law for these properties. To explain why S is angry at t, we explain the occurrence of x, y, and z, tokens of the mental element-types that make up anger, as effects of previous mental events and/or current extramental stimuli. (Perhaps we shall also need to explain why conditions were propitious for the combination of x, y, and z into anger. Compare: many chemical combinations require a fair amount of heat, or a catalyst, to take place.)

The project, then, was to discover the fundamental and introspectively unanalyzable elements of consciousness and to formulate the principles of combination whereby these elements are synthesized into the complex and familiar experiences of ordinary life. Every compound mental state or process was to be explained componentially, the characteristics of the whole derived from the characteristics of the parts and mode of combination. The central part of this program—analyzing familiar conscious experiences—was an extension of the project Locke initiated in Book Two of his *Essay Concerning Human Understanding*. The simple ideas of sense and reflection became the extramentally produced mental elements and the mentally produced elements, respectively, and the principle of association became the central law of mental bonding. Not surprisingly, introspectionists spoke of mental valences, of mental equilibrium, and of mental elements neutralizing each other. The fundamental issues of mental analysis and synthesis never got significantly beyond Locke, however, and the reason is fairly clear. There were simply no well-focused experimental procedures; nothing like qualitative analysis in chemistry. Lacking anything comparable to titration, centrifuges, solubility, tables, oxidation, and analytical balances, the student of introspective psychology was, in the end, simply left to compete with Locke at more or less well disguised linguistic analysis. Notable improvements on the *Essay* were not forthcoming.

So the analysis of consciousness bogged down for lack of analytical tools. But the correlative project to explain the elements as responses did not, for here introspectionists had an experimental paradigm in the Weber-Fechner experiments. Here is Titchener's textbook description of a typical variation.

Method—to find the numerical expression of Weber's law for noise—An ivory ball is let fall from two different heights, upon a hardwood plate. The difference of intensity between the two sounds (i.e., the difference between the two heights of fall) must be slight. The two sounds are given in irregular order in different experiments (to avoid the influence of expectation) and the subject is required to say, in each case, whether the second is louder than the first. In 100 experiments, he will give a certain number of right answers, and a certain number of wrong.

The method assumes that if the two sounds are just noticeably different in intensity, the subject will give about 80% right and 20% wrong answers. This proportion is calculated by what mathematicians call the 'law of probability'. Now suppose that a certain difference gave 70 right and 30 wrong answers in 100 experiments. We could calculate, by aid of the integral calculus, how much larger the difference must have been to give 80 right and 20 wrong—i.e., to be just noticeable. The calculated difference (difference of height of fall) is the numerator, and the original intensity (height of fall) of the weaker sound, the denominator, of the fraction which expresses Weber's law. (Titchener, 1897, pp. 81-82)

Here is a different and more fundamental description:

Suppose that we are investigating the intensity of noise. We shall begin with a stimulus of moderate intensity: say, the noise made by the fall of an ivory ball upon a wood plate from a height of 90 cm. We will call the intensity of this sensation 1. If we gradually increase the height of fall, we shall reach a point at which the noise of the fall is just noticeably greater than the original noise. We may call the intensity of this second sensation 2. If we further increase the height of fall we shall presently get a noise, 3, which is just noticeably louder than 2; and so on. Now what are the different heights of fall—i.e., intensities of stimulus—necessary to arouse sensations of the intensities 2, 3, 4, etc.?...An addition of 30 cm. suffices to raise the intensity of sensation from 1 to 2; but if we are affected by the stronger stimulus 120 cm., we must add more than 30 to it to change intensity 2 to intensity 3. In other words: change in the intensity of sensations does not keep even pace with change in the intensity of the stimuli which occasion them.

Experiment enables us to replace this general statement of the relation of sensation intensity to stimulus intensity by a definite scientific law. If sensations are to increase in intensity by equal amounts, their stimuli must increase by relatively equal amounts.
(Titchener, 1897, pp. 79-80)

This sort of experiment, begun by Weber (1834) and refined and expanded by Fechner (1860), led to Fechner's law (mistakenly called Weber's law by Titchener and Fechner—compare Gregory, 1981, pp. 500-505). Here was a law—a transition law—and a procedure for testing and refining it. Does the law hold for all sensation? What are the values of the constants of proportionality for each case? Work on these matters proceeded apace.

But this apparent bright spot in the program proved to be an Achilles heel. Although the analysis of consciousness languished, the fact was that there were no generally accepted or clearly articulated canons for the evaluation of analytic explanations, especially as applied to consciousness. Hence there was no way to determine whether the trouble was merely practical or deeply conceptual. On the other hand, there did exist well-articulated and generally accepted canons for the evaluation of the sort of explanations Fechner's law was used to construct, since the idea is to explain an event in consciousness—e.g., a discernible difference in loudness—as an effect of an external cause—a change in stimulus strength.

We can bring out the problem in a few lines. The canon requiring independent access to causes and effects applies to the Weber-Fechner experiments as follows. Suppose we find a subject whose responses don't fit the law? Is the subject (A) misdescribing his/her sensations, or (B) psychologically idiosyncratic? For that matter, how do we know that a subject responding normally is not in fact (C) psychologically idiosyncratic but systematically misdescribing experience, rather than (D) psychologically normal? We cannot compare a subject's descriptions with what is supposed to be described. Our only access to the sensations of the subject are (i) inference from established connections with responses, including verbal reports, and (ii) inference from established connections with stimuli. Obviously we cannot establish connections of the sort required unless we can, at least sometimes, distinguish (A) from (B) and (C) from (D). But we cannot make these distinctions unless we can establish the connections.[2] Since the accuracy of introspective observation cannot be checked, it seems it cannot play the role required of scientific observation. With introspection ruled out of court, the only way to measure sensation intensity is to measure stimulus intensity and then calculate sensation intensity using Fechner's law, but if we do this, we are using Fechner's law to define sensation intensity, and we cannot then turn around and pretend to explain the intensity of a sensation by appeal to stimulus intensity and Fechner's law. Once introspection is disqualified, we have no access to sensation intensity other than the very law that is supposed to explain it.

Thus it was that introspectionists became vulnerable to a powerful methodological attack. The methodology of explanation by causal subsumption had been well entrenched by Bacon, Berkeley, Hume, Mach, and Mill. The empiricist doctrine was (and is) that causal laws have no explanatory power unless the causes and effects they subsume are knowable independently of each other. It is ironic that this doctrine was so fatal to introspectionism, for introspection was held to be the only avenue to noninferential knowledge by the very empiricist philosophers who developed the line of argument that killed introspectionist psychology. Locke's inverted spectrum problem returned to haunt those who attempted to pick up where Book Two of the *Essay* left off. The inverted spectrum problem turned on the conceptualist assumption that linguistic training would inevitably disguise sufficiently systematic "psychological" differences. This is just the possibility raised by (A) and (C).

Whatever we may think of this critique,[3] the mere certainty that it could be formulated was eventually enough to kill any psychology based on introspection. And just as Berkeley's critique of representational realism seemed to point inevitably to a single alternative (if we know about tables and can know only about ideas, then tables are ideas), so this critique of introspectionism seemed to point inevitably to a single alternative. Introspection isn't genuine observation, so the Weber-Fechner law cannot be about consciousness. It is obviously about something, though, for experimenters certainly observe and record something in the sort of experiment described by Titchener in the passages above. What? Since experimenters record what their subjects say, the Weber-Fechner law must correlate stimuli with "introspective behavior." This is verbal behavior, *per accidens*, in the usual experimental set-up, but button pressings would do just as well.

This bit of diagnostics would probably have sufficed to produce behaviorism eventually, but a number of other factors conspired to guarantee a quick takeoff. Two, I think, are especially worthy of note. First, pragmatism was in vogue in United States philosophical circles, and pragmatists emphasized the importance of understanding connections with action to the understanding of traditional philosophical problems involving mental states and processes. According to Dewey, for instance, the central mistake of empiricists and rationalists alike is the supposition that knowledge and belief can be understood independently of action, and treated as antecedent conditions to be investigated in their own right. Refuting this supposition was always a central theme in Dewey's writings. It isn't a long step from this to the doctrine that talk of mental states is just shorthand for talk of identifiable behavioral patterns. Watson was a student at Chicago at a time when the influence of Dewey's pragmatism was very strong there.

The other significant factor was Pavlov's discovery of stimulus substitution. This was an important discovery in its own right, but in the intellectual climate I have been describing, it had a special significance, for it seems to account for the sort of phenomenon generally attributed to the association of ideas, without recourse to ideas. Someone taken with the empiricist critique of introspection, and the pragmatist treatment of doxastic states, could hardly have failed to conclude that Pavlov had shown that it was stimuli and responses, not ideas, that were associated. This had to seem a major breakthrough for association was the only principle of learning on hand.

Introspectionist psychophysics was accessible to methodological scrutiny because it was a case of subsumption, an explanatory strategy the structure and logic of which were clear and familiar to Introspectionist and

Behaviorist alike. In sharp contrast, the analytical strategy, though regularly applied, had not been clearly articulated as a strategy. The logic at the core of Titchener's program was, therefore, largely invisible, hence inaccessible to serious scrutiny. With the benefit of hindsight, however, we can see that Titchener's central explanatory program was naive in the extreme, and this gives us a measure of the extent to which the structure of property theories was tacit (at best) rather than explicit.

Titchener conceived the objects of psychological analysis as processes. This was the standard Wundtian line, a line that appears to go back to the empiricist conception of mind as a series of events in consciousness. Processes can be analyzed, of course, and so can properties of processes, just as chemical compounds and their properties can be analyzed. Chemists sought to discover what water is—namely H_2O—and this is naturally thought of as an analytic question: When we analyze water, what do we get? This, I think, is how Titchener and the introspectionists generally tended to think of their questions. They asked, "What is anger made of?"

Analysis of a compound, though often intrinsically interesting, is of theoretical interest mainly as a step toward explaining the dispositional properties of the compound. Similarly, Titchener's interest in analyzing anger was that such an analysis would be (hopefully) a necessary and useful step toward explaining the properties of anger—most importantly, its dispositional properties. But Titchener must have lost sight of this fact, or at least failed to note its implications. To see this, recall the schematic illustration of a Titchenerian explanation of anger. We are to imagine the feeling of anger analyzed into mental elements x, y, and z, and each of these explained as effects of previous disturbances. Now, the obvious question is: How does throwing x, y, and z together make anger? Is their mere coincidence in consciousness enough? This is surely naive in the extreme, for the whole program would then depend on the assumption of a "bushel basket" theory of consciousness (i.e., the theory that consciousness is simply a receptacle for elements that do not interact). On the other hand, if there is serious mental bonding, so that x, y, and z form a mental compound, then (i) we need an account of the capacities of x, y, and z to form this compound, and/or of the capacity of the mind to so compound them, and (ii) we can no longer seriously suppose that we can discern the components in the compound simply by "looking" at it. (Imagine someone supposing that one could tell what was in a chemical compound simply by examining it: this is surely a confusion between compounds and mixtures.)

Introspection as a method thus sorts ill with the explanatory goal of the theory. This goal was to explain analytically the properties of the complex contents of consciousness, and perhaps the capacities of the mind required for it to have such contents. Since introspection is at best a form of observation it *can* hope to yield *data*—the properties of conscious contents—but it *cannot* hope to yield analyses. The elements and their properties will not be "visible" when compounded unless we assume that there is no serious composition at all. If we assume a bushel basket theory of consciousness, a step even Hume did not take, then the properties of anger, say, will simply be the union of the properties of the elements in consciousness when one is angry. Analysis in such a case is morphological analysis at best. Morphological analysis certainly has significant applications, but the idea that the properties of complex, conscious experiences can be explained by morphological analysis is implausible in the extreme. "Implausible" is not the same as "false" or even "disconfirmed," of course, but if this implication of the program had been clearly perceived, the program would surely have looked much less attractive than in fact it did. It is evident that neither Titchener nor his critics were aware of this implication. It was thus the introspectionists' use of subsumption (in their psychophysics) that was vulnerable to well-focused criticism, and since their use of psychophysics was peripheral to their main project, the net result was more or less unsatisfactory to everybody concerned. As it happens, however, it is the use of introspection that is critical in both cases—though for different reasons—and therefore the general conclusion that introspectionist psychology fails because it is (exclusively) introspectionist is basically sound. Notice, however, that a critique of the role of introspection in the analytical part of the project does not point to a behaviorist alternative, as does the critique of the use of introspection in the subsumptivist part of the project. In the former case, the problem is not at all that introspective reports are unverifiable but rather that (i) the objects of introspection (the mental compounds) are not what need to be observed (the mental elements), and (ii) introspection, being a sort of observation, isn't an analytical tool (looking at anger won't take it apart any more than looking at sugar will take it apart). Now that verificationism is out of style, reliance on introspection in psychophysics seems less problematic than its role in Titchenerian analysis.

University of Colorado, Boulder

NOTES

[1] Since 'in the mind' meant 'in consciousness' for Titchener, as it did for Locke, this emphasis on perception as the source of mental elements is hardly surprising.

[2] Appeal to abnormal ear structure is tempting but ultimately only a delaying tactic: how do we establish the needed connections between ear structure and sensation?

[3] Perhaps the experimenter could avoid this difficulty by using him/herself as subject. Since the objects of introspective "observation" are events, one cannot look at the same object twice; hence as Wittgenstein pointed out (1953, par. 244-315), there is no guarantee that one is describing consistently over time. One simply has to assume that the same (type of) stimulus produces the same (type of) experience. With this assumption, one can check oneself and others for consistency. What about the assumption? It certainly isn't safe as a general rule: experience and response vary with other factors besides stimulus conditions. But it seems one might remember: "Yes, that's just what it was like last time." Wittgenstein seems to have thought that this sort of thing isn't a genuine memory report because there is no way to check it. But have someone prick you in the back with a pin, then touch you with a finger, and ask yourself whether, so far as you can remember, the second felt like the first.

REFERENCES

R. Gregory, *Mind in Science* (Cambridge: Cambridge University Press, 1981).

G. Fechner, *Elemente der Psychophysic* (Leipzig: Breitkopf and Hartel, 1860). Trans. H. Adler, *Elements of Psychophysics* (New York: Holt, Reinhart, and Winston, 1966).

E. Titchener, *An Outline of Psychology* (New York: Macmillan, 1897).

E. H. Weber, *De Tactu* (Leipzig, 1834). Available in translation as *The Sense of Touch*, trans. E. Ross and D. Murray (New York: Academic Press, 1978).

L. Wittgenstein, *Philosophical Investigations* (New York: Macmillan, 1953).

JOHN R. SEARLE

ANALYTIC PHILOSOPHY AND MENTAL PHENOMENA

1. Introduction: The Behaviorist Background

Throughout most of its history analytic philosophy has exhibited a curious prejudice against the mental. Many, perhaps most, analytic philosophers have felt that there was something especially puzzling about mental processes, states, and events, and that we would be better off if they could be analyzed away or explained in terms of something else or somehow eliminated. One sees this attitude, for example, in the persistent use of pejorative adjectives, such as 'mysterious' and 'occult', that analytic philosophers from Ryle to Rorty use to characterize mental phenomena naively construed.

I first became aware of the pervasiveness of this attitude when I tried to extend my analysis of speech acts to intentional states. No one doubts the existence of promises, statements, apologies, and commands, but when the analysis is extended to beliefs, fears, hopes, desires, and visual experiences, suddenly philosophers raise a host of "ontological" doubts. I think that thinking and other mental processes and events, like linguistic processes and events, are biologically based and are as real as digestion, conversation, lactation, or any other of the familiar biologically based processes. This seems to me so obviously true as to be hardly worth arguing, but I am assured that it is a minority opinion in contemporary philosophy.

During the positivist and verificationist phase of analytic philosophy the reason for the urge to eliminate the mental was not difficult to see: if the meaning of a statement is its method of verification and if the only method of verification of statements about the mental is in the observations of behavior, at least where "other minds" are concerned, then it would appear that some sort of behaviorism is an immediate logical consequence of verificationism. Statements about the mental are equivalent in meaning to statements about behavior.

Most philosophers today regard behaviorism as dead; yet, as I shall argue, many of them hold views that suffer from the same sort of defects as behaviorism. So, let us begin by examining what is wrong with behaviorism. Sometimes in reading the literature on this subject one gets the

J-C. Smith (ed.), Historical Foundations of Cognitive Science, 243–265.
Reprinted from *Midwest Studies in Philosophy* **6** (1981), pp. 405-423, with permission of the University of Notre Dame Press.

impression that behaviorism is at fault because of some more or less technical reason: the behaviorist never quite got a satisfactory account of the notion of a disposition, or their analyses suffered from some sort of circularity, or they never could give a satisfactory formulation of the antecedent clauses in the analysanda, or they had weak answers to the "perfect actor" or simulation arguments. Or some such. I want to suggest that these defects are not the basic flaws in behaviorism; at most they are surface symptoms of the underlying problem. If we remember that in its material mode version, behaviorism is the view that mental phenomena just are patterns of behavior, then it just seems obviously false. To me at least, its falsity seems clear as soon as I ask myself what it is like to have some conscious mental phenomenon such as a pain and contrast that with what it is like to engage in certain sorts of behavior appropriate to pains. For example, I now have a stomachache, not a violent one, but a nagging pain at the back of my stomach. The fact that I have this stomachache is quite a different fact, and indeed a different sort of fact altogether, from any facts, including conditional facts, about my behavior. Having a stomachache is one thing; engaging in various sorts of behavior appropriate to having a stomachache is something quite different.

I do not present this objection as an *argument* against behaviorism; if so, it would be question begging since the facts I am reminding myself of are stated in a way that simply asserts the falsity of behaviorism. What I want ultimately is not a refutation of behaviorism but an understanding of the underlying motives that gave rise to such an unplausible thesis in the first place.

Well, one might say, Rylean behaviorism was always troubled by the analysis of sensations such as pains. But it is much more satisfactory for beliefs and desires. It is not at all obvious that behavioristic analyses of beliefs and desires are false. Perhaps having beliefs and desires, especially in the right combination,[1] just is being disposed to behave in certain ways.

Here again I want to argue that behaviorism is false, though its falsity is not as obvious in this case as in the case of sensations. Suppose I now have a desire to drive to my office; suppose I want to drive to my office, as they say, more than I want to do anything else right now. Suppose, furthermore, that I have a set of beliefs about my abilities to drive to my office, about my car and its functioning, about the proper route, and so on. Now, doesn't that all entail that I will, other things being equal, engage in certain sorts of behavior, namely behavior describable as driving to my office, or at least trying to drive to my office? And furthermore, doesn't the appropriate kind of behavior, such as driving to my office, itself manifest or express my beliefs and desires in such a way that given the appropriate formulations, we could say that the behavior is at least a sufficient condition of having a certain range of beliefs and desires? At this point another objection to behaviorism arises: If the project of behaviorism is to analyze mental notions in terms of behavioral notions, then the project fails because the notion of behavior in question, the kind of behavior described in the analysans of the behaviorists' analysis, is itself *intentional* behavior, and therefore the behavior in question is still, in the relevant sense, mental. If by "behavior" we mean human action, then behavior is more than a set of bodily movements or muscle contractions. The bodily movements count as actions only if they are caused by intentions, and the intentions in question are as much mental as the beliefs and desires that they are being invoked to analyze.[2] Rylean behaviorism about beliefs and desires is therefore confronted with a dilemma: either the behavior in question is full-blown human behavior, that is to say, human actions performed intentionally, or it is not. If the former, then the analysis of the mental into the behavioristic is the analysis of the mental into the mental, and we are still left with the mental component in the notion of behavior. If the latter, then we have to construe behavior as bodily movements described solely as bodily movements; and no analysis of statements about beliefs and desires and intentions into statements about bodily movements will ever be adequate because the movements are not yet human actions, and the notion of a bodily movement by itself is inadequate for any analysis of mental states. A human body, for example, might be wired up in such a way that it would undergo all the bodily movements characteristic of driving a car to an office even though it was totally unconscious or even dead. If, in short, "behavior" means action, then behavior is mental. If it means bodily movements, then it is not behavior, and the analysis fails. At this point, the objection to analyzing beliefs and desires in terms of bodily movements (including dispositions to bodily movements) is much like the objection to analyzing sensations such as pain in terms of behavior (including dispositions to behavior): the distinction between mental states and sheer physical bodily movements is as stark as the distinction between pain and behavior.

I want to keep in mind these two objections to behaviorism—the non-identity of sensations and behavior and the mentalistic element in human actions—in what follows. I now want to introduce a distinction between what I will call ascriptions of *intrinsic mental phenomena* on the one hand and *observer relative mental ascriptions* on the other hand. This distinction can be made clear using unproblematic and uncontroversial examples. Suppose I now say that I have a stomachache or that Reagan believes he can win the election. When I make such statements I am ascribing intrinsic mental phenomena to myself and Reagan. But suppose I say that the expression "il pleut" in French means it is raining or that my pocket calculator adds and subtracts but does not divide. In such cases I am not ascribing any intrinsic mental phenomena to the expression "il pleut" or to my pocket calculator. Such statements are in part shorthand statements for the intrinsic mental phenomena of French speakers or users of pocket calculators. They are, roughly speaking, shorthand for saying such things as that by convention people who speak French use the sentence "il pleut" to mean "it is raining" and that I am able to use my calculator for addition and subtraction but not division. There are not two kinds of mental phenomena, intrinsic mental phenomena and observer relative mental phenomena; rather, there are ascriptions of mental predicates where the ascription does not ascribe an intrinsic mental phenomenon to the subject of the ascription. In such case the only mental phenomena in question are in the minds of the observers (or users), not in the subject of the ascription. Thus there are two kinds of *ascriptions* but only one kind of mental phenomena.

2. Carburetor Functionalism

I want now to turn to functionalism, the most influential of the current forms of the rejection of the mental. I will argue that functionalism suffers from very much the same sort of difficulties as behaviorism and that, in addition, it rests on a confusion between observer relative and intrinsic mental ascriptions.

First, a few words about the recent history of the philosophy of mind to explain how functionalism evolved out of physicalism. In the early days of physicalism, there were a series of objections from more or less dualistically inclined philosophers to the physicalists' view that mental states were identical with brain states. Most of these objections were versions of the argument from Leibniz's Law (e.g., my stomachache is in my stomach but my brain state is in my head, so my stomachache cannot be identical with my brain state), and Smart[3] and his colleagues justifiably felt that they could answer them (e.g., it is not my stomachache, but my experience of having a stomachache, which is identical with a state of my central nervous system). But there were two classes of arguments that were more troublesome.

First Objection. Suppose physicalism is true as a contingent identity thesis. Then all mental states are identical with some physical states of the brain. But then some physical states of the brain are also mental states, and some other states of the brain are not mental states. What is the difference between them? Well, the obvious difference is that the physical states that are mental states have mental properties and the others do not. But now we seem to be left with a dualism of properties in place of our original dualism of entities; so physicalism is a species of the very dualism it was designed to replace. In place of "mysterious," "occult" mental entities, we are left with "mysterious," "occult" mental properties.

Smart answered this objection by saying that it was possible to describe these entities in a language that did not mention any mental properties, in a language that was "topic neutral." For example, instead of saying, "I am having an orange afterimage," one could say, "Something is going on in me which is like what goes on when I see an orange." But that is no answer. The fact that one can mention an object that has mental properties without mentioning that it has those properties is as irrelevant to its possession of those properties as the fact that one can mention a locomotive without mentioning the fact that it is a locomotive is irrelevant to the fact that it is a locomotive. One could refer to a locomotive in a "topic neutral" way, as, e.g., "A certain item belonging to the Southern Pacific Railroad." But that does not make it any less of a locomotive, and the fact that one can talk of mental states in a veiled, topic-neutral vocabulary does not make them any less mental. This, by the way, is a recurring fallacy in analytic philosophy, the confusion between features of the language we use to describe a phenomenon and features of the phenomenon.

Second objection. Early physicalists believed that if two people were in the same type of mental state, they must be in the same type of corresponding neurophysiological state, that, for example, if you and I both now believe that it is snowing, we must both have the same type of neurophysiological state which realizes that belief. But for a variety of reasons it seems most implausible to suppose that this is true. Even if mental states are physical states, it can hardly be true that type identical mental states can be identified with type identical neurophysiological states.

The way out of this objection is to notice that many sorts of things can be type identical at one level of description even though each instance is token identical with some object at a lower level of description and those objects at the lower level of description are not type identical. Clocks are the same type of thing at the level of description "clock," even though each clock is token identical with some physical realization, say, a set of mechanically powered gears and wheels, or a set of electrically operated quartz oscillators; and those physical realizations may be type different. And if this works for clocks and carburetors, why not for mental states as well? Thus, you and I can each have the same belief that it is snowing, even though our neurophysiologies may not be type identical, just as your car and my car can each have a carburetor, even though mine is steel and yours is brass. But what is it then that the mental states have in common that makes them type identical? I believe the obvious answer is that mental states are type identical because of their common mental features, whether features of consciousness or of intentionality or both or some further sorts of mental features. But such is the reluctance of many analytic philosophers to take mental phenomena at their face value that this commonsense answer was not the one proposed.

According to the functionalists, just as clocks and carburetors are identified by their functions, not by how those functions are realized in the physical structure, so mental states are identified by their functions, and not by how those functions are realized in the brain. Mental states are functional states in a sense that is supposed to be made clear by analogy with clocks and carburetors, and so on. And now the answer to the second objection has, *en passant*, also answered the first. For the "mysterious," "occult" mental properties turn out to be metaphysically harmless functional properties. In a word, type-type identity theories gave way to token-token theories, and with them came carburetor functionalism. Functional materialism turned out, more or less inadvertently, to be a kind of eliminative materialism because the functional analysis eliminated any problems about irreducibly mental properties.

An immediate and obvious objection to functionalism is that it seems it could not be right because the mental states in question are intrinsic and functions are always observer relative. The ascription of a function to a system or to an element of a system is always made relative to some goal, purpose, or objective; functions are never just causes, they are causes relative to a teleology. My carburetor intrinsically causes many things, only some of which are its functions; for example, it exerts pressure on the engine block, it makes a hissing sound, it supports the air cleaner, it accumulates dirt, and it mixes air and gasoline. When we say that its *function* is to mix air and gasoline, we are saying more than that it causes the mixing of air and gasoline; we are also saying that *its purpose* is to mix air and gasoline relative to the overall purposes of the system. But such purposes are never intrinsic; they are assigned by us. If I decide to use my carburetor solely for supporting my air cleaner, then its function changes even though there are no changes in any of the intrinsic causal features of the system. The observer relativity of functions, by the way, is also true of systems that are not artifacts. For example, the heart causes the circulation of the blood, a thumping noise in the chest, and pressure on the lung. To say that its *function* is to pump blood is to assign observer relative intentionality to it. A linguistic clue to these facts is that as soon as we assign a function to some causal element, a whole intentionalistic vocabulary becomes appropriate which is inappropriate for causal relations *tout court*; thus we can speak of "malfunction," "breakdown," "functioning properly," and so on.

Now because mental states are intrinsic and functional states are observer relative, it can never be constitutive of mental states that they are functional states. Even if mental (=brain) states always had the same causal relations and even if we always assigned the same functions to these states, still, the features of these states that make them *mental* states are intrinsic and so could not be constituted by any observer relative stance we might take. To put the same point differently, if we try to define mental states in terms of their functions, then something else might serve just those functions, even though it lacked the intrinsic features we were trying to define. The very feature that made carburetor functionalism so appealing, namely, the analogy between the functional level and the physical level of descriptions of beliefs, etc., on the other hand, is fatal in the end to carburetor functionalism, because the mental level of description is intrinsic and not functional.

I believe that most contemporary functionalists would feel that this is not a serious objection to their view because, so they might argue, the notion of a function is not really essential to functionalism. Ultimately they want to define mental states in terms of their causal relations where the causes are intrinsic, even if the functions are not. The argument would then turn on whether the causal features are sufficient to define intrinsic mental features. And I will come to this point shortly.

3. Turing Machine or Organizational Functionalism: Putnam and Dennett

Let us continue with our story. Carburetor functionalism was soon replaced by Turing machine functionalism. On this view, mental states are indeed functional states, but not just any kind of functional state; rather, they are logical states of a computer, and thus they are *intrinsic* states at least at the level of description of the computer program. Turing machine functionalism is also superior to carburetor functionalism in that it seems to promise a richer theory of the mind, according to which mental processes are computational processes; and this theory seems to be consistent with and supported by current work in cognitive psychology and artificial intelligence. I will consider the work of two authors in this tradition, Hilary Putnam and Daniel Dennett.

I begin with Putnam even though he claims to have abandoned Turing machine functionalism, at least in its cruder forms (Putnam, 1975, pp. 298-99). I believe he abandoned it for the wrong reasons. The two reasons he gives for abandoning it are first, that a Turing machine can only be in one state at a time whereas a human can be in many psychological states at the same time, and second, some psychological states such as jealousy are related to other psychological states in a holistic fashion that is unlike the relations among Turing machine states. But this would still permit him to hold that psychological states are matters of functional organization, and it is this aspect of Turing machine functionalism that I want to discuss, not the specific limitations of the computer analogy that have to do with the scope and discreteness of machine states. So perhaps 'organizational functionalism' would be a better term than 'Turing Machine functionalism' for the view I am criticizing.

Putnam introduces the notion of a "Description" of a system, where a "Description" is defined as a specification of the states and input-output relations of the system as determined by the machine table (the computer program) for that system. He then states a version of organizational functionalism about conscious mental states such as pains.

> The hypothesis that "being in pain is a functional state of the organism" may now be spelled out more exactly as follows:
>
> (1) All organisms capable of feeling pain are Probabilistic Automata.
>
> (2) Every organism capable of feeling pain possesses at least one Description of a certain kind (i.e., being capable of feeling pain *is* possessing an appropriate kind of Functional Organization).
>
> (3) No organism capable of feeling pain possesses a decomposition into parts which separately possess Descriptions of the kind referred to in (2).
>
> (4) For every Description of the kind referred to in (2), there exists a subset of the sensory inputs such that an organism with that Description is in pain when and only when some of its sensory inputs are in that subset.
>
> (Putnam, 1975, p. 434)

It seems to me that this analysis suffers from the same sort of defects as classical behaviorism: being in pain is one thing; satisfying all these conditions or anything like these conditions is something quite different. I will argue for that conclusion by showing that a system could satisfy all these conditions and not feel anything at all, that these conditions are not constitutive of nor sufficient for having sensations such as pains. Since (2) and (4) are the heart of the analysis, I will dispose of (1) and (3) briefly.

(1) "All organisms capable of feeling pain are Probabilistic Automata."

Since everything is a probabilistic automaton under some description or other, as Putnam concedes, this statement is not much help. It does not serve to discriminate organisms capable of feeling pain from anything else. Furthermore, it is important to emphasize that 'organism' in the statement does not mean organism in any biological sense. It just means system. It is essential to functionalism of the sort that both Putnam and Dennett adhere to that any system at all capable of having the right functional organization is capable of pain. It need not be an organism in the sense that plants and animals are organisms. Indeed, as we noted earlier, one of the original motivations for functionalism was the recognition that an indefinite number of different types of system could be functionally isomorphic. Robots, for example, could count as "organisms" for Putnam's purposes. So, in what follows, I will use the word 'system' since systems are what is being talked about.

(3) "No system capable of feeling pain possesses a decomposition into parts which separately possess Descriptions of the kind referred to in (2)."

The reason for putting in this condition is that many systems can consist of subsystems and the subsystem may be capable of feeling pain even when the system is not capable of feeling pain. Suppose, for example, we have a system composed of many people working together in a large bureaucracy. Such a system can realize a certain machine table without it being the case that the system suffered any pain in addition to the pain suffered by the individual people, and in order to eliminate these sorts of counterexamples Putnam has inserted condition (3). The elimination of a class of potential counterexamples by simple fiat is generally the sign of a theoretical weakness, and we will shortly see that Putnam's analysis is still subject to related counterexamples.

(2) and (4). "Every system capable of feeling pain possesses at least one Description of a certain kind (being capable of feeling pain *is* possessing an appropriate kind of Functional Organization). For every Description of this kind there exists a subset of the sensory inputs such that a system with that Description is in pain when and only when some of its sensory inputs are in that subset."

What (2) and (4) together amount to is this: Beings capable of pain are those and only those that instantiate a certain computer program; the computer program specifies a set of transition probabilities between input, output, and internal states, and the system will actually be in pain when the right input activates the system according to the program.

There is a way of trivializing Putnam's view so that it implies that a system is in pain if and only if it is in pain. That, I take it, is not his intent. His idea is rather that just as there are specific machine tables for pain, so there are specific inputs for pain. We do not need to know exactly *what they are* but simply *that there are some*, and we can define functionalism in terms of them whatever they are. Furthermore, the use of the expression 'sensory' in "sensory input" is as potentially misleading as the use of the term 'organism'. 'Sensory' in this context does not mean sensory in the biological sense, since any system such as, e.g., a robot could have all sorts of sensory inputs without having any of the familiar biological apparatus for the senses. Otherwise, it is impossible to interpret Putnam's various remarks about robots.

Now once you lay bare the real character of the claims made in (2) and (4), you can easily see that it is possible for systems or organisms to satisfy the conditions and not have any pains. I think the claim is as obviously false as behaviorism, and I will present two related arguments to try to make its falsity apparent.

The argument from anesthesia.

Let the system in question be me. By hypothesis I am capable of feeling pain because I instantiate a certain machine table. Let us suppose that the machine table is upset and that this is done by anesthetizing the relevant portion of my central nervous system, so that I cannot feel any pains. I now no longer instantiate the machine table. But suppose further that I make up for this by simply memorizing the steps in the machine table, so that I can now go through the steps in my mind. On Putnam's account there is no reason why my own mind could not be the element of the system that instantiates the machine table. Whenever somebody gives me the relevant sensory input, say he punches me in the nose, I do not feel anything because I am anesthetized, but I "look up" in the machine table what I am supposed to do next, and I follow through all the steps until I reach the output and then I say or print out "Ouch!" I cannot feel anything, but I still have the relevant sensory inputs. I still instantiate the machine table, and I still have the right input-output "transition probabilities." I satisfy all of Putnam's conditions (1)-(4), but I do not feel any pains at all.

This argument may seem rather swift, but I think in fact the counter-example is decisive. Machine table plus inputs could not be constitutive of nor sufficient for pain because a human agent could in his own mental processes instantiate any machine table you like and could have any inputs you care to induce in him and still not feel the relevant sensations if his specific neurophysiological states are not appropriate. This suggests that functional organization is not the feature of the neurophysiology that really matters, and this suggestion will be pursued in the next objection. It is also important to emphasize that though this objection is stated in terms of machine tables, it would apply to other sorts of functional organization. It is directed at organizational functionalism and not just at the Turing machine version of organizational functionalism.

The argument from biology.

Insofar as we know anything about the causal basis of pains and other sensations, we know that they are specific to quite definite sorts of animal and human nervous systems. Humans and many species of animal are capable of suffering pains; but trees, shrubs, and flowers are not, much less stones, waterfalls, and mountains. It is an amazing consequence of the claims of Turing machine functionalism that *any sort of substance whatever* is capable of feeling pain, provided only that it instantiates the machine table of principle (2) and has the sort of receptors described in principle (4). Though functionalists recognize and even embrace this consequence, its implausibility is masked from us by the fact that they seldom go through any detailed examples, so let us now try such an example. In my office is a computer console hooked up to a routine garden-variety computer called a PDP-10. Now let us suppose, in line with Putnam's number (2), there is a quite specific machine table which human beings instantiate and which enables them to feel pain. I know of no reason at all to suppose it is true, but let us give it to Putnam for the sake of the argument. Let us now program the PDP-10 with exactly this machine table. Notice that the machine table is purely formal; it has nothing to do with the specific material in which it is instantiated. As Putnam repeatedly insists, two systems can have the same Functional Organization, while having quite different constitutions, and to repeat, "Being capable of feeling pain *is* possessing a certain kind of Functional Organization" (ibid., p. 434). So on Putnam's definition, the PDP-10 is now capable of pain, presumably all the way from mild discomfort to the most frightful agony. We now add a "sensory input." We introduce a transducer which is such that whenever I slam the door of my office or hit the console with my fist this produces an input to the computer as programmed for pain, and the "transition probabilities" are such that on my automatic typewriter the machine prints out, "Ouch!" "You are giving me a terrible stomachache!" "Please stop!" and so on.

We have now satisfied all of Putnam's conditions. We have a probabilistic automaton; by fiat we can give it any machine table or other sort of functional organization that you like; it has no subsystems with the same Description; and it has sensory inputs that trigger the appropriate pain-responses. (And if I have chosen the wrong inputs, it does not matter because the argument remains the same no matter what sensory inputs you put in it.) Now is there any reason at all to suppose that the computer is in pain? Real intrinsic pain? Notice that the functionalist thesis is not that for all we know the computer *might* be in pain, but that the computer *must* be in pain because it is in a functional state equivalent to the functional state that human beings are in when they are in pain; and that is all that pain is: a functional state. I think it is obviously empirically false that the computer is in pain, because, to put it as an understatement, it is made of the wrong kind of material and is the wrong sort of system altogether to have pains. To have pains it would have to have a nervous system with neurons, biologically specific pain receptors, etc., or at least it would have to have something that was *causally* and not merely *formally* equivalent to an animal nervous system.

Some other sort of system besides animal nervous systems might be capable of having pains; that is an empirical question. But in order to be capable of having pains the system would have to have the relevant causal powers equal to the causal powers of animal nervous systems, and merely instantiating the same program would never be sufficient to guarantee that. Suppose by some incredible miracle my PDP-10 were in frightful agony. Even if that were true, it would still be no help to Turing machine functionalism, because if it were true it would have to be because the specific causal structure of the hardware somehow duplicated the causal powers of animal nervous systems, and that, as far as functionalism is concerned, is quite accidental. Even if my PDP-10 had, unknown to us, an electrochemical structure capable of feeling pains, the thesis of functionalism is that structure does not matter; what matters is functional organization, and that is defined by purely formal considerations quite independently of the structure. Even if the PDP-10 is in pain, we can then put the same program in a set of water pipes, wind machines, or any crazy Rube Goldberg apparatus you might care to construct, provided only that it is stable and enduring enough to carry the program. We could, e.g., make the system out of windmills, old beer cans, and paper clips, and according to organizational functionalism, any such system must be in pain. Against this I am arguing that from everything we know about physiology, it seems incredible to suppose that these other systems *could be* in pain, much less that they *must* be in pain.

If we put these two arguments together, we get a sequence of steps to show that both the main claims of organizational functionalism are false. Anybody who wanted to defend this sort of functionalism would have to show that something is wrong with this argument.

The two claims are:

a. For a human to be in pain or some other mental state is just to instantiate a certain sort of functional organization (such as a computer program) with the right inputs and outputs.
b. Any system such as a robot that was functionally equivalent to a human, i.e., that had the right organization and the right input and output relations, would also have to be in pain in virtue of those facts.

Against these I argue:

1. Humans and at least some animals have nervous systems that are causally sufficient to enable them to feel pains and have other sorts of mental states (this is an empirical assumption, shared by both sides in the dispute, about the causal basis of mental phenomena).
2. Any system capable of having pains and other mental states must have the relevant causal powers equivalent to human and animal nervous systems (this is a trivial consequence of 1).
3. Instantiating a certain functional organization such as a program could never by itself be sufficient for mental states, because it would be possible for an agent to instantiate the program and not have the relevant mental states (by the argument from anesthesia). This entails the negation of *a*.
4. Therefore, a system could not have the relevant causal powers described in 2 solely in virtue of instantiating a certain organization with certain input-output relations (by 2 and 3).
5. Human and animal nervous systems have a causal capacity to produce mental states which is not constituted by their having a certain functional organization with certain input-output relations (this is a consequence of 1 and 3).
6. Other systems such as robots could not have mental states solely in virtue of having a functional organization with the right input-output relations but would have to have other causal powers equivalent to those of the brain (a consequence of 2 and 4). This entails the negation of *b*.

Throughout both these arguments I have been supposing, as a concession for the sake of argument, that my machine table must have some relevant role to play in my feeling of pains, and that might even be true, but there is so far not the slightest reason to suppose it is true. What I am doing here is arguing against the view that organizational functionalism gives us a sufficient condition for the mental. I am conceding for the sake of argument that it might give us a necessary condition, though I do not believe there is any argument for that view, and once you see that it does not give a sufficient condition, most of the motivation for supposing it might give a necessary condition is removed.

I now turn to Dennett, whose opposition to the mental is even more explicit than Putnam's. Typical of Dennett's remarks about mental phenomena is the claim that "beliefs, desires, pains, mental images, experiences—as all these are *ordinarily* understood" are not "good theoretical entities" (Dennett, 1978, p. xx). The difficulty with this claim is that there are quite different ways in which something can fail to be a "good theoretical entity." Witches and goblins fail to be good theoretical entities because such things do not exist; but tables and chairs, and mountains and beaches, fail to be good theoretical entities not because such things do not exist, but because the laws that describe the course of nature do not use these terms; they do not mention objects under the description "chair," "table," "mountain," or "beach." In that sense, these sorts of things are not good theoretical entities, but so what? At most, that is a remark about which sorts of terms figure in the construction of scientific theories and not about the *existence* of chairs, tables, mountains, and beaches. Now, does Dennett think mental phenomena are like witches or like mountains? Can it be that Dennett really thinks pains and beliefs are like witches and goblins in that they do not exist at all "as ordinarily understood?" (How many non-ordinary ways are there to understand, say, the stomachache I referred to earlier?) He discusses how difficult it is to "convince someone that there are no pains or beliefs" (ibid., p. xx), and he declares that it will have to be done piecemeal. He concludes one such effort with the sentence "Then, since I am denying that any entity could have the features of a pain or a thought, so much the worse for the ontological status of such things" (ibid., p. 38). And in a subsequent discussion (Dennett, 1980, p. 76), he says he is indeed "claiming that there are no such things as pains, although people do feel pain." He does not find this view self-contradictory (I do), but for present purposes our concern is not with its consistency but with the argument that "strictly speaking there could be no such things as pain" (ibid., p. 76). The argument is as follows:

> But if, as I have claimed, the intuitions we would have to honor were we to honor them all do not form a consistent set, there can be no true theory of pain, and so no computer or robot could instantiate the true theory of pain which it would have to do to feel real

pain. Human beings and animals could no more instantiate the true theory of pain (there being none), which lands us with the outrageous conclusion that no one ever feels pain. (Dennett, 1978, 228)

This argument contains, in my view, something more outrageous than the conclusion, and that is its logic. From the fact that the set of intuitions we have about a class of objects is inconsistent, and therefore the class is such that none of the objects could satisfy *all* of the intuitions, it simply does not follow that no such objects exist. Compare: the intuitions we have about chairs do not form a consistent set; therefore, there is no true theory of chairs; therefore, chairs do not exist. Or compare: the set of intuitions people have about Jimmy Carter do not form a consistent set; therefore, there is no true theory of Jimmy Carter; therefore, Jimmy Carter does not exist. With the best will in the world there is simply no way you can rescue the formal structure of Dennett's argument that "strictly speaking" pains do not exist, and the objections I am making to it are at least as old as Wittgenstein's discussion of "game" and other family resemblance notions.

But even assuming that Dennett's arguments against the existence of mental phenomena such as pains are inadequate, what about his positive thesis concerning the mental, about those states which presumably are left over after the nonexistent ones are eliminated? I now turn to his "Cognitive Theory of Consciousness" (ibid., pp. 149-173).

He describes, in some detail, a flow chart for a system that has consciousness. He then asks (Thomas Nagel's question), "What is it like to be an entity that instantiates the flow chart?" His supposition is that there is something it is like if the entity is conscious, and if not, not. Here is the answer: "Suppose I put forward the bold hypothesis that you are a realization of the flow chart and that it is in virtue of this fact that it seems to us—and to you—that there is something it is like to be you" (ibid., p. 165). Having consciousness is thus on his account "a matter of having a certain sort of functional organization." He then challenges us, "Can you give good grounds for denying the hypothesis and if so, what are they? What personal access do you have and to what?" (ibid., p. 165).

One feels one must be misunderstanding the questions because the answer seems too obvious. Yes, one can give good grounds for denying the hypothesis. Indeed, taken quite literally, it seems obviously false and false for the same sort of reason that old-fashioned logical behaviorism is false. Having conscious states such as bodily sensations and visual experiences is one thing, and instantiating the flow chart is something quite different. Bodily sensations such as tickles and itches are constituted by the way they intrinsically feel and are realized in nervous systems. Flow charts are constituted by satisfying certain functional descriptions and can be realized in any substance you like. They have no essential connection with either feelings or nervous systems. If it is not obvious on the face of it that instantiating the flow chart is neither constitutive of nor sufficient for the possession of the relevant conscious states, the same two arguments I marshalled against Putnam can again be invoked here. First, the flow chart is totally independent of the forms of its realization. This means that all sorts of substances can instantiate the flow chart even though they are inappropriate for having visual experiences, tickles, itches, and so on. The flow chart can be instantiated by our fancy Rube Goldberg machine made entirely of beer cans, paper clips, and windmills.

And second, the argument from anesthesia that we used against Putnam also applies to Dennett, though in a slightly modified form. The flow chart will have to have variations for different sorts of conscious states, say, itches, visual experiences, and so on. Let a person whose eyes are closed go through the steps in the flow chart that are specific to vision. He can use braille or any other method to get through the flow chart. He will indeed have conscious states, he is conscious of going through the flow chart, but he has the wrong conscious states since he has his eyes closed and has no visual experiences. He cannot see anything, even though the flow chart in question is specific for vision.

I said earlier that functionalists showed a tendency to confuse intrinsic mental states with the causal features that might warrant observer relative ascriptions of mental states. In Dennett's case this shift from intrinsic to observer relative ascriptions of mental phenomena is made a matter of principle with his introduction of the notion of "intentional systems" and "the intentional stance." On his account both humans and computers with certain sorts of program are "intentional systems," and an intentional system is just one where we find it appropriate to adopt "the intentional stance." In the adoption of the intentional stance, "One predicts behavior...by ascribing to the system *the possession of certain information*, and supposing it to be *directed by certain goals* and then working out the most reasonable or appropriate action on the basis of these ascriptions and suppositions" (ibid., p. 6). It is then, he says, but a small step to describe this information and goals as beliefs and desires. But we must not ask if these really are beliefs and desires because "the definition of intentional systems I have given does not say that intentional systems really have beliefs and desires but that one can explain and predict their behavior by ascribing beliefs and desires to them, and the decision to adopt this strategy is pragmatic and is not intrinsically right or wrong" (ibid., p. 7).

Well, one can agree that it is open to anybody to adopt any strategy he finds useful, but the question remains, what is the status of the ascription of the mental states, whether the ascription is of information and goals, or beliefs and desires? Even if we have not defined intentional systems in such a way that we say that they really have beliefs and desires, there will still be a difference between those intentional systems that really have beliefs and desires and those that do not, and in the case of those that do, the ascriptions have entirely different interpretations from the case of those that do not. For example, in the case of those that do, the beliefs and desires play causal roles determined by their specific intrinsic features. Dennett thinks the case of the computer and the human being are the same; in both cases we find it useful to adopt the intentional stance, and it is up to anybody to adopt some other strategy if he finds it more useful. But there is an enormous difference between my attributing a "desire" to a chess-playing computer to castle on the strong side and my saying I have a desire to drink a glass of cold beer. In the computer case, it is just a useful shorthand for describing how the system functions. There are no intrinsic mental phenomena in question at all. But in my own case I do not attribute to myself a desire for a beer because I find it useful in predicting my behavior, but because I want a beer. In my own case, I am stating facts about intrinsic mental phenomena, and whether or not people find it useful to adopt "the intentional stance" toward me is quite irrelevant to what the facts really are. In short, whether a system really has beliefs and desires is quite independent of whether or not we find it useful to make observer relative ascriptions of beliefs and desires. Furthermore, we only understand the observer relative attributions in a way that is parasitic on the use of the same expressions to specify intrinsic mental states. We understand the metaphorical non-literal use of 'belief' and 'desire' as applied to computers because we see these attributions as based on an analogy with systems such as human beings who literally have intrinsic beliefs and desires.[4]

I think I can make clear my sense of the strangeness of Dennett's project and of functionalism in general with the following analogy. Suppose that there were a group of philosophers who were puzzled by the existence of hands; suppose there was a long tradition of being worried about the ontological status of hands. And suppose that a functionalist view of hands became fashionable. On one version of this view, we are told that we do not have to worry about the existence of hands because it is all a matter of adopting the "manual stance" toward certain systems which we will describe as "manual systems." To paraphrase Dennett: "the definition of a manual system does not say that manual systems really have hands but that one can explain and predict their behavior by attributing hands to them, and the decision to adopt this strategy is pragmatic and not intrinsically right or wrong." I think that the intentional stance approach to understanding beliefs and desires is about as useful as the manual stance approach would be to understanding hands. In each case the question of analyzing the intrinsic features of mental states (or hands) gets replaced by a different question: under what conditions do we find it useful to talk *as if* a system had mental states (or hands)? And to ask that question amounts to changing the subject while still deluding ourselves into thinking we are addressing the original philosophical issues.

In sum, I have made three objections to Dennett's account: first, that the argument to show that pains, etc., do not really exist is invalid; second, that the cognitive theory of consciousness is subject to counterexamples; and third, that the notion of the intentional stance conceals but does not get rid of the crucial distinction between observer relative mental attributions and attributions of intrinsic mental phenomena.

Most of my discussion of functionalism has been addressed to organizational functionalism because it seems to me the richest and most interesting of the versions I have seen, but I want to conclude this part of the argument by showing how some of the same worries apply quite generally to other sorts of functionalist accounts. Consider the following functional definition from Grice which he calls a "first shot" at a definition of belief. "*X* believes that p just in case *x* is disposed, whenever *x* wants (desires) some end E, to act in ways which will realize E given that p is true rather than in ways which will realize E given that p is false" (Grice, 1975, p. 24). Now compare that with an analogous definition of hands: "*X* has a hand just in case *x* is disposed whenever *x* wants (desires) some graspable and retrievable object O_{gr} to act in such a way that will bring it about that x grasps and retrieves O_{gr} rather than that he does not grasp and retrieve O_{gr}."

The problem I have with such definitions is not just that one can always think up counterexamples, but that the enterprise of this sort of definition seems so dubious in the first place. Shoemaker (1980) tells us that on his version of functionalism it is the view that "mental states can be defined in terms of their relations, primarily their causal relations, to sensory inputs, behavioral outputs and other mental states." Analogously, a functionalist view of hands would be that "hands can be defined in terms of their relations, primarily their causal relations, to manual inputs, behavioral outputs, and other features of the body including mental states." Now let us suppose that after years of ingenious philosophical effort the functionalist dream of satisfactory definitions was realized for both hands and mental states; suppose we had such definitions and nobody could think up a good counterexample. What would we have achieved? In the case of our definition of hands, not very much. Because our definitions still leave out the intrinsic properties of hands, all they tell us about is the functional role of hands. Analogously, our definitions of mental states do not tell us the intrinsic properties of mental states, only their functional role. But it might be replied that the very thesis of functionalism is that mental states do not have any intrinsic properties except their functional role, so that is hardly an objection. But now I want to pick up the thread of our earlier discussion of the intrinsic/observer-relative distinction. We left off where the question was: are the acknowledged causal properties of mental states sufficient to define their intrinsic mental properties? And by now it should be clear that my answer to that question is no, and for the same sorts of reasons that the acknowledged behavioral manifestations of pains are not sufficient to define pains and the acknowledged causal features of hands are not sufficient to define hands. Indeed, all the arguments I have presented so far are just ways of trying to exhibit this basic intuition. The intrinsic features of, e.g., beliefs, pains, and desires are just that: intrinsic features. The strategy of the functionalist, like that of the behaviorist, is to get us to take a third person observer relative stance, in which we agree that mental states do indeed stand in causal relations; just as we agree with the behaviorist that they have behavioral manifestations. But the arguments that I have presented, such as the argument from anesthesia and the argument from biology, are designed to remind us that something could have all the right causal relations and still not have the right mental properties.

4. Diagnosis and Conclusion

I believe that functionalism, like the behaviorism which preceded it, requires not so much refutation as diagnosis. Why would anyone hold such an unplausible view? Why would anyone believe, for example, that one's own pains, tickles, and itches, as well as one's beliefs, fears, hopes, and desires, were constituted by the fact that one was the instantiation of a flow chart or a computer program? And more generally, why does this phase of analytic philosophy seem to exemplify a longer-term reluctance to take mental phenomena at their face value?

I believe, in fact, that there are many reasons, more than I can hope to discuss in this article, and some of those that I will not be discussing should be mentioned at least in passing. There are arguments, such as those of Quine on indeterminacy, to show that where intentional states are concerned there are no facts of the matter beyond certain dispositions to behavior. And there are a series of confusions about the relations between intentionality with-a-t and intensionality with-an-s. Many philosophers believe that the propositional contents of beliefs and desires belong to a class of mysterious entities, called intensions, and that we would be better off to avoid believing in such entities. But I think these are not the arguments that have really mattered in the long run. On my diagnosis there are three major reasons for the appeal of such views as functionalism and for the long-run prejudice against the mental in analytic philosophy.

(1) *Verificationism*. Positivism and verificationism are officially dead, but they live on in certain verificationist tendencies that one finds in functionalist authors. In analyzing pains and beliefs Putnam and Dennett do not try to describe the intrinsic properties of such mental phenomena; rather, they are concerned with such questions as: "When would we attribute such mental states to a system? Under what conditions do we think ourselves justified or not justified in making such intentionalistic attributions?" But these questions are not the same as, "What are the intrinsic properties of mental states?" Notice that in both authors the approach is almost entirely from a third-person point of view, and indeed the confusion between observer relative and intrinsic attributions of the mental derives largely from this third-person approach. As long as we think of mental states as something to be ascribed from outside, it can be made to seem as if what matters is the adoption of an "intentional stance" and the "pragmatic" usefulness of treating something as an "intentional system." No one would think it of much use in the philosophy of mind to worry about these "stances" if he were concentrating on his own mental states. Think about what it is actually like to have a pain in the stomach or a passionate desire for a cold beer and ask yourself if you are talking about real mental phenomena or just adopting a stance. The cure, in short, for the verificationism of Putnam, Dennett, and others and for their persistent confusion of the question "What is it to have mental states?" with such questions as "What is the functional role of the ascription of mental states?" is to insist at least at some point on the first-person point of view. It reveals a fundamental confusion to suppose that we can get clear about mental states entirely by examining the functioning of third-person attributions of the mental.

(2) *Cognitive Science*. The second reason is peculiar to functionalism, but I think it is important. Many philosophers have the conviction that they are utilizing the results of something called "cognitive science" and that somehow the computer has given us remarkable new insights into the operation of the mind. On the surface, this manifests itself in a great deal of more or less intimidating technical jargon. There is much talk

about Turing machines, finite state automata, analog and digital computers, machine tables, flow charts, transducers, and so on. The illusion is thus conveyed that philosophical problems about the mental have been converted into technical questions in automata theory capable of solution by the cooperative efforts of philosophers and cognitive scientists, including both psychologists and computer scientists. Underlying this surface appearance is a fairly serious fallacy. From the fact that at some level of description the brain is a digital computer (or, if you like, a set of digital computers) together with the fact that the programs of the brain can be instantiated by any other sort of computer sufficiently rich and complex to carry the program, the fallacious inference is drawn that what the brain does in producing mental phenomena is nothing more than what these other sorts of computers would be doing if they instantiated the program. And this fallacious inference leads naturally to the thesis of Turing machine functionalism. Being in a certain mental state just is instantiating a certain sort of computer program. To put it another way, the fallacy is in moving from the true premise that the brain is a computer, in the sense that it instantiates computer programs, to the false conclusion that all that the brain does which is relevant to the production of the mental is to instantiate computer programs, that the only psychologically relevant feature of brain states is that they are logical states of a computer. The premise could hardly be false: since everything is a digital computer at some level of description, brains are too. But the conclusion is obviously false, as I have claimed in both the argument from biology and the argument from anesthesia. It is false both because no program by itself guarantees the causal powers that are specific to the biochemistry of the brain and because a system such as a human agent could instantiate the program and still not have the relevant mental states.

(3) *Fear of Cartesianism*. I believe that the deepest source of the prejudice against the mental is the fear that to admit mental entities at their face value would necessarily involve us in the worst excesses of Cartesianism, that we will be postulating a class of mysterious and occult entities, inhabiting another metaphysical realm beyond the reach of scientific investigation, that we will be left with the self or the soul, with privileged access and incorrigibility, and all the rest of it. Or, perhaps worst of all, that we will be left with a "mind-body problem."

Indeed it is a curious fact about the history of behaviorism and functionalism that neither was *independently* motivated. No one examined his own pains and discovered them to be patterns of behavior or Turing machine states; rather, these theories were proposed as solutions to other problems in philosophy, such as the problem of other minds and especially the "mind-body problem." But suppose there is not any such problem as the "mind-body problem" anymore than there is such a problem as the "digestion-stomach problem." Suppose that, as I believe to be the case, thinking and perceiving are as much natural and biological phenomena as digestion or the circulation of the blood. Suppose that mental phenomena are both caused by and realized in the structure of the brain. If so, we can give up not only dualism but "monism" and "physicalism" as well, for the antidualist jargon only has a point if we accept the dualist categories, such categories as kinds of substance and relations between kinds of substances. No one feels that he has to choose between monism and dualism where digestion is concerned; nor does he have to choose between epiphenomenalism, interactionism, etc., nor to postulate the existence of a metaphysical self or a transcendental ego that digests. Why should cognition be treated any less naturalistically? The most effective way to answer behaviorism and functionalism is not, as I have done, by making "objections" to them, but by removing the philosophical picture that motivated them in the first place.

University of California, Berkeley

NOTES

[1] It is sometimes correctly objected to behaviorism that in order to analyze desires one would have to assume fixed beliefs, and conversely. But again, this is an "in house" objection that does not expose the underlying absurdity of the behaviorist approach.

[2] For a discussion of the mental element in actions, see Searle (1979).

[3] For a discussion of these issues see the articles by Smart, Shaffer, and Cornman in Rosenthal, (1971).

[4] Also, when I adopt the intentional stance, is that supposed to be intrinsic or not? Do I really have an intentional stance, or is it just a case of adopting an intentional stance to my intentional stance? If the former, we are left with intrinsic intentionality; if the latter, it looks like we are in a vicious regress.

REFERENCES

Daniel C. Dennett, *Brainstorms: Philosophical Essays on Mind and Psychology* (Montgomery, VT: 1978).

Daniel C. Dennett, "Reply to Professor Stich," *Philosophical Books* **21**, no. 2 (1980): 73-76.

H. P. Grice, "Method in Philosophical Psychology," *Proceedings and Addresses of the American Philosophical Association*, vol. 48, 1974-75.

Hilary Putnam, *Mind, Language, and Reality: Philosophical Papers*, vol. 2 (Cambridge: 1975).

James W. Cornman, "The Identity of Mind and Body" in David M. Rosenthal (ed.), *Materialism and the Mind-Body Problem* (Englewood Cliffs, NJ: 1971), 73-79.

John R. Searle, "The Intentionality of Intention and Action," *Inquiry* 22 (1979), 253-280.

Jerome Shaffer, "Mental Events and the Brain" in David M. Rosenthal (ed.), *Materialism and the Mind-Body Problem* (Englewood Cliffs, NJ: 1971), 67-72.

Sydney Shoemaker, "The Missing Absent Qualia Argument."

J. J. C. Smart, "Sensations and Brain Processes" in David M. Rosenthal (ed.), *Materialism and the Mind-Body Problem* (Englewood Cliffs, NJ: 1971), 53-56.

LANCE J. RIPS

INTUITIVE PSYCHOLOGISTS: MENTAL ACTIVITIES AND THEIR PARTS

If there's one assumption that comes naturally to cognitive scientists, it's that mental activities can be divided into parts. When these psychologists try to explain some complex activity—say, reading or problem solving—they typically do so by specifying the components that make it up. An analysis of reading might consist of a description of individual steps like word recognition, parsing, and interpretation, together with an account of how these steps are temporally and causally combined. An analysis of problem solving might proceed by breaking the process into subparts such as encoding of the given information and the goal, assessing differences between givens and goal, and applying operators to reduce the differences. This type of analysis might be repeated with respect to the parts. For example, we might decompose word recognition into letter encoding, memory access, and decision. At some point in this successive partitioning we reach the level of atomic mental activities—"elementary information processes," as Newell and Simon (1972) call them—which cannot themselves be decomposed within a cognitive framework. Exactly which activities one takes as atomic depends on one's theory, but they might include simple operations of adding a new piece of information to a location in memory, comparing one piece of information to another, and issuing motor commands.

This familiar partitioning of mental activities may well be the right way to do cognitive science, but it is not an inevitable way of conceiving of these activities. The cognitive scientist's notion of a mental part is a special one, involving certain functional relationships. Mental parts, in this sense, are subprocesses that accept specified input, carry out a cohesive sequence of internal actions, and produce output as a result. However, there are other ways of thinking about mental activities and their parts. Imagine, for example, a certain type of mental activity—call it 'mentating'—that is homogeneous with respect to functional components, something like steadily attending to a stationary dial on a control panel. Even though mentating has no subroutines, it might still have parts of a different sort. If nothing else, it will have temporal segments like its beginning, middle, and end. The beginning of mentating is a part of mentating, in one reasonable sense of 'part', and the same is true for the beginnings of ordinary mental activities like reading and problem solving.

J-C. Smith (ed.), Historical Foundations of Cognitive Science, 267–292.

There is another type of part, though, that we can use in connection with mental activities and that tells us more about them. This is related to the sense of 'part' that we get in expressions like "Having acne is part of being a teenager" or "Looking glamorous is part of being a starlet." We might paraphrase these examples by saying that having acne is a characteristic of (many) teenagers and looking glamorous is a characteristic of starlets; i.e., they're typical properties of these individuals. Similarly, we can talk about criterial or characteristic properties of activities as parts of those activities. To say, for example, that exerting oneself is part of exercising means that exerting oneself in a certain manner is a criterial property of exercising. Exerting oneself is not a subprocess of exercising in the way that performing a push-up might be, since exerting oneself *encompasses* the exercising. What's important about this for our purposes is that this use of 'part' also extends to mental activities. Just as we can say that exerting oneself is part of exercising, we can say, for instance, that exerting oneself is part of solving difficult problems. This truism isn't supposed to convey a subprocess analysis of problem solving; it merely gives one property that characterizes the activity. We could equally well say that exerting oneself is part of mentating, even though (by hypothesis) mentating has no subprocesses.

Let's use the term 'attributional part' to refer to the notion of a mental part as a characteristic property and use 'procedural part' for the notion of part as subprocess. What I'd like to claim is that in people's ordinary way of conceiving mental parts these parts tend to be attributional rather than procedural. I assume that this is not because these intuitive psychologists have no concept of a procedural part or subprocess, since they are certainly familiar with subprocesses of physical activities. They know that kneading the dough is a procedural part of bread making and that taking off the lugs is a procedural part of changing a flat (Schank and Abelson, 1977).[1] In special cases, they probably recognize procedural parts of mental activities, especially if the activities are extremely complex ones such as reading a novel. For many mental activities, however, procedural parts are not readily apparent, either because the activities themselves seem homogeneous (like mentating) or because the components of the activities are too variable to qualify as distinct procedural components. If these activities have parts at all, the parts will be attributional.

The main evidence for this contrast between scientific and folk psychology comes from the results of experiments—to be described later on—in which subjects judged whether particular mental activities were parts of others. The best way to explain the results of these experiments, I believe, is on the hypothesis that the subjects based their decisions on the attributional rather than the procedural sense of 'part'. Before looking at the experiments, though, it might be helpful to consider the way in which the idea of a procedural part surfaces in scientific psychology. Section 1 deals with these background matters. The second section takes up the experimental evidence, highlighting some unusual features of judgments about mental parts. The most important of these is a converse relationship between parts and kinds in this domain: Subjects will often agree that $activity_1$ is a part of $activity_2$ if $activity_2$ is a kind of $activity_1$. In the final section, I look at this evidence from the standpoint of what Smith and Medin (1981) call "Probabilistic models" of concepts. What is interesting for us about these models is that characteristic properties are at the heart of the way they represent categories. I argue that attributional parts are very natural in the context of these models and that they can therefore give a satisfying explanation of the experimental findings.

One more thing before pushing on: Many cognitive scientists (especially hard-headed psychologists) become fidgety about studying people's concepts of mental activities, even though they are quite positive about similar studies about physical activities (e.g., changing a flat or going to a restaurant). Maybe this is because everyday ideas about psychological processes strike them as trivial beside the more intricate ones they are developing. Or perhaps they view such an endeavor as an attempt to build a theory of mental processes by quick and dirty methods. But, of course, the point of studying people's concept of reading, say, isn't to construct a scientific theory of reading any more than studying people's concept of going to a restaurant is an attempt to put together a science of restaurant going. In both cases the aim is to find out something about the mental representation of a class of (physical or mental) events.

Moreover, although people's intuitive notions of mental activities are probably a lot less beefy than their scientific counterparts, there are still several reasons for taking them seriously. Here's a sampling:

(a) There is evidence from studies of "metacognition" that the way students represent activities like mathematical problem solving (Schoenfeld, 1983) and writing (Scardamalia, Bereiter, and Steinbach, 1984) affects the quality of their performance in these tasks.

(b) A number of theories in artificial intelligence suggest that the ability to represent one's own inference processes is essential for reasoning in an efficient manner (Davis and Buchanan, 1977; de Kleer, Doyle, Steele, and Sussman, 1977; Weyhrauch, 1980; Moore, 1982).

(c) The ability to deal with embedded propositional attitudes (e.g., "Susan wanted Jeff to recognize that she intended him to learn that she knew he hadn't come home the previous night") is essential to understanding even simple utterances (e.g., Susan's pointed remark, that she hadn't heard the key turn in the lock last night). But on a cognitive analysis, understanding these attitudes entails being able to represent mental processes or states (which can represent other processes or states...). In Dennett's (1983) view, this "higher-orderedness of intentional characterization is a deep mark...of intelligence."

(d) Mental notions permeate other conceptual domains, such as criminal categories, speech acts, personality traits (D'Andrade, 1987), and even biological notions in children's world view (Carey, 1985). So understanding intuitive psychology may be prerequisite for making sense of other culturally important systems.

(e) We might have an easier time conveying to students our pet scientific theories if we had a better idea of the initial psychological concepts these students bring with them.

These applications aren't taken up in this paper, but I hope their prospects will persuade psychologists and other cognitive scientists that this domain is worth a closer look.

1. Mental Parts in Early Experimental Theory

To clarify the distinction between procedural and attributional parts, I'd like to consider a picture of mental experience that comes from the early history of experimental psychology. I'm going to attribute this picture to Wundt (1873/1904, 1896/1907), although I am not concerned with historical accuracy. The point of bringing it up in this context is that it provides a convenient framework for thinking about different aspects of inner activities. Wundt's picture is a synoptic one, allowing us to see how emphasis on these different aspects can lead to different conceptions of the activities' parts.

In its recent guise, scientific theorizing about mental processes is the outcome of computational analysis, in which the processes are construed as algorithms or subroutines. Mental processes were also the center of interest among the early experimentalists (e.g., James, 1890; Titchener, 1902; Wundt, 1896/1907), although the processes were conceived in a different way. Wundt took immediate experience as given and defined psychology as the scientific study of this experience. In this view, psychology seeks an "account of processes which are evinced by our own consciousness" (Wundt, 1873/1904, p. 1). As Wundt says elsewhere, "the contents of this science are exclusively *processes*, not permanent objects" (1896/1907, p. 24). This emphasis on processes probably reflects a fear of dualism that could be motivated by a psychology based on mental substance or mind-stuff. Perhaps it is also the result of a "brain metaphor" for the mind, as in current connectionist models where brain activity is featured over static representations. But in any case, the most important aspect of this orientation, and the part that seems closest to cognitive theory, is the idea that mental processes could be studied scientifically by breaking them into subprocesses. In Wundt's picture, the contents of mental experience are always *composite* processes, and a main task of psychology is therefore to analyze these composites into their elementary constituents.

We have, then, at least two levels of analysis for mental processes. At the level of immediate experience, there are "psychical compounds"—compound mental processes—each of which is "marked off from other contents of this experience by characteristics peculiarly its own, in such a way that it is recognized as a relatively independent unity and is, when practical necessity demands it, designated by a special name" (Wundt, 1896/1907, p. 99). Below this level is the realm of "psychical elements"—elementary mental processes—which are "absolutely simple and irreducible components of psychical phenomena" (ibid., p. 31). Since the contents of experience are composite, we can discover the elementary processes only by analyzing and abstracting over the different composites in which they occur. I have tried to indicate these two levels in Table 1, where at the level of compound processes we have the perception of an object and at the level of elementary processes we have sensations like redness. The elementary processes will fall into two general types: sensations, such as redness or warmth, and feelings, such as pleasantness.

Processes at these different levels are associated with different characteristics. Elementary processes always possess a qualitative character and a degree of intensity. Attributes of compound processes are not, in general, the same as the attributes of their elementary processes but also depend on the manner in which the elementary ones combine. Wundt held,

> It is a general principle valid for all psychical compounds, whether they are composed of sensations only, or feelings only, or of combinations of both sensations and feelings, that *the attributes of psychical compounds are never limited to the attributes of the elements which enter into them.* It is true rather that new attributes, peculiar to the compounds themselves, always arise as a result of the combinations of elements. (1896/1907, p. 100)

Table 1
Wundt's Conception of Mental Processes and Their Parts

	Mental Processes	Attributes of Processes	Classes of Processes
Compound	Perceiving a particular object	3-dimensionality Shape Spatial location Distance	Perceiving
Simple	Sensations of color	Color tone Saturation Brightness	

In Table 1, for example, the perception of an object will have attributes such as dimensionality that arise from the fixed arrangement of its elementary components with respect to one another and with respect to its background. In Wundt's (1896/1907) words,

> The actual content of psychical experience always consists of various combinations of sensational and affective elements, so that the specific character of a given psychical process depends for the most part, not on the nature of its elements, so much as on the union of these elements into a composite psychical compound. (p. 32)

There is disagreement in the literature about how seriously we should take the "mental chemistry" metaphor as applied to Wundt (see Blumenthal, 1985; Mischel, 1970), but it is uncontroversial that in his system the properties of the elements don't by themselves yield the properties (specific characters) of the compounds.

Since compound mental processes and their properties are more obvious than the elementary ones, natural language terms for classes of mental activities are likely to be based on the former. In the *Principles*, Wundt says that "language brings against us an array of concepts like 'sensibility', 'feeling', 'reason', 'understanding'--a classification of the processes given in internal perception against which, bound down as we are to the use of these words, we are practically powerless" (1873/1904, p. 18). Although he acknowledges that these terms might have value for psychological analysis if they are based on actual distinguishing characteristics (1896/1907, p. 99), it is plain that he regards these classes as superficial.

In the first place, they fail to indicate the more important breakdown of compound processes into elementary ones. Moreover, terms like 'feeling' and 'understanding' smack of faculty psychology, which Wundt (1873/1904) saw as a taxonomy masquerading as an explanatory scheme:

> Here, therefore, in the first formation of psychological concepts, we have the germ of that confusion of classification with explanation which is one of the besetting sins of empirical psychology....As a matter of fact, one can quite well conceive of a natural science of the internal experience in which sensibility, memory, reason, and understanding should be conspicuous by their absence. For the only things that we are directly cognizant of in internal perception are individual ideas, feelings, impulses, etc.; and the subsumption of these individual facts under certain general concepts contributes absolutely nothing toward their explanation. (p. 19)

Table 1 represents this subsumption relation in a rough way by placing the term for the class of the mental processes ('perceiving') next to the attributes that it categorizes.

Cognitive science differs from this Wundtian picture of mental processes in the relative importance of activity types, such as perceiving, reasoning, and understanding. In the current view, these types are not simply rudimentary categories that can be eliminated in a completed theory. It is these types, rather than individual mental experiences, that are subject to subprocess analysis. In fact, what makes these types respectable is that their powers can be explained as a function of their subprocesses. Perceiving, reasoning, and understanding are not merely postulated as the causal source of specific mental skills—which is the practice that bothered Wundt—but are unpacked into simpler, more comprehensible components. What mattered for Wundt were elementary sensations and feelings, whereas what matters for a cognitive scientist are the process kinds and their components. But despite this shift in explanatory strategy from the left side to the right side of Table 1, the dimensions of this picture should be relatively familiar. For both cognitivist and Wundtian, the real scientific task is subprocess decomposition. The attributes of mental processes, at least at the top level, are taken as obvious: they are the starting points, not the constituents, of a psychological theory.

What I'd like to emphasize about this picture, however, is that it presents us with two ways of factoring mental activities. On one hand, we can proceed in a vertical direction to break compound processes into subprocesses (and their modes of combination). Wundtians and cognitive psychologists agree that this strategy generally requires experimental methods and offers the best chance of a true scientific explanation. On the other hand, we can proceed along the more intuitive horizontal path, which takes into account the surface properties of the compound processes and sorts them into kinds. These two analytic dimensions correspond quite closely to the two types of parts that we met earlier: The vertical analysis gives us procedural parts, whereas the horizontal analysis gives us the attributional parts of the same mental activities. Thus, if there is anything to this general picture, attributional parts should be the ones closest to people's experience. Procedural parts have a more obscure relation to their wholes, since they depend on theoretical and experimental analysis. It is certainly possible to question aspects of this overall view, but there may nevertheless be enough that's right about it to provide clues to some puzzling facts about mental parts.

2. Some Evidence on Mental Parts and Kinds

One way to investigate conceptions of mental parts is simply to ask subjects to judge whether a given activity is part of another. By surveying these judgments, we might be able to induce the general principles that underlie the subjects' decisions. Of course, comprehending data of this sort isn't always an easy matter. Even for relatively small numbers of subjects and target activities, the sheer size of the dataset can exceed our ability to assess it. To make matters worse, the data are likely to contain noise—due to subjects' inattention, misunderstandings, and other slips—that can obscure the crucial relationships. For these reasons, it's very helpful to have a technique for displaying data like these in a more comprehensible format, and we will employ a method that Frederick Conrad and I have developed for this purpose (Rips and Conrad, 1988). This technique produces a partial order that is, in some sense, as close as possible to the subjects' raw judgments and that is much easier to apprehend. The inferences that we draw from these orderings will be indirect, of course, since they will depend on the assumptions of our method. But there are payoffs: The derived orderings highlight some nonobvious properties of the mental-activities domain, and they also allow us to predict the results of further studies.

2.1. *Partial orderings for parts and kinds*

In the basic experiment subjects evaluated the truth of sentences of the form '*X-ing* is a part of *Y-ing*' and, for contrast, '*X-ing* is a kind of *Y-ing*', e.g., "Reasoning is a part of conceptualizing" or "Reasoning is a kind of conceptualizing." The mental activity terms in these sentences came from a set of 30 that we had previously collected from other subjects who had been asked to write down as many examples of mental activities as they could. Our aim in compiling the set in this way was to ensure that subjects were familiar with the terms and believed they referred to mental activities. The sentences in the main study resulted from substituting all possible ordered pairs of these activity terms for *X* and *Y* in the two sentence frames just mentioned. Twenty subjects evaluated the "part-of" sentences, and another twenty the "kind-of" sentences. Both groups saw the sentences in a randomized list and simply indicated whether each was true or false. In our instructions we emphasized that they should respond "yes" to a "part-of" sentence only if every instance of *Y* had an instance of *X* as a part and "yes" to a "kind-of" sentence only if every instance of *X* was an instance of *Y*. The instructions illustrated this with an analogy to physical activities: putting a foot forward is a part of walking, and strolling is a kind of walking. In this stage of the investigation, we hoped that subjects would interpret "part-of" in the procedural sense, and we picked the illustration to underscore this idea.

The subjects in this study (and in the studies described later) were undergraduates at the University of Chicago who were not psychology majors. Their opinions about mental activities are probably similar to those of most intelligent laymen, though of course they may be more interested in mental activities of an academic sort, such as reasoning and conceptualizing.

The results from the study, then, consist of the subjects' binary judgments about the activity pairs. To get a handle on these results, it's useful to make some simplifying assumptions about subjects' underlying beliefs. Let's suppose, in particular, that intrinsic to subjects' concepts of *part-of* and *kind-of* is the idea that these relations are irreflexive, asymmetric, and transitive. So if *X*, *Y*, and *Z* are mental activities, then (a) *X* is not a part of (kind of) *X*; (b) if *X* is a part of (kind of) *Y* then *Y* is not a part of (kind of) *X*; and (c) if *X* is a part of (kind of) *Y* and *Y* is a part of (kind of) *Z* then *X* is a part of (kind of) *Z*. Of course, in a study like this one these properties may sometimes be violated because of memory limits, misinterpretations, and other factors; but we will assume that under ideal conditions subjects' judgments obey these properties.[2] If so, then *part-of* and *kind-of* should partially order the set of mental activities, since a transitive, asymmetric, irreflexive relation is by definition a (strict) partial order. This is consistent with formal theories of mereology (Goodman, 1966, ch. 2; Simons, 1987) and taxonomy (Thomason, 1969), though such theories usually place additional constraints on these relations.

Under these assumptions, we can attempt to recover the underlying structure in our data by searching for a partial order that comes as close as possible to the subjects' actual judgments. There are alternative ways of specifying the sense of 'as close as possible' in this context, but one

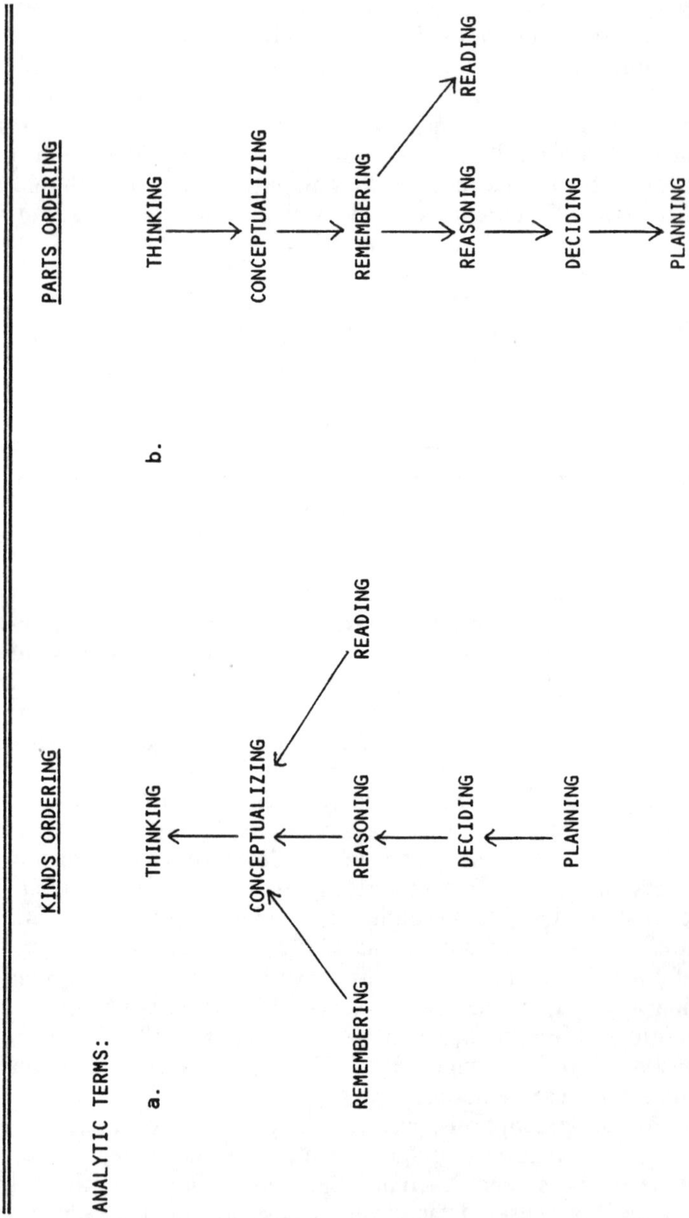
ANALYTIC TERMS:
a.
KINDS ORDERING
THINKING
CONCEPTUALIZING
REMEMBERING
READING
REASONING
DECIDING
PLANNING
b.
PARTS ORDERING
THINKING
CONCEPTUALIZING
REMEMBERING
READING
REASONING
DECIDING
PLANNING

Figure 1

NONANALYTIC TERMS:

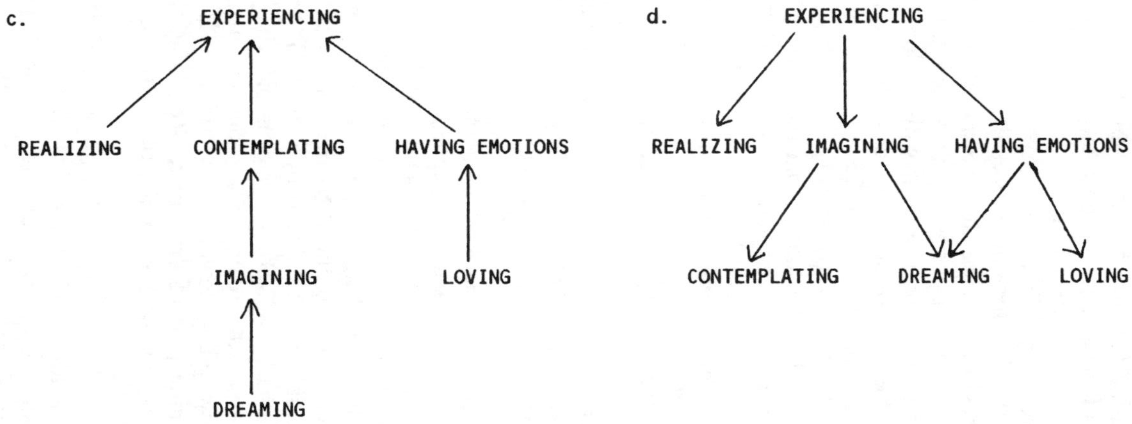

Figure 1. Nearest partial orders for analytic terms (Graphs a and b) and nonanalytic terms (Graphs c and d). The left-hand graphs represent ordering by <u>kind-of</u>, and the right-hand, ordering by <u>part-of</u>.

method is this: Let matrix S represent the subjects' actual decisions, with each entry s_{ij} indicating the proportion of subjects who agree that mental activity i is a part of activity j. Second, let **T** represent a partial order defined over the same activities, with $t_{ij} = 1$ if activity i is a part of j in the partial order and $t_{ij} = 0$ if it is not. We can then define the 'nearest partial order' to be that **T** which minimizes the sum of squared deviations, $\Sigma_{ij}(s_{ij} - t_{ij})^2$. In practice, finding the nearest partial order is a difficult matter because the number of possible orders increases exponentially with the number of elements. For instance, with 5 elements to be ordered (e.g., 5 mental activities), there are 4,231 distinct partial orders to choose from; with 6 elements, there are 103,023 partial orders; and with 7 elements, there are 6,129,859 (Evans, Harary, and Lynn, 1967). It's known that the problem of finding the nearest partial order is NP-complete; i.e., the problem is equivalent in computational complexity to the Traveling Salesman Problem and others for which there is no known algorithm that takes less time than some exponential function of the number of elements (see Rips and Conrad, 1988).

This complexity effectively squelches our chances of finding the nearest order for the full set of 30 mental activities in our experiment, but it is possible to find nearest orders for subsets of these items. Figure 1 shows the nearest partial orders for two subsets: one composed of what we will call *analytic* activities (thinking, conceptualizing, remembering, reasoning, reading, deciding, and planning), and the other composed of *nonanalytic* activities (experiencing, realizing, imagining, having emotions, contemplating, dreaming, and loving). These groupings are informal ones and are not meant to imply an essential distinction between them. The items within each group were selected because they seem intuitively related and therefore more likely to yield informative partial orders.[3] Graphs a and c in Figure 1 are the orders induced by the *kind-of* relation, and Graphs b and d are the partial orders induced by *part-of*. In the Kind orderings, the arrows point upward, indicating, for example, that conceptualizing is a kind of thinking and that dreaming is a kind of imagining. In the Part orderings, the arrows point downward, so thinking is a part of conceptualizing and imagining part of dreaming. Reversing the direction of the arrows in this way highlights the similarities between the Kind and Part graphs.

The Kind graphs in Figure 1 seem, for the most part, sensible representations of both the analytic and nonanalytic items. As we would expect, thinking and experiencing are the most general activities in these taxonomies, all of the remaining items being kinds of these two. More complex activities, such as planning, loving, and reading, tend to appear at the base of the graphs. There is room for doubt about some of the specific relations: Is imagining really a kind of contemplating or deciding a kind of reasoning? In most cases, however, it's easy to see how these links arise from particular ways of interpreting the activity terms. Imagining is a kind of contemplating if contemplating means something like mentally inspecting, and deciding is a kind of reasoning if deciding means deliberating.

The Part graphs, however, present more of a puzzle. In the first place, the *part-of* links are hard to interpret on the procedural sense of 'part'. For example, although having emotions is a part of loving according to Graph d, it's not clear what it would mean to say that having emotions is a subprocess of loving. The same goes for experiencing as a part of realizing or thinking as a part of reasoning. If experiencing is a subprocess of realizing, what are realizing's *other* subprocesses?

The second piece of the puzzle is that the Part graphs tend to be the converse of the corresponding Kind graphs. If activity_i is a part of activity_j then activity_j tends to be a kind of activity_i. In Graphs b and d, 28 pairs of activities are linked by *part-of*, counting both direct links (e.g., the one from *thinking* to *conceptualizing* in Graph b) and indirect links implied by closure under transitivity (e.g., the one from *thinking* to *reasoning* in the same graph). Of these relations, all but 6 are linked in the reverse direction by *kind-of*. Similarly, if activity_i is a kind of activity_j then activity_j tends to be a part of activity_i. In Graphs a and c, 24 pairs of activities are linked by *kind-of*, and all but two are linked in the reverse direction by *part-of* in Graphs b and d. Most of the exceptions involve the activities contemplating and remembering. Contemplating has dreaming and imagining as kinds according to Graph c, but contemplating is not a part of dreaming or imagining. Remembering is part of reasoning, reading, deciding, and planning in Graph b, but the latter activities are not kinds of remembering. It's important to recall that separate groups of subjects evaluated kinds and parts; so this near-converse relationship is not the result of the two types of judgment influencing each other.

What makes this converse relationship especially mysterious is that it clearly *doesn't* hold for parts and kinds of physical objects. A stem is part of a daffodil, but a daffodil is not a kind of stem. A daffodil is a kind of flower, but a flower is not a part of a daffodil. We'll come back to the interpretation of the Part graphs in the third section of this paper. But first, it might be wise to assure ourselves that these graphs correctly represent people's beliefs about the activities—that they are not merely artifacts of the algorithm we used to derive them. For this reason, I'll describe two additional studies of the same activities and try to show that the results can be predicted by the position of the activities within the Part orderings. (For further evidence, see Rips and Conrad, 1988).

2.2. *Ratings of importance and of "possible minds."*

According to the Part graphs of Figure 1, thinking and experiencing are parts of the remaining activities in their domains, and we might therefore expect people to view these two as more important and less dispensable than the rest. In fact, the graphs impose a scale of what we might refer to as *part centrality*, where one activity has greater part centrality than another if it is a part of more activities. In Graph b, for example, thinking has the highest centrality, while planning and reading have the lowest. In Graph d experiencing has highest centrality, with realizing, contemplating, dreaming, and loving tied for lowest.

If these graphs are an accurate depiction of people's beliefs about the activities, then part centrality ought to predict other judgments about them. As an example, consider the possibility of creatures who possess some of these activities but not others. It's easy, for instance, to imagine a creature who is unable to dream but is nevertheless able to experience, realize, imagine, have emotions, contemplate, and love; but it's much more difficult to envision a creature who is unable to experience and yet can realize, imagine, have emotions, love, contemplate, and dream. In general, the least "dispensable" mental activities in this sense should be those that are part of many others, whereas the most dispensable ones should be part of relatively few. In other words, the greater the part centrality of an activity, the less dispensable that activity should be and the more difficulty subjects should have imagining a mind without it. In the same way, activities with high part centrality ought to be ones that subjects should view as more important in the way the mind works. An activity that is part of many others should be more important in this respect than an activity that is part of only a few.

As a test of these predictions, two groups of subjects rated the activities. The first group read descriptions of hypothetical creatures who could engage in six of the seven analytic or nonanalytic activities but not in the remaining one. For example, one creature was described as being able to engage in experiencing, realizing, imagining, having emotions, contemplating, and loving, but not in dreaming. Each of the 14 activities in Figure 1 served as the missing item for one of the creatures. We asked subjects to rate the descriptions in terms of "how likely it is that such a creature could exist, given what you know about the different activities mentioned." The second group of subjects received a list of the same 14 activities intermixed with 16 other "filler" activities. These subjects rated

Table 2
Mean Likelihood and Importance Ratings for Mental Activities

Activity	Part Centrality	Mean rating of Likelihood of Creature Lacking Activity	Mean Rating of Importance
Analytic			
thinking	6	2.5	8.0
conceptualizing	5	4.2	7.0
remembering	4	3.1	7.2
reasoning	2	3.7	7.7
deciding	1	3.2	6.1
planning	0	6.5	5.8
reading	0	9.8	6.0
Nonanalytic			
experiencing	6	1.4	6.5
having emotions	2	4.5	7.3
imagining	2	6.1	7.0
realizing	0	5.2	5.6
contemplating	0	6.2	5.8
dreaming	0	7.1	4.8
loving	0	8.1	6.2

"the importance of these activities in how the mind operates." The mean ratings from the two groups appear in Table 2. High numbers for the first group indicate that subjects believed a creature missing that activity was possible, and low numbers, that the creature was impossible. For the second group, high numbers mean that the activity was important, and low numbers that it was unimportant.

We are interested in comparing these ratings with part centrality, and for this purpose Table 2 also contains a simple count of the number of other activities that a target activity is part of. In terms of the **T** matrix defined earlier, this measure of part centrality for activity$_i$ is just the sum of the *i*th row; i.e., $\Sigma_j t_{ij}$. Thus, thinking has a score of 6, conceptualizing a score of 5, and so on. A comparison of the first two columns in Table 2 shows that, in general, as part centrality increases, the rated likelihood of a creature missing that activity decreases. In statistical terms, the correlation between part centrality and rated likelihood is -.78, which is highly significant. Along the same lines, an increase in part centrality is accompanied by an increase in the rated importance of the activity. This trend is somewhat less clear for the nonanalytic activities, since the importance of experiencing is lower and the importance of loving higher than we would anticipate on the basis of the centrality measure. This may mean that subjects were taking into account criteria of importance other than the way these activities support each other. Nonetheless, the correlation between centrality and importance is still .70 and again highly reliable.

Given the reciprocal relationship between *part-of* and *kind-of* in the partial orders, it would not be surprising if the ratings also correlated with a comparable measure of centrality defined on the Kind graphs. One such measure, for example, is the number of subkinds that a target activity possesses. Thus, experiencing has 6 subkinds, contemplating 2, realizing 0, and so on, according to Graph c. This measure is, of course, highly correlated with part centrality ($r = .85$). It also predicts both the likelihood and importance ratings, though in both cases the correlations are somewhat lower than in the case of part centrality ($r = -.70$ for likelihood and $r = .50$ for importance).

These results suggest that the Part graphs are an accurate reflection of people's beliefs about these activities. The same conclusion follows from additional experiments that Conrad and I have reported (Rips and Conrad, 1988): Subjects identify the Figure 1 graphs as more reasonable than ones in which the terms are permuted; they remember the Figure 1 graphs more accurately than the permuted ones; and when they misremember the position of an item in the permuted graphs, they tend to displace the item to the position it occupies in Figure 1. This last result holds even though the subjects had never actually seen Graphs a-d. I'll assume, then, that the Part graphs reveal something of the way people think about relations among mental activities. But this still leaves us with the puzzle that we noticed earlier. How are we to understand the notion of *part-of-a-mental-activity* that is embodied in them?

3. The Mental Representation of Mental Parts

What seems odd about the Part graphs is that subjects identify as parts of a mental activity other *more general* activities. Subjects believe that thinking is a part of reasoning, experiencing is a part of loving, and so on. At the same time, they attest to the fact that reasoning is a kind of thinking and loving is a kind of experiencing. But if one activity is a species of another, how can the genus be a part of that species? If we look at the activities in set-theoretic terms, it would seem that the part links in Figure 1 should point in exactly the opposite direction: If reasoning is a kind of thinking, the set of activities that we can describe as reasoning is a subset of thinking activities. It is possible to say, loosely speaking, that a subset is a part of its superset. But this would make reasoning a part of thinking, whereas what subjects tell us is that thinking is a part of reasoning. We might try to dismiss the Part graphs as reflecting idiosyncratic judgments of just a few subjects. However, at least 45% of the subjects in our first experiment affirmed each of the links in (the transitive closure of) these graphs, and we have just seen that these links predict other aspects of subjects' conceptions of mind.

3.1. *An explanation in terms of procedural parts.*

As I've already hinted, I think the best way to understand the subjects' intuitions is to assume that they interpreted 'part' (as applied to mental activities) in attributional rather than procedural terms. It certainly seems unlikely at first glance that a more general activity could be a procedural part or subprocess of a more specific one. Still, there may be ways of reconciling subjects' judgments with the procedural interpretation, and one such possibility comes to light if we consider the following situation. Imagine that you are a subject with no special knowledge of psychology and that you are given the task of enumerating the procedural parts of reasoning. It might occur to you that an initial part of reasoning is trying to sort out the given information and that the final part is arriving at some conclusion. But what comes in between? What would you call the procedure that takes you from the premises to the conclusion? The only natural response seems to be that this middle part is the "reasoning" itself or "reasoning proper" or some paraphrase like "getting from the premises to the conclusion." There seems to be a lexical gap at just the crucial point, and to fill it you might reach for a more general term like 'thinking'. After all, this middle part of the reasoning process is a kind of thinking, so why not call it that?

These considerations would then lead you to say that thinking was a part of reasoning. However, once we get this far, we can see that this result doesn't depend on the presence of a lexical gap, though the gap makes it apparent why one might want to say that "thinking" is a part of reasoning, rather than using a more specific term. Mental activities like sorting-out-the-premises or arriving-at-the-conclusion also seem to be instances of thinking, and since these activities are clearly part of reasoning they also permit us to say that thinking is part of reasoning. To put it a bit more formally, for each instance of reasoning, there's an instance of thinking—for example, arriving at the conclusion—that is a procedural part of the reasoning. Since this seems to be exactly what it means for one activity type to be a part of another—it's the way we defined the relation for subjects—thinking is indeed a part of reasoning.

On this account, what makes the result seem odd at first is merely that these relationships often fail to hold for categories of physical objects with which we may be more familiar. The parts of an object are usually not members of the same kind as the object itself. For example, the parts of a daffodil, such as its stem or leaves, are not themselves flowers, so flowers are not parts of daffodils. But with mental activities, the procedural parts do happen to be instances of the superordinate kinds, and it's this that permits more general activities to be parts of more specific ones. To get the parallel result with physical objects, we have to go to a more abstract level. Although daffodil stems and daffodil leaves aren't flowers, they are physical objects. Hence, we should be able to say that physical objects are parts of daffodils, and this does seem marginally acceptable.

I think there is something to be said for this explanation of the parts data, but there is at least one aspect of it that looks suspect. Notice that it isn't invariably true that procedural parts of an activity are members of all the same kinds as that activity. In fact, in cognitive theories this seems more the exception than the rule. On the standard cognitive analysis, for example, recognizing a printed word involves procedural parts such as sampling visual features, accessing long-term memory for representations of candidate words, determining the fit between features and representations, and drawing some conclusion about the best candidate. But although recognizing a word is a kind of recognizing, none of the procedural parts just mentioned is. No one would contend, for instance, that sampling visual features is itself a kind of recognizing. To take another example that is closer to the terms we used in the experiment, a part of deciding may be multiplying outcome probabilities and utilities, and deciding may be a kind of problem solving. But the multiplying in this case isn't itself a kind of problem solving.

Now it isn't an accident that procedural parts in cognitive theories have these properties. If our purpose is to explain a mental activity by analyzing its parts, then it is usually not much help if one of the parts is of the same proximal kind as the original activity (cf. Cummins, 1983, chap. 2). Suppose Professor Wheeze suggests that one of the components of recognizing a word is itself a kind of recognizing, perhaps recognizing the individual letters that the word contains. Although this is legitimate, Wheeze's theory won't be satisfying for long unless he or she gives us some idea of how this other form of recognizing works. Analyzing one kind of activity in terms of another of the same kind has its place, but only as a temporary step on the way to a fuller analysis. Of course, this needs to be qualified since we can always view the analyzed activity and its procedural parts as members of the same kind if we go to a high enough level of abstraction (just as we could classify a daffodil and its stem as joint members of the category of physical objects). Recognizing a word and sampling visual features are both kinds of mental activities, but joint membership in that kind doesn't preclude a meaningful analysis. Trouble comes as the kinds in question approach each other in the abstraction hierarchy, and in the limit we get the sort of vacuous analysis that was Wundt's bugbear—recognizing a word explained simply as the result of the word recognition procedure.

In order to explain the results in Figure 1 on the above interpretation, subjects would have to believe that the unnamed procedural parts of the activities are members of the same proximal kinds. For deciding to be a part of planning, there must be a procedural part of planning that is a kind of deciding; for having emotions to be a part of loving, there must be a procedural part of loving that is a kind of having emotions; and so on. It's not inconceivable that subjects believe this, but if they do, the procedural parts must be quite a bit different from the parts to which cognitive theory has accustomed us. An activity and its procedural parts must be so similar in subjects' conceptions that the purported subprocesses must begin to approach a redescription of the whole. Of course, it isn't surprising that subjects with little knowledge of psychology can't give a procedural analysis that's as detailed as the ones cognitive scientists offer; it would be embarrassing for cognitive science if they could. Nevertheless, subjects have had no lack of experience with the activities in question. If we have to assume that their notion of parts as coarse as it must be in order to explain the data in procedural terms, then we might want to consider the possibility that procedural parts aren't natural units in the subjects' representations of mental activities.

3.2. *An explanation in terms of attributional parts.*

Here's another way of thinking about mental activities' intuitive structure. Dreaming, reasoning, remembering, and the rest just happen; they don't have parts in the sense of discernible subprocesses. Of course, they don't happen all at once. Reasoning, for example, may go on for an appreciable time. But when we try to break it up, all we get are episodes of the same sort, not simpler subprocesses. In reasoning our way through a problem, we might solve one lemma after another, but all of these are just further segments of the reasoning and not something else. Sure, there may be some stage-setting at the beginning and some recognition that the process is over at the end. However, these aren't procedural parts of the process itself; they're causes or consequences. The real mental activity—the stuff that comes in the middle—is nondecomposable. We might want to allow for exceptions when the activity becomes extremely complex, as in the case of reading *The Ambassadors* or planning a polar expedition. But for any garden-variety mental activity, the sort that takes on the order of a few seconds, there's only a single procedure in play at a time.

This isn't to say that the activities are nondescript. Reasoning may be strenuous, dreaming may be frightening, remembering may be pleasurable. Sometimes these properties hold for only a portion of the activity's duration, and so they can be said to divide the activity into parts (e.g., the frightening part of a dream). Sometimes the properties are typically true of the activity, and in this case we can talk about the property as a characteristic part of the activity (e.g., strenuousness is a part of reasoning). The latter are what we called attributional parts at the beginning of this paper. But in neither case do these property-induced parts amount to procedural parts in the sense discussed earlier.

What's appealing about this conception, for present purposes, is that it lends itself to a systematic explanation of the Part results from our main experiment. Suppose that subjects represent each type of mental activity in terms of its attributional parts (i.e., characteristic properties). For example, reasoning might have attributes p_1, p_2,..., p_k, where these attributes will be internally represented by corresponding mental predicates. More general mental activities, such as thinking, will presumably have fewer attributes (thinking has $p_1, p_2,..., p_i$, where $i < k$), since reasoning should have all the attributes of thinking plus some distinctive ones of its own. We can imagine that subjects represent these attributes in simple lists of predicates or in more complex relational structures (frames or schemas). The predicates themselves may also be simple or highly relational ones. Perhaps they are also weighted for importance with respect to the activities with which they are connected. I'll ignore these details since they're irrelevant for our current goals. What's important is that, given this set-up, subjects can make decisions about which mental activities are kinds of others by comparing their stored predicates. For example, subjects believe that reasoning is a kind of thinking because their representation of reasoning has all of the stored predicates of thinking, or at least some criterial number of them. These assumptions are, of course, commonplace in the psychological literature on categorization. They're just what Smith and Medin (1981) call a "Probabilistic model." Needless to say, there are rival models of categorization, but the Probabilistic model has some interesting implications for the data at hand.

If we assume that subjects represent mental activities along these lines, how should they respond to questions about whether one activity is a part of another? Well, if 'part' means procedural part, then they should always say "no." But there is another way of dealing with the question. We've already noticed that it's possible to think of the attributes that are associated with these representations as attributional parts. One reason for this is that the corresponding predicates are, quite literally, parts of the representation of the activities, and it would be a very small step to thinking of the attributes as parts of the activities themselves. It would be surprising if subjects *failed* to think this. But then if all or most of the attributional parts of one activity are included among those of another, it might also seem that the first activity is itself part of the second. For instance, all (or almost all) of the attributional parts of thinking are attributional parts of reasoning, so thinking is a part of reasoning. On this account, the same comparison of predicates that establishes that $activity_1$ is a kind of $activity_2$ also establishes that $activity_2$ is a part of $activity_1$.

We can visualize how this works in terms of Euler circles like these:

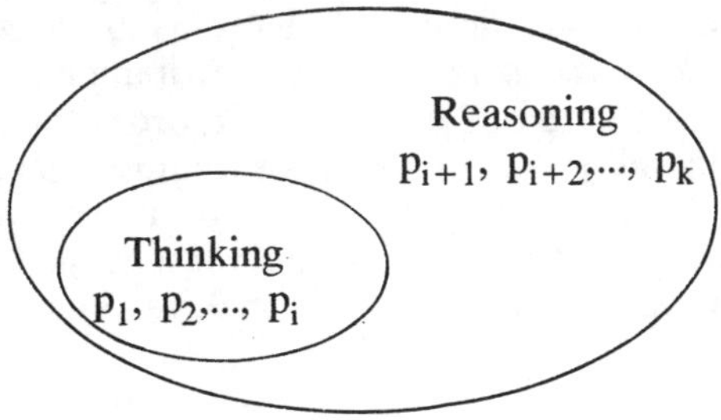

Since reasoning has attributional parts $p_1, p_2, \ldots, p_k$, it has all of the parts of thinking and is therefore a kind of thinking. Conversely, since certain of the parts of reasoning (namely, $p_1, p_2, \ldots, p_i$) constitute thinking, thinking is a part of reasoning. According to this account, the converse relation between *kind-of* and *part-of* is very similar to the usual converse relation between concepts' extension and intension.

It's important to keep in mind that this theory does not imply that during an act of reasoning two different activities or procedures take place, one that does the thinking and another that does the reasoning. What we have instead is one procedure with two different labels.[4] This helps explain why it is that *kind-of* and *part-of* aren't converses when they relate other sorts of entities. For example, although a stem is part of a daffodil, it isn't true that daffodils are a kind of stem. Probabilistic models represent categories of physical objects in the same way that we've proposed representing mental activities. So why don't we get the same reciprocal part-kind relationship? The best answer, I think, is that because a physical whole and its part are conceived as two different objects, the whole doesn't automatically inherit the attributes of the part. Daffodil stems have the attributes of being less than a centimeter in diameter and entirely green, but daffodils are neither less than a centimeter in diameter nor entirely green. Since not all of the attributes of stems are attributes of daffodils, Probabilistic models don't classify daffodils as stems. On the other hand, since thinking and reasoning are not distinct entities, the attributes of the part are attributes of the whole, and the converse relation follows.

4. Conclusions

There are a couple of reasons, I think, for preferring the attributional over the procedural explanation of the Part data. One is that the attributional approach fits well with Probabilistic models of categorization. These models do a fairly good job in predicting results from other types of experiments (see Smith and Medin, 1981, for a review of this evidence), and the fact that they are needed for these independent reasons may be grounds for supposing that similar mechanisms are at work in the present studies. A second point in favor of the attributional explanation is that the alternative is such a bloodless affair. In order for the Part judgments to reflect procedural parts, these subprocedures have to be extremely shallow ones, not much different, in fact, from the procedures that they are supposed to characterize. For the reasons sketched earlier, these procedural parts are a far cry from the analysis Wundt was up to or from the subroutines we're familiar with in cognitive science. Finally, the attributional approach is more systematic. According to this explanation, part-kind converses are a function of the ordinary representation of these activities. The procedural explanation, by contrast, has to posit a new procedural part for each pair of activities in which the converse relation obtains.

These three reasons certainly aren't decisive ones, however. Although the weight of evidence seems slightly on the attributional side, further experiments might easily tip the scale toward procedural parts.[5] As it is, the attributional approach may have to acknowledge the existence of intuitive subprocedures for certain kinds of mental events. For example, in order to explain why subjects said that remembering is part of reading but reading is not a kind of remembering (see Figure 1), we might well assume that subjects were treating remembering as a true procedural part.

This suggests that it might be worthwhile to look for a compromise between these positions. Perhaps what the data are telling us is that there is really not much difference between extremely shallow procedural parts and groups of characteristic attributes. Suppose there is an aspect of dreaming, for example, which we could describe as a kind of imagining, just as we can describe dreaming itself as a kind of imagining. Should we count this aspect as a distinct subactivity in its own right, or does it merely reflect those properties of dreaming that allowed us to label dreaming as a type of imagining in the first place? Maybe we could answer this question if we knew more about the causal role that this aspect plays, but this causal information may be the kind of thing that's hard for our commonsense psychologists to get. As both Wundtians and cognitivists might say, determining causal roles is what lab experiments are supposed to be for.

It's good to remind ourselves that what we've been after is not the truth about mental activities, but the truth about people's ordinary representations of them. However, I don't want to quit on a tone of scientific hauteur. What the results seem to show is that intuitive psychologists view mental activities as part of a systematic structure—a structure that orders the domain in such a way that activities which are parts of many others (or those which have many others as kinds) are seen as more central aspects of mind. Scientific psychology can offer more detailed or theoretical analyses of some of these activities, but it is hard to see how it could resist the general outlines of the intuitive view.[6]

University of Chicago

NOTES

1 There is some disagreement in the psychological literature about how people represent routine events like changing a flat. It is unclear, for example, whether the mental representation of these events is structured as a linear sequence of episodes as in the original Schank and Abelson (1977) "script" proposal or in some more hierarchical fashion. (For recent evidence, see Abbott, Black, and Smith, 1985; Barsalou and Sewell, 1985; Galambos and Rips, 1982; Haberlandt and Bingham, 1984.) Abelson (1981) and Schank (1982) also seem to have changed their minds on this issue. But no one doubts that people have access to information about the procedural components of these routines, however they happen to store them.

2 There is some evidence in the psychological literature for systematic intransitivities in subjects' judgments of whether one class of objects is a kind of another. Hampton (1982) found, for example, that subjects will agree that canoes are vehicles and that vehicles are machines, but deny that canoes are machines. However, these intransitivities are probably best explained either as the result of equivocation (e.g., two senses of 'vehicle') or misinterpretation of the question ("Are vehicles typically machines?" rather than "Are all vehicles machines?"). For similar rebuttals to purported examples of intransitivity of *part-of*, see Cruse (1979), Macnamara (1986, chap. 8), and Winston, Chaffin, and Herrmann (1987).

3 One might also question whether all of these items are in fact activities. Activities are usually taken to imply some sort of change (cf. Miller and Johnson-Laird, 1976, e.g.); but a few of our items (having emotions, for example) don't involve an obvious change and thus might better be conceived as mental states. Our justification for grouping these items with "true" mental activities is that this is how the subjects themselves classified them in our preliminary study. It seems unlikely that any of our substantive conclusions would be changed by omitting these borderline items.

4 There's an echo here of rival analyses in the philosophy of action (e.g., Davidson, 1963; Goldman, 1970; Thalberg, 1977). If a player moves his foot rapidly, thereby kicking a football and scoring the winning point, how many actions are performed? From one point of view, only a single action takes place, variously described. From another, several actions have taken place, including moving the foot, kicking the football, scoring the point, and winning the game. The multiple-action theory is consistent with the notion of procedural parts (winning the game is an action with distinct sub-actions, including moving the foot), and the single-action theory with attributional parts (winning the game and moving the foot are the same action, picked out by distinct attributes). But of course, the question in the philosophy of action (how best to analyze activities, including mental ones) is different from the one we are asking here (how to reconstruct the commonsense conception of mental activities). The answer to one of these problems doesn't foreclose an answer to the other.

5 One might suppose, for example, that the issue could be settled simply by asking subjects to list the parts of mental activities and inspecting these lists to see whether the parts were procedural or attributional. But a pilot study of this kind turned out to be equivocal. Many of the items subjects listed were in fact superordinate activities (e.g., "thinking" was listed as a part of reasoning). This replicates the results from the main experiment but doesn't help settle the procedure-attributes issue. Other responses could be classified as attributes (e.g., "being logical or rational" for reasoning), subprocesses ("find practical examples"), and near synonyms ("identify implication"). In still other cases, subjects failed to list any parts. Asked for the parts of imagining, one subject wrote, "There are no parts to this; one just thinks about something to dream about and does it." Perhaps these results should be seen as support for a compromise position, as discussed below.

[6] I'm very grateful to Rob Chametzky, Fred Conrad, Ron McClamrock, Byron Sebastian, Edward Smith, J-C. Smith, Barbara Tversky, and Rita Walter for their help on this project. The experiments reported here were supported by NIMH Grant MH393633.

REFERENCES

R. Abelson (1981), "Psychological Status of the Script Concept," *American Psychologist* **36**, pp. 715-729.

V. Abbott, J. B. Black, and E. E. Smith (1985), "The Representation of Scripts in Memory," *Journal of Memory and Language* **24**, 179-199.

L. W. Barsalou and D. R. Sewell (1985), "Contrasting the Representation of Scripts and Categories," *Journal of Memory and Language* **24**, 646-665.

A. L. Blumenthal (1985), "Wilhelm Wundt: Psychology as the Propaedeutic Science" in C. E. Buxton (ed.), *Points of View in the Modern History of Psychology* (San Diego, CA: Academic Press).

Susan Carey (1985), *Conceptual Change in Childhood* (Cambridge, MA: MIT Press).

D. A. Cruse (1979), "On the Transitivity of the Part-Whole Relation," *Journal of Linguistics* **15**, pp. 29-38.

Robert Cummins (1983), *The Nature of Psychological Explanation* (Cambridge, MA: MIT Press).

R. D'Andrade (1987), "A Folk Model of Mind" in D. Holland and N. Quinn (eds.), *Cultural Models in Language and Thought* (New York: Cambridge University Press).

Donald Davidson (1963), "Actions, Reasons, and Causes," *Journal of Philosophy* **60**, pp. 685-700.

R. Davis and B. G. Buchanan (1977), "Meta-level Knowledge: Overview and Applications," *Proceedings of IJCAI 77*, pp. 920-927.

J. De Kleer, J. Doyle, G. L. Steele, Jr., G. J. Sussman (1977), "AMORD: Explicit Control of Reasoning," *SIGART Newsletter* **64**, pp. 116-125.

D. C. Dennett (1983), "Intentional Systems in Cognitive Ethology: The 'Panglossian Paradigm' Defended," *Behavioral and Brain Sciences* **6**, pp. 343-390.

J. W. Evans, F. Harary, and M. S. Lynn (1967), "On the Computer Enumeration of Finite Topologies," *Communications of the ACM* **10**, pp. 295-313.

J. A. Galambos and L. J. Rips (1982), "Memory for Routines," *Journal of Verbal Learning and Verbal Behavior* **21**, pp. 260-281.

Alvin I. Goldman (1970), *A Theory of Human Action* (Princeton, NJ: Princeton University Press).

Nelson Goodman (1966), *The Structure of Appearance*, 2nd ed. (Indianapolis, IN: Bobbs-Merrill Company, Inc.).

K. Haberlandt and G. Bingham (1984), "The Effect of Input Direction on the Processing of Script Statements," *Journal of Verbal Learning and Verbal Behavior* **23**, pp. 162-177.

J. A. Hampton (1982), "A Demonstration of Intransitivity in Natural Categories," *Cognition* **12**, pp. 151-164.

William James (1890), *Principles of Psychology* (New York: Holt & Co.).

J. Macnamara (1986), *A Border Dispute: The Place of Logic in Psychology* (Cambridge, MA: MIT Press).

G. A. Miller and P. N. Johnson-Laird (1976), *Language and Perception* (Cambridge, MA: Harvard University Press).

T. Mischel (1970), "Wundt and the Conceptual Foundations of Psychology," *Philosophy and Phenomenological Research* **31**, pp. 1-26.

R. C. Moore (1982), "The Role of Logic in Knowledge Representation and Commonsense Reasoning," *Proceedings of AAAI* **82**, pp. 428-433.

A. Newell and H. A. Simon (1972), *Human Problem Solving* (Englewood Cliffs, NJ: Prentice Hall, Inc.).

L. J. Rips and F. G. Conrad (1988), *The Folk Psychology of Mental Activities* (ms. under review).

M. Scardamalia, C. Bereiter, and R. Steinbach (1984), "Teachability of Reflective Processes in Written Composition," *Cognitive Science* **8**, pp. 173-190.

Roger Schank (1982), *Dynamic Memory: A Theory of Reminding and Learning in Computers and People* (New York: Cambridge University Press).

R. Schank and R. Abelson (1977), *Scripts, Plans, Goals, and Understanding* (Hillsdale, NJ: Erlbaum Associates, Inc.).

A. H. Schoenfeld (1983), "Beyond the Purely Cognitive: Belief Systems, Social Cognitions, and Metacognitions as Driving Forces in Intellectual Performance," *Cognitive Science* **7**, pp. 329-363.

P. Simons (1987), *Parts: A Study in Ontology* (Clarendon Press).

E. E. Smith and D. L. Medin (1981), *Categories and Concepts* (Cambridge, MA: Harvard University Press).

I. Thalberg (1977), *Perception, Emotion, and Action: A Component Approach* (New Haven, CT: Yale University Press).

R. H. Thomason (1969), "Species, Determinates, and Natural Kinds," *Nous* **3**, pp. 95-101.

E. B. Titchener (1902), *An Outline of Psychology* (New York: Macmillan Publishing Co.).

R. W. Weyhrauch (1980), "Prolegomena to a Theory of Mechanized Formal Reasoning," *Artificial Intelligence* **13**, pp. 133-170.

M. Winston, R. Chaffin, and D. Herrmann (1987), "A Taxonomy of Part-Whole Relations," *Cognitive Science* **11**, pp. 417-444.

W. M. Wundt (1904), *Principles of Physiological Psychology*, vol. 1, trans. by E. B. Titchener, (Swan Sonnenschein). (Original work published in 1873.)

W. M. Wundt (1907), *Outlines of Psychology*, trans. by C. H. Judd (Engelmann). (Original work published 1896.)

INDEX

PHILOSOPHICAL STUDIES SERIES IN PHILOSOPHY

1. JAY F. ROSENBERG, *Linguistic Representation,* 1974.
2. WILFRID SELLARS, *Essays in Philosophy and Its History,* 1974.
3. DICKINSON S. MILLER, *Philosophical Analysis and Human Welfare.* Selected Essays and Chapters from Six Decades. Edited with an Introduction by Lloyd D. Easton, 1975.
4. KEITH LEHRER (ed.), *Analysis and Metaphysics.* Essays in Honor of R. M. Chisholm. 1975.
5. CARL GINET, *Knowledge, Perception, and Memory,* 1975.
6. PETER H. HARE and EDWARD H. MADDEN, *Causing, Perceiving and Believing.* An Examination of the Philosophy of C. J. Ducasse, 1975.
7. HECTOR-NERI CASTAÑEDA, *Thinking and Doing.* The Philosophical Foundations of Institutions, 1975.
8. JOHN L. POLLOCK, *Subjunctive Reasoning,* 1976.
9. BRUCE AUNE, *Reason and Action,* 1977.
10. GEORGE SCHLESINGER, *Religion and Scientific Method,* 1977.
11. YIRMIAHU YOVEL (ed.), *Philosophy of History and Action.* Papers presented at the first Jerusalem Philosophical Encounter, December 1974, 1978.
12. JOSEPH C. PITT, *The Philosophy of Wilfrid Sellars: Queries and Extensions,* 1978.
13. ALVIN I. GOLDMAN and JAEGWON KIM, *Values and Morals.* Essays in Honor of William Frankena, Charles Stevenson, and Richard Brandt, 1978.
14. MICHAEL J. LOUX, *Substance and Attribute.* A Study in Ontology, 1978.
15. ERNEST SOSA (ed.), *The Philosophy of Nicholas Rescher: Discussion and Replies,* 1979.
16. JEFFRIE G. MURPHY, *Retribution, Justice, and Therapy.* Essays in the Philosophy of Law, 1979.
17. GEORGE S. PAPPAS, *Justification and Knowledge: New Studies in Epistemology,* 1979.
18. JAMES W. CORNMAN, *Skepticism, Justification, and Explanation,* 1980.
19. PETER VAN INWAGEN, *Time and Cause.* Essays presented to Richard Taylor, 1980.
20. DONALD NUTE, *Topics in Conditional Logic,* 1980.
21. RISTO HILPINEN (ed.), *Rationality in Science,* 1980.
22. GEORGES DICKER, *Perceptual Knowledge,* 1980.

23. JAY F. ROSENBERG, *One World and Our Knowledge of It,* 1980.
24. KEITH LEHRER and CARL WAGNER, *Rational Consensus in Science and Society,* 1981.
25. DAVID O'CONNOR, *The Metaphysics of G. E. Moore,* 1982.
26. JOHN D. HODSON, *The Ethics of Legal Coercion,* 1983.
27. ROBERT J. RICHMAN, *God, Free Will, and Morality,* 1983.
28. TERENCE PENELHUM, *God and Skepticism,* 1983.
29. JAMES BOGEN and JAMES E. McGUIRE (eds.), *How Things Are, Studies in Predication and the History of Philosophy of Science,* 1985.
30. CLEMENT DORE, *Theism,* 1984.
31. THOMAS L. CARSON, *The Status of Morality,* 1984.
32. MICHAEL J. WHITE, *Agency and Integrality,* 1985.
33. DONALD F. GUSTAFSON, *Intention and Agency,* 1986.
34. PAUL K. MOSER, *Empirical Justification.* 1985.
35. FRED FELDMAN, *Doing the Best We Can,* 1986.
36. G. W. FITCH, *Naming and Believing,* 1987.
37. TERRY PENNER, *The Ascent from Nominalism.* Some Existence Arguments in Plato's Middle Dialogues, 1987.
38. ROBERT G. MEYERS, *The Likelihood of Knowledge,* 1988.
39. DAVID F. AUSTIN, *Philosophical Analysis.* A Defense by Example, 1988.
40. STUART SILVERS, *Rerepresentation.* Essays in the Philosophy of Mental Rerepresentation, 1988.
41. MICHAEL P. LEVINE, *Hume and the Problem of Miracles. A Solution,* 1979.
42. MELVIN DALGARNO and ERIC MATTHEWS, *The Philosophy of Thomas Reid,* 1989.
43. KENNETH R. WESTPHAL, *Hegel's Epistemological Realism.* A Study of the Aim and Method of Hegel's *Phenomenology of Spirit,* 1989.
44. JOHN W. BENDER, *The Current State of the Coherence Theory.* Critical Essays on the Epistemic Theories of Keith Lehrer and Laurence Bonjour, with Replies, 1989.
45. ROGER D. GALLIE, *Thomas Reid and 'The Way of Ideas',* 1989.
46. J-C. SMITH (ed.), *Historical Foundations of Cognitive Science,* 1990.
47. JOHN HEIL (ed.), *Cause, Mind, and Reality.* Essays Honoring C. B. Martin, 1989.
48. MICHAEL D. ROTH and GLENN ROSS (eds.), *Doubting.* Contemporary Perspectives on Skepticism, 1990.